D0215870

Process Improvement and CMMI® for Systems and Software

Process Improvement and CMMI® for Systems and Software

Ron S. Kenett
Emanuel R. Baker

CRC Press
Taylor & Francis Group
Boca Raton   London   New York

CRC Press is an imprint of the
Taylor & Francis Group, an **informa** business

AN AUERBACH BOOK

* Architecture Tradeoff Analysis Method, ATAM, Capability Maturity Model, Capability Maturity Modeling, Carnegie Mellon, CERT, CERT Coordination Center, CMM, CMMI, FloCon, and OCTAVE are registered in the U.S. Patent and Trademark Office by Carnegie Mellon University.

SM CMM Integration, COTS Usage Risk Evaluation, CURE, EPIC, Evolutionary Process for Integrating COTS-based systems, Framework for Software Product Line Practice, IDEAL, Interim Profile, OAR, OCTAVE, Operationally Critical Threat, Asset, and Vulnerability Evaluation, Options Analysis for Reengineering, Personal Software Process, PLTP, Product Line Technical Probe, PSP, SCAMPI, SCAMPI Lead Appraiser, SCE, SEPG, SoS Navigator, T-Check, Team Software Process, and TSP are service marks of Carnegie Mellon University.

Auerbach Publications
Taylor & Francis Group
6000 Broken Sound Parkway NW, Suite 300
Boca Raton, FL 33487-2742

© 2010 by Taylor and Francis Group, LLC
Auerbach Publications is an imprint of Taylor & Francis Group, an Informa business

No claim to original U.S. Government works

Printed in the United States of America on acid-free paper
10 9 8 7 6 5 4 3 2 1

International Standard Book Number: 978-1-4200-6050-8 (Hardback)

This book contains information obtained from authentic and highly regarded sources. Reasonable efforts have been made to publish reliable data and information, but the author and publisher cannot assume responsibility for the validity of all materials or the consequences of their use. The authors and publishers have attempted to trace the copyright holders of all material reproduced in this publication and apologize to copyright holders if permission to publish in this form has not been obtained. If any copyright material has not been acknowledged please write and let us know so we may rectify in any future reprint.

Except as permitted under U.S. Copyright Law, no part of this book may be reprinted, reproduced, transmitted, or utilized in any form by any electronic, mechanical, or other means, now known or hereafter invented, including photocopying, microfilming, and recording, or in any information storage or retrieval system, without written permission from the publishers.

For permission to photocopy or use material electronically from this work, please access www.copyright.com (http://www.copyright.com/) or contact the Copyright Clearance Center, Inc. (CCC), 222 Rosewood Drive, Danvers, MA 01923, 978-750-8400. CCC is a not-for-profit organization that provides licenses and registration for a variety of users. For organizations that have been granted a photocopy license by the CCC, a separate system of payment has been arranged.

Trademark Notice: Product or corporate names may be trademarks or registered trademarks, and are used only for identification and explanation without intent to infringe.

Library of Congress Cataloging-in-Publication Data

Kenett, Ron S.
    Process improvement and CMMI* for systems and software / Ron S. Kenett, Emanuel R. Baker.
        p. cm.
    Includes bibliographical references and index.
    ISBN 978-1-4200-6050-8 (hardcover : alk. paper)
    1. Capability maturity model (Computer software) 2. Software engineering. I. Baker, Emanuel R., 1934- II. Title.

QA76.758.K47 2010
005.1--dc22
                                            2009048731

Visit the Taylor & Francis Web site at
http://www.taylorandfrancis.com

and the Auerbach Web site at
http://www.auerbach-publications.com

To Sima, my wife; our children, Dolav, Ariel, Dror, Yoed, and their spouses; and our grandchildren Yonatan, Alma, Tomer, and Yadin. They are an essential part in everything I do, including this book.

RSK

To my wife, Judy; to my sons, Todd and Russ; and to our grandchildren, Evan, Ean, Caleb, Margot, and Isabelle. They represent our future.

ERB

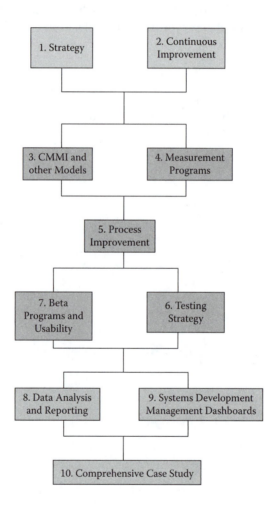

Contents

SECTION II ASSESSMENT AND MEASUREMENT IN SYSTEMS AND SOFTWARE DEVELOPMENT

SECTION III IMPROVEMENTS AND TESTING OF SYSTEMS AND SOFTWARE

5 System and Software Development Process Improvement................. 161

6 System and Software Testing Strategies 191

SECTION IV MANAGING AND REPORTING DATA, AND A COMPREHENSIVE CASE STUDY

Foreword

Systems in general and computing systems in particular have been extensively used in almost all aspects of our daily lives, ranging from operating and managing critical infrastructures to improving productivity in offices and factories, to enhancing research and education, and to controlling home appliances. All these systems rely on dependable and efficient software. However, despite decades of development in software and systems engineering technologies, many challenging issues to produce high-quality systems still need to be addressed due to the following challenges.

First, the pervasiveness of software requires systems to operate in increasingly heterogeneous and dynamic environments. Nowadays, software systems run on various computing devices ranging from tiny sensors, handheld and embedded devices, and personal computers to enterprise servers over various network infrastructures. Different devices and network infrastructures have unique characteristics of available resources and operational environments.

Second, the scale of software systems has grown from stand-alone systems running on single computers to globally networked systems consisting of thousands of geographically distributed computers. Furthermore, large-scale software systems are often a "system of systems," that is, the integration of available software systems.

Third, with the increasing demands and complexity, systems need to be more secure and dependable. Current systems are vulnerable to various security threats to the software that runs them, malicious attacks (such as viruses, worms, and denial of services), and often lack effective mechanisms in their design to handle unexpected system and network failures.

Fourth, systems need to be situation-aware and autonomous. Due to the dynamics of operating environments and available resources for many time-critical applications, software systems need to be aware of changes in user activities, system status, and operating environments, and autonomously adapt themselves to continuously provide satisfactory quality for the users.

Fifth, customers often have requirements on multiple quality aspects, such as real-time performance, availability, and security, and hence systems need to be able to satisfy the requirements on multiple quality aspects simultaneously. However, satisfying the requirements on one quality aspect may need to compromise another quality aspect. This requires adaptive trade-offs among various quality aspects.

Finally, systems need to have excellent composability and interoperability to improve the ability to reconfigure systems and accelerate systems development. The "system of systems" view of large-scale systems requires the rapid integration of available systems. Although the software technologies, such as component-based software development and service-oriented computing, have greatly improved software reusability, it is still difficult to realize the "system of systems" view due to the poor composability and interoperability of current systems.

Developing new system and software technologies, such as requirements modeling techniques, new software architecture, programming languages, and testing techniques, is not sufficient to address these challenges. Human participation in systems and software development remains one of the most important factors in developing high-quality systems. Many defects and quality issues are due to human errors, instead of technical barriers. Hence, process is as important, if not more important, as software and systems development technologies in order to meet the challenges for developing high-quality systems. Over the past 40 years, researchers and practitioners have developed many process models and established various international standards for software and systems engineering processes. In this book, the authors have presented the state-of-the-art in this area with special focus on planning, implementation, and management in system and software processes. The authors have also presented important basic concepts on strategic planning and process improvement, introduced CMMI® and various other software and systems process models, and discussed various concepts and approaches to system and software measurement. Various testing strategies and techniques for system and software improvement have also been presented. The authors have discussed how data related to systems and software development can be collected, analyzed, and reported. They have also presented a case study in software process quality management and control to illustrate the models and techniques. This book will help readers understand software quality and process improvement and provide an effective tutorial for the planning, implementation, and management of system and software projects.

Stephen S. Yau
Professor of Computer Science
Arizona State University
Past President, IEEE Computer Society
Past President, American Federation of Information-Processing Societies (AFIPS)

Preface

Process has been a major part of human life since the beginning of time. The biblical Book of Genesis tells the story of creation, showing that a process existed to take this vast mass of diffuse matter floating around in order to create a universe. First, a differentiation between night and day was created. Quality control was part of the process: "And G-d saw every thing that he had made, and, behold, it was very good" (Genesis 1:31), a first example of testing and quality control. On the second day, a separation of the waters between the oceans and the sky was created. Again, quality control was imposed to make sure that the results were what were intended. On the third through the sixth days, the process continued until we had a world essentially like what exists today, consisting of dry land and oceans, and life in the form of flora and fauna. At the end of each day, G-d implemented self-control to check on his work.

So, we see that *process* has been with us since earliest times.

Process is an essential part of human endeavor, particularly in the creation of products and services. Imagine a shop floor with all the piece parts of an automobile (from chassis to the smallest screw) strewn all over it. Next, imagine trying to assemble all those piece parts into an automobile that can be safely driven. Could it be done without a process that defines the sequence of things to put together and an accompanying set of procedures and work instructions to do it? Hardly. Yet, we often vigorously attack problems, or engage in creating something, without the foggiest notion of how to proceed. Software development, for instance, has been notorious in many quarters for developing applications by immediately starting to code without trying to first solidify the requirements or perform design.

Processes are an essential element of being able to deliver products and services that make our customers happy and bring them back for repeat business. Processes are the link between past, present, and future activities. We believe that following a highly capable process does make a difference. Achieving such processes requires intelligent development of it, as well as continuous improvement of it.

What are the barriers to establishing a process or implementing process improvement? A nonexhaustive list includes

- *Lack of data:* Many organizations have no idea what their current levels of performance are. They have never implemented effective measurement programs and have very little idea if they are performing well, or if they are in deep trouble. For instance, they don't know if the defect rates in their delivered systems are at, above, or below industry norms, or how long it really takes to deliver a given set of capabilities. For many companies that do collect data, the data is never used, or is not used properly.
- *Extreme focus on time to market:* Working on process improvement is perceived as getting in the way of getting new products (or updates to existing products) out the door. While process improvement would be nice to have, "we just can't spare the people or the time."
- *Cost:* While there is often money to acquire new tools or computing equipment, there is usually little money allocated to process improvement. It's not perceived as something that requires a capital investment.
- *Managers who don't understand process:* Many people can recognize the importance of process in activities such as car manufacturing or chemical processing, yet don't recognize the fact that it plays an important role in software *and* hardware development. For example, many software development organizations are populated with "developers" who know how to hack out code, but have never had to work in an environment where development was performed in anything other than a chaotic, fire-drill environment. These people later become software project leaders or software development managers. Why would they require their developers to follow a process as managers, when they never did that themselves as developers? As another example, many system development organizations consider the hardware and software development activities as two separate and distinct enterprises to be done in separate corners of their facilities with little or no interaction between the two. Systems engineering is not part of the process.
- *The educational pipeline:* Few schools teach software engineering or systems engineering as an undergraduate program. The emphasis is on learning how to do specific pieces of the effort (like designing a circuit or developing a compiler), with little or no emphasis on how one develops a system to perform as a system.
- *Passive management:* Many senior managers who decide to implement process improvement believe that it can be done by "decree." They often do not hold their middle managers accountable for the success or failure of process improvement. Many manage process improvement by "remote control." The middle managers, left to their own devices and agendas, will concentrate on getting product out the door. Without upper management setting process improvement goals to measure middle management performance with middle management becomes a black hole when it comes to process improvement.

Undoubtedly, there are other factors, but it doesn't have to be so hard. Process improvement can be implemented and managed effectively. It takes careful planning and monitoring, but it can be done. *Process Improvement and CMMI® for Systems and Software: Planning, Implementation, and Management* presents a methodology for establishing the current status of a development or maintenance process and laying out a reasoned plan for process improvement. The book identifies practical ways of implementing measurement programs used to establish current levels of performance and baselines against which to measure achieved improvements.

The ten chapters of the book provide details and examples of measures and metrics that can be used to establish a baseline, as well as for use in monitoring process improvement projects. The book concludes with a realistic case study to illustrate how all the methods, tools, and techniques discussed in the book fit together.

Process Improvement and CMMI® for Systems and Software: Planning, Implementation, and Management was designed to be practical but not necessarily exhaustive. Our objective was to provide the reader with a workable approach for achieving cost-effective process improvement.

Acknowledgments

The authors would like to acknowledge the contributions of so many people who helped us by providing input and advice and by reviewing portions of the text for us. We would like to thank Dr. Di Cao, Schlumberger, Beijing, China; Prof. Fevzi Belli, University of Paderborn, Paderborn, Germany; Mr. Sumit Popli, Tata Consulting Services, Gurgaon Haryana, India; Prof. Bai Xiaoying, Tsinghua University, Beijing, China; Mrs. Hongyu Pei-Breivold, ABB, Vasteras, Sweden; Dr. Jun-Jang Jeng, IBM Research Center, Yorktown Heights, New York, United States; Prof. Tad Gonsalves, Sophia University, Tokyo, Japan; Dr. Rajesh Subramanyan, Siemens, Princeton, New Jersey, United States; Prof. Joao Cangusu, University of Texas, Dallas, Texas, United States; Prof. Stephen Yau, Arizona Stat University, Tempe, Arizona, United States; Mr. Marek Stochel, Motorola, Krakow, Poland; Mr. Sebastiano Lombardo, Creativitas.net, Oslo, Norway; Dr. Yossi Raanan, KPA, Ra'anana, Israel; Prof. Paolo Giudici, University of Pavia, Pavia, Italy; Dr. William Hefly, Institute for Software Research, Carnegie Mellon University, Pittsburgh, Pennsylvania, United States; Ms. Angela Tuffley, Griffith University, Nathan, Queensland, Australia; Mr. Terry Rout, Griffith University, Nathan, Queensland, Australia; Mr. Frank Koch, Process Strategies, Walpole, Maine, United States; and Mr. John Marciniak, Upper Marlboro, Maryland, United States.

Finally, we would like to thank John Wyzalek, Jessica Vakili, Pat Roberson, Prudy Board, and all the people at Taylor & Francis associated with this book for their patience. This book has been in the works for some time, and they never gave up on us.

About the Authors

Ron S. Kenett, PhD, is CEO and senior partner of KPA Ltd., an international management consulting firm, research professor at the University of Torino, Torino, Italy and international research professor, Gerstein Fisher Center for Finance and Risk Engineering, New York University Poly, New York, USA. He has more than 25 years of experience in restructuring and improving the competitive position of organizations by integrating statistical methods, process improvements, supporting technologies, and modern quality management systems. For ten years he served as Director of Statistical Methods for Tadiran Telecommunications Corporation and, previously, as researcher at Bell Laboratories in New Jersey. His 150 publications are on topics in industrial statistics, software quality, and quality management. Ron is co-author of four books including *Modern Industrial Statistics: Design and Control of Quality and Reliability* (with S. Zacks), Duxbury Press, 1998, Chinese edition, 2004 and *Software Process Quality: Management and Control* (with E. Baker), Marcel Dekker Inc., 1999, which is based on a course he taught at Tel Aviv University School of Engineering. He is editor-in-chief of the Encyclopedia of Statistics in Quality and Reliability (Wiley, 2007) and of the international journal *Quality Technology and Quantitative Management*. He is past president of ENBIS, the European Network for Business and Industrial Statistics and has been a consultant for HP, EDS, SanDisk, National Semiconductors, Cisco, Intel, and Motorola. Professor Kenett's PhD is in mathematics from the Weizmann Institute of Science, Rehovot, Israel, and a BSc in mathematics with first class honors from Imperial College, London University.

Emanuel R. Baker, PhD, is a principal owner of Process Strategies, Inc., an internationally recognized software and systems engineering consulting firm based in Los Angeles, California and Walpole, Maine. He and Dr. Ron Kenett are co-authors of the book, *Software Process Quality: Management and Control*, published by Marcel Dekker, Inc. in 1999. He has written numerous papers and other publications on the topics of process improvement and software quality. Dr. Baker is an SEI-certified High Maturity Lead Appraiser for the SCAMPI appraisal methodologies for the CMMI-DEV, CMMI-SVC, and CMMI-ACQ models, and instructor for the SEI's introduction to the CMMI-DEV and CMMI-ACQ supplement courses. He has held a number of management positions and was manager of the Product Assurance Department of Logicon's Strategic and Information Systems Division. In that capacity, along with his duties of managing the department, he had responsibility for the contract to develop the former DoD software quality standard, DoD-STD-2168. His current interest is to help organizations in implementing process improvement and assisting them in their implementation of the CMMI®.

STRATEGY AND BASICS OF QUALITY AND PROCESS IMPROVEMENT

I

Chapter 1

The Strategic Role of Systems and Software Quality

Synopsis

> Perfect quality, perfect delivery, perfect reliability, perfect service—
> these are achievable. ... The quality system that will be embraced by
> any particular organization who takes the subject very seriously will
> aim for those goals and be measurable by the appropriate and dedicated
> use of the statistical systems that are now readily available.
>
> **—Robert W. Galvin**
> *Former Chairman of the Board,*
> *Motorola Inc.*

This quote from the foreword of the *Encyclopedia of Statistics in Quality and Reliability* [11] presents, in clear terms, the vision of a top executive who decided to embark his organization on the never-ending journey of quality improvements. Galvin set up a goal of perfection that he operationalized in the form of Six Sigma in the early 1980s. In 1995, Jack Welch, Chairman and CEO of General Electric, mandated each GE employee to work toward achieving Six Sigma. Six Sigma was eventually adopted all over the world by small and large companies, not-for-profit and educational organizations, government services organizations, and health-care organizations. Since its introduction, and subsequent widespread acceptance, a number of variants have emerged, such as Lean Sigma, Lean Six Sigma, and Design for Six Sigma (DFSS).

Six Sigma is an outgrowth of the insights of W. Edwards Deming [9] and Joseph M. Juran [14] on how organizations can be managed to achieve the vision of Galvin. Six Sigma was preceded by other quality initiatives, such as Total Quality Management, that had varied success.

A key focused element in striving to build quality into the systems we acquire, use, or develop are the processes we use to specify, design, build, test, and install new versions or updates. The systems and software development community has also recognized that "the quality of a system or product is highly influenced by the quality of the process used to develop and maintain it" [8]. A sound process is needed to ensure that end-user requirements are accurately specified for the systems they acquire. The challenge is in making sure that the development effort results in a product that meets those needs. Moreover, the processes used to deploy, operate, and maintain the systems must not degrade the system's quality. Such systems should be the driving force behind quality and customer satisfaction. This book is about implementing and managing with quality the processes needed to cover all aspects of a system's life cycle, from concept to retirement.

The book is divided into four parts:

Part I: Strategy and Basics of Quality and Process Improvement
Part II: Assessment and Measurement in Systems and Software Development
Part III: Improvements and Testing of Systems and Software
Part IV: Managing and Reporting Data, and a Comprehensive Case Study

Part I consists of two chapters. In Chapter 1, we set the stage by addressing the strategic role of systems and software quality. Chapter 2 then moves into a discussion of the importance of process improvement relative to systems and software quality, and addresses various approaches for implementing process improvement.

Part II also consists of two chapters. Process improvement is difficult to achieve without using a model as a framework. These models are discussed in Chapter 3. Chapter 4 discusses measurement and measurement programs, which are essential for determining whether process and quality improvement has been achieved, and for determining the return on investment for process improvement.

Part III consists of three chapters. Chapter 5 discusses how the process improvement effort is structured. Chapters 6 and 7 move into topics related to demonstrating whether process improvements have provided the intended results. Chapter 6 discusses testing programs and Chapter 7 discusses beta testing and usability programs.

Part IV consists of three chapters. Chapter 8 focuses on data reporting from testing and usability programs, as well as data reporting from the user community. In Chapter 9, we discuss the use of systems development management dashboards, a useful took for decision making. The book concludes with a case study in Chapter 10.

We begin this chapter with a brief overview of strategic planning, the first step in setting up effective, efficient, and sustainable process improvement initiatives.

1.1 Basic Strategic Planning

At the 30th Anniversary Congress of the Asian Productivity Organization, Hideo Segura, former chairman of Honda Motor Company, explained the roles of senior management and strategic planning better than anyone (see [13]). He described four "sacred obligations" of management:

1. To develop a clear vision of where the company is going; this must be clearly stated and communicated to every member of the organization in language he/she understands
2. To provide a clear definition of the small number of key objectives that must be achieved if the company is to realize its vision
3. To translate these key objectives throughout the entire organization so that each person knows how performing his/her job helps the company achieve the objectives
4. To communicate a fair and honest appraisal so that each and every employee, subcontractor, and supplier knows how his/her performance has contributed to the organization's efforts to achieve the key objectives, accompanied by guidance as to how the individual can improve this performance

A. Blanton Godfrey [13], the past chairman of the Juran Institute and a world expert on quality management, lists seven milestones that delineate a roadmap for the top management of organizations planning their journey toward quality improvement. With some modifications, these milestones are:

1. *Awareness* of the competitive challenges and your own competitive position.
2. *Understanding* of the new definition of quality and of the role of quality in the success of your company.
3. *Vision* of how good your company can really be.
4. *Plan* for action: clearly define the steps you need to take to achieve your vision.
5. *Train* your people to provide the knowledge, skills, and tools they need to make your plan happen.
6. *Support* actions taken to ensure changes are made, problem causes are eliminated, and gains are held.
7. *Reward and recognize* attempts and achievements to make sure that quality improvements spread throughout the company and become part of the business plan.

Quoting Joseph M. Juran [14], the "Father of Quality" who passed away on February 28, 2008, at the advanced age of 103:

> For most companies and managers, annual quality improvement is not only a new responsibility; it is also a radical change in style of management—a change in culture … All improvement takes place project by project and in no other way.

The message is clear: (1) Management has to lead the quality improvement effort, and (2) any improvement plan should consist of stepwise increments, building on experience gained in initial pilot projects, before expanding horizontally to cover all subprocesses. The idea is to explore all the ramifications of the improvement in a pilot project before promulgating the change to the organization as a whole. In other words, go deep before going wide.

A generic plan for driving a software development organization toward continuous process improvement and a case study implementation of the plan was presented by Kenett and Koenig [16] and Kenett [17]. Although originally developed for a software development organization, an expanded adaptation of this plan for systems development consists of the following five steps:

1. *Define* an integrated systems development process.
2. *Support* this framework with an automated development environment, where possible.
3. *Identify key areas* for process improvements.
4. Within these areas, *assign ownership, determine metrics, create feedback loops, provide training,* and *establish pilot projects.*
5. *Support and fortify* the continuous improvement efforts.

The first three steps should be carried out by an interdisciplinary team of experts from the various systems development activities. The last two steps must involve management and working groups centered around the main development subprocesses. Step 3 sees the phasing-out of the team of experts and the phasing-in of improvement projects and middle management's direct involvement. The whole effort requires a dedicated "facilitator" and management's commitment and leadership. The facilitator function is sometimes carried out by the Systems Engineering Process Group (SEPG) (see Chapter 2). The global objective is to include every member of the systems development organization, including its suppliers and customers, in the continuous process improvement effort.

While reading a *Scientific American* article on the relative efficiency of animals, Steve Jobs of Apple was struck by how ordinary human beings are. We are no better than somewhere in the middle of the pack in converting energy to speed. But, put a human on a bicycle, and the human far exceeds the efficiency of any animal. Jobs' vision for Apple became a "bicycle for the mind."

The key role of "vision" for organizations and cultural groups was recognized long ago. In the Old Testament it is written, "Where there is no vision, the people perish" (Proverbs 29:18).

Some examples of vision statements are listed below:

- *Aetna:* We will achieve *sustained, superior profitability,* and *growth* by developing and delivering commercial property/casualty insurance products and services that meet the needs of small businesses.
- *General Electric: Be number one or number two in every market we serve.* (This vision statement was not intended to limit an organization to finding comfortable niches where they could easily achieve a number-one or -two status, but rather was intended to make organizations stretch into other markets and strive for number-one or -two status in whatever markets they entered).
- *Eastman Chemical: Be the best international chemical company.*

The vision of an organization is translated by its management into *strategic initiatives.* As an example, the strategic initiatives of Eastman Chemical are:

1. Achieve higher customer satisfaction.
2. Create shareholder value.
3. Be a trusted member of the community.
4. Join alliances to gain competitive advantage.

If we zoom in on "Achieve higher customer satisfaction" and study topics drawn from Pareto analysis of customer satisfaction survey responses, we identify the following as first choices for improvements:

1. Improve reliability of supply by 50%.
2. Improve responsiveness to product changes to 70% of the time.
3. Demonstrate concern for the customer by doubling customer follow-ups.
4. Improve feedstock on-time delivery by 80%.
5. Improve billing accuracy 95%.

By studying customer complaints we find that the main causes for unreliable supply will be addressed if we launch improvement projects in three areas:

1. Improve tanker supply.
2. Decrease equipment downtime.
3. Revise scheduling process.

The *vision* and *strategic initiatives* of an organization serve as background to the short-term and long-term decision-making process. In running a strategic analysis of a new project or business proposition, several dimensions should be considered. Examples of criteria for evaluating such projects include:

■ Fit with business or corporate strategy (low, medium, high)
■ Inventive merit (low, medium, high)
■ Strategic importance to the business (low, medium, high)
■ Durability of the competitive advantage (short term, medium term, long term)
■ Reward based on financial expectations (modest to excellent)
■ Competitive impact of technologies (base, key, pacing, and embryonic technologies)
■ Probabilities of success (technical and commercial success as percentages)
■ R&D costs to completion (dollars)
■ Time to completion (years)
■ Capital and marketing investment required to exploit (dollars)
■ Markets or market segments (market A, market B, etc.)
■ Product categories or product lines (product line M, product line N, etc.)
■ Project types (new products, product improvements, extensions and enhancements, maintenance and fixes, cost reductions, and fundamental research)
■ Technology or platform types (technology X, technology Y, etc.)
■ Organizational culture (employee satisfaction and attitude surveys)

In this book we show how organizations developing and integrating software and system products can launch and implement concrete action plans that will help them achieve their vision. The idea is to implement SMART goals (i.e., goals that are Specific, Measurable, Achievable, Relevant, and Time-bound). We begin with a brief description of several basic strategic planning models.

1.2 Strategic Planning Models

Basic strategic planning is facilitated by the application of models that have proven effective for a large number of organizations. In this section we discuss these models and how they are applied and utilized.

A number of models for strategic analysis have been proposed, over time, by both consultants and academics [1, 7, 19]. Some of the most popular ones include:

■ The Boston Consulting Group market position model (see Figure 1.1)
■ The Booz innovation model (see Figure 1.2)
■ The Arthur D. Little life-cycle model (see Figure 1.3)
■ The risk-reward model (see Figure 1.4)

We provide a brief introduction to these four models and suggest that a combined analysis, using several such models, is the effective approach to strategic analysis.

Each model brings its own perspective. Integrating several points of view provides synergies that have proven useful in practice.

The *Boston Consulting Group market position model* (Figure 1.1) is a portfolio planning model developed by Bruce Henderson of the Boston Consulting Group in the early 1970s. It is based on the observation that a company's business units can be classified into four categories based on combinations of market growth and market share relative to the largest competitor—hence the name "growth-share." Market growth serves as a proxy for industry attractiveness, and relative market share serves as a proxy for competitive advantage. The matrix thus maps a company's activities within these two important determinants of profitability. The horizontal axis represents relative market position and the vertical axis represents market attractiveness. The market attractiveness and market position scales are divided into three parts, with the highest third being in the most favorable position, and the lowest third in the most unfavorable position. On this map, projects in the top third of attractiveness and market position are "winners," and projects in the bottom third of attractiveness and market position are "losers." Projects that are high in relative market position but low in attractiveness can produce profits, but organizations may choose not to pursue such projects because they may not provide new business opportunities in the long run. Projects that are medium on both scales provide only average opportunities, while those that are low in relative market position but high in attractiveness may bear watching for future business opportunities.

Figure 1.1 shows the positioning of four projects labeled P1 through P4. The circle circumference is proportional to the addressable market of each project's product. By positioning the projects within the matrix, management has the capability to evaluate and develop a focused strategy, decide where to expand, what projects are cash cows that produce revenues without substantial investments, and what areas the organization should consciously decide to abandon. Products or projects in the "Winner" corner are sometimes called "Stars." Profit producers are

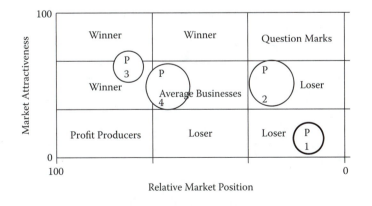

Figure 1.1   The Boston Consultants Market Position Model.

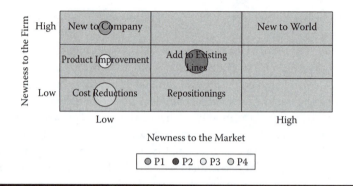

Figure 1.2   The Booz innovation model.

also known as "cash cows" and losers as "dogs." Cash cows are current bread for the company, so they are constantly fed; however, cash cows probably may turn into dogs in the future.

The *Booz innovation map* is based on an assessment of the status of the project technology in terms of newness to the firm conducting the project and newness to the market (Figure 1.2). The circle areas here can represent, for example, forecasted research and development expenditures. Projects involving mature technology that is new to the firm require aggressive benchmarking and investments in knowledge acquisition and hiring. Projects that are not new to the market and where the organization has accumulated significant experience should be motivated by cost-reduction initiatives in order to sustain the market.

The *Arthur D. Little (ADL) model* maps projects by their life-cycle maturity and relative market position (Figure 1.3). Here again, the circle areas can represent forecasted R&D expenditures or requested budgets. The horizontal axis represents relative market position, and the vertical axis represents the positioning of the technology in its natural life cycle. The model identifies four main strategies

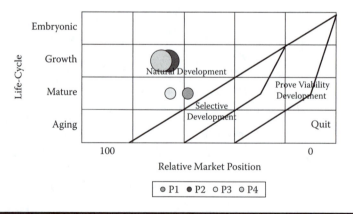

Figure 1.3   The ADL life-cycle model.

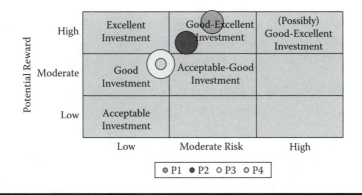

Figure 1.4 The risk reward model.

that are demarcated by the solid lines on the map: (1) natural development, (2) selective development, (3) projects requiring a viability study, and (4) projects that the organization should simply stop. The four strategies are determined by the two-dimensional positioning of specific projects.

The *risk reward model* assesses technical and marketing risks versus reward potential. Figure 1.4 maps the various combinations of risk and reward in specific projects into specific strategy qualifications. Projects with low risk and high potential reward are excellent investments. Projects with low risk and low reward are acceptable. Projects with high risk and low potential reward are unacceptable in responsible organizations.

With a vision, a clear strategy and a mapping of current and future project positions, a company is well set to develop and deliver the systems or software that will make it successful. However, to execute effectively and efficiently, the company needs to have in place a comprehensive quality system. The next section describes how to plan and implement such a system.

1.3 Elements of System and Software Quality

Strategic planning, without a corresponding approach to ensure quality in the products and services, can result in efforts that bear minimal fruit. An organization may have a great vision, but if its quality system is ineffective, it will have little success in achieving its goals. And, of course, a quality system depends on the definition of quality. In this section we discuss the elements of a quality program that help ensure that the organization will achieve its goals.

Development of a system or a software product should be supported by an appropriate quality plan to implement quality in processes for the development of systems and software. A *system/software quality program* (SQP) is the overall approach to influence and determine the level of quality achieved in a system or

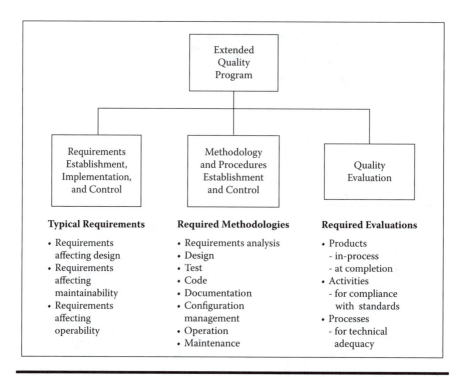

Figure 1.5 Elements of a systems quality program. Kenett, R.S. and Baker, E.R., *Software Process Quality: Management and Control*, New York: Marcel Dekker, Inc., 1999.

software product (see Figure 1.5). The concept of the SQP was first articulated in 1982 [2], and has recently been extended to include the development of systems [5]. It consists of the activities necessary to:

■ Establish and maintain control of requirements for the quality of a system or software product.
■ Establish, implement, and put into practice methodologies, processes, and procedures to develop, operate, and maintain the software or system.
■ Establish and implement methods, processes, and procedures to evaluate the quality of a system or software product and to evaluate associated documentation, processes, and activities that impact the quality of the product.

The foundation of an SQP is neither how well one can measure product quality nor the degree to which one can assess product quality. While these are essential activities, they alone will not attain the specified quality. Quality can only be built in during the development process. System quality cannot be tested, audited, evaluated, measured, or inspected into the product. This is a difficult concept to understand and implement in systems development organizations, whether these are organizations that develop systems for direct sales, develop systems for a third party under a contractual arrangement, or just develop systems for in-house use. Furthermore, it is not sufficient to pay attention to only the development aspects of the systems enterprise. Once the quality has been built in, the operating and maintenance processes must not degrade it. It is that understanding that lays the basis for the system and software quality program.

The foundation of the SQP stems from the definition of system quality. It is the concept that product quality means, in effect, achieving its intended end use, as defined by the user or customer. Is the system doing what it is supposed to do? In other words, does it correctly meet requirements that accurately capture what the user or customer wants? These requirements include the system's functional and performance requirements; requirements for maintainability, portability, interoperability; and so on. The significance of this concept is that product requirements are, in reality, the quality requirements. And they must *accurately* capture and reflect the way that the user/customer wants to use the system.

Figure 1.6 illustrates the interaction of the elements of the system quality program and how it affects system quality. The interaction of the SQP with the other parts of the project elements, as depicted in the figure, is necessarily complex. That involvement is at all levels of the project organization and takes place throughout every project's life. In some cases, the SQP directs activities; in other circumstances, it can only influence activities. In any case, all the project activities, in some way, affect system quality.

The SQP includes both technical and management activities. For example, the element of the SQP concerned with establishing and implementing methodologies for systems development is a management activity, while the specification of the

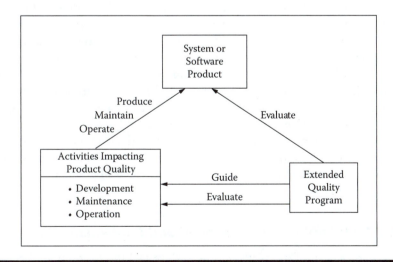

Figure 1.6   Interaction of the elements of system quality. Kenett, R.S. and Baker, E.R., *Software Process Quality: Management and Control*, New York: Marcel Dekker, Inc., 1999.

methodologies is a technical activity. We cover such elements in the next sections, starting with setting requirements establishment and control.

1.3.1 Requirements Establishment and Control

The first element of the SQP concerns the establishment of the requirements for the system to be produced on a project. These include the technical requirements as well as the requirements for its use in service. Technical requirements, as used here, refer to the functional, performance, security, safety, "ilities" (i.e., reliability, maintainability, etc., characteristics), and other related attributes that the system should possess. Requirements also include constraints imposed on the development of the system by considerations such as external standards (e.g., interface protocol standards) and regulatory agency regulations (e.g., those imposed by the Food and Drug Administration [FDA]).

As previously pointed out, the requirements for the system are, in fact, the requirements for the quality of the system. Consequently, the requirements must accurately reflect the functionality, performance, etc., that the customer or user expects to see in the system.

This activity includes not only defining the requirements, but also baselining (formalizing) them. This means that once the definition of the requirements is reasonably stable, they must be formally documented and placed under the control of a centralized authority. They should not be changed by any individual acting alone; rather, they should be changed only after a centralized function performs an impact assessment of the proposed change. To accomplish that, a formalized change control process must exist that assesses impact, feasibility, and assignment of resources to implement the change. Representatives of other systems or software that interface with the system for which the change is contemplated should participate in order to concur that (1) there is no impact on them, or (2) if there is an impact, that it is an acceptable one. However, merely defining and formalizing the requirements is insufficient. The agreed-to requirements must be implemented and the developers must not be allowed to ignore, unilaterally deviate from, or accept changes in requirements that have not been properly reviewed, analyzed, and agreed to by the change authority. Unilateral deviations assume that the developer is better qualified than the customer, the user, or the change authority to know what is needed or when to change the requirements. Such assumptions lead to systems that do not meet customer needs.

1.3.2 Methodology Establishment and Implementation

The second element of the SQP pertains to the establishment of, implementation of, and putting into practice the methodologies for the development, operation, and maintenance activities for the system. These are the processes by which the system will be developed, operated, and maintained. There is a very strong link between system quality and the processes used to develop it. If the processes in use by a systems development organization are not well defined or organized, the quality of those systems will not be predictable or repeatable from project to project. In contrast, consistency in the quality of the end product can be achieved by having well-defined and -organized processes. By having well-defined and -organized processes, and having the organization follow them as a matter of course, consistency in the quality of the end product can be achieved. These processes may not necessarily produce finely tuned products of the highest quality, but the quality will be consistent from project to project. Furthermore, by having stable, consistent processes, we can begin to acquire a quantitative understanding of how well the processes are performing. As will be discussed in Chapter 2, an organization at Maturity Level 3 on the CMMI® maturity rating scale is able to produce products of consistent quality, but will have little quantitative knowledge of how good its processes are performing. At the higher maturity levels, the organization knows quantitatively how well it is performing, and at Maturity Level 5 has enough quantitative knowledge on processes to assess quantitative benefits of focused improvements.

The dependency of system or systems quality on processes has been characterized by the Software Engineering Institute (SEI) at Carnegie Mellon University in the Capability Maturity Model IntegrationSM (CMMI®) [8]. Baker, Fisher, and Goethert [5] relate the elements of the SQP to the practices elaborated in the CMMI, and demonstrate how the concepts upon which the SQP are based are very consistent with the CMMI. The original Software Quality Program was formulated well before the CMMI.

1.3.3 Quality Evaluation

The third element pertains to those activities necessary to evaluate the systems development process and the resultant products. This element is sometimes referred to as the system or software quality evaluation (SQE) program. SQE is a set of assessment and measurement activities performed throughout systems development and maintenance to evaluate the quality of systems and to evaluate associated documentation, processes, and activities that impact the quality of the products. "Assessment" pertains to qualitative evaluation while "measurement" pertains to quantitative evaluation.

Measurement encompasses all quantitative evaluations—specifically, tests, demonstrations, metrics, and inspections. For these kinds of activities, direct measures can be recorded and compared against preestablished values to determine if the system meets the requirements. Accordingly, unit-level testing, systems integration, and systems performance testing can be considered measurement activities. Similarly, the output of a compare program or a path analyzer program for software can also be considered measurement.

Measurement also includes the use of metrics and measures that can be used to directly determine the attainment of numerical quality goals. Measures of reliability such as the number of faults per thousand lines of source code (faults/KSLOC) or faults per function point (faults/FP) are examples of such measures for software. Mean time to failure (MTTF) is a measure that can be used for systems, as well as for software.

Statistical analyses can also be used to determine the attainment of quality. The use of control charts, run charts, or process performance models, for example, are statistical and probabilistic procedures that can be used to quantitatively assess whether the required levels of quality are being maintained.

On the other hand, any evaluative undertaking that requires reasoning or subjective judgment to reach a conclusion as to whether the system meets requirements is considered an assessment. It includes analyses, audits, surveys, and both document and project reviews.

The "set of assessment and measurement activities" refers to the actions, such as reviews, evaluation, test, analysis, inspections, and so on, that are performed to determine (1) that technical requirements have been established; (2) that products and processes conform to these established technical requirements; and ultimately, (3) system quality. The focus of these activities varies as a function of the stage of the development effort. For instance, peer reviews conducted during the systems requirements analysis activities will focus on the adequacy of the evolving system requirements. On the other hand, peer reviews conducted during detailed design will focus on how well the design of a hardware component or a unit of software is implementing the requirements allocated to it.

These evaluations may be performed by a number of different organizations or functions, some or all of which may be within the project organization. Furthermore, evaluations may be performed by different organizations or functions. As an example, the evaluators of a systems requirements specification for a flight control system may include a flight controls engineer (to ensure that all the technical requirements have been implemented), a systems engineer (to ensure that the requirements, as specified, can be implemented in the system), a test engineer (to ensure testability), and QA (to ensure that the overall requirements for the content of the document, as well as its quality, have been addressed). The "set of assessment and measurement activities" to be performed are generally documented in project plans, systems development plans, project-specific quality procedures, and/or company quality plans and related quality procedures.

Determination of system quality can be performed by comparing the product and its components against pre-established criteria. However, evaluation of product quality may be difficult if the criteria have not been established quantitatively.

The evaluation program also includes assessment and measurement of the systems development and maintenance processes, and the activities/procedures comprising them. It may be that the process is being properly implemented but the products are not attaining the desired level of quality. (We discuss this aspect of quality evaluation in more detail in Chapter 4 where we discuss quantitative project management.) These evaluations constitute a check of the existing process against the initial analysis performed prior to the start of the development, that is, during the methodology selection process discussed above. A principle of quality management is that product quality can be improved through continuous improvement of the processes used to produce it. Continuous improvement is achieved by focusing on the processes and using product evaluations, as necessary, as indicators of process adequacy. This evaluation may be implemented by examining interim system products, such as initial specifications, materials presented at walk-throughs or inspections, or the artifacts (products) that result from the process itself, such as a document or drawing. Such measurements are aimed at evaluating the quality of the content of these products as the determinant of the process quality.

In general, the basis for determining which process to implement has been to look at a candidate process or set of processes and evaluate the proposed process on the basis of its track record. The assumption is that if the process has been shown to produce "high-quality systems" in the past, then proper implementation of the process will result in a high-quality product. This argument is somewhat misleading. In Chapter 3 we discuss the CMMI and levels of organizational process maturity that can be achieved by following a standard process. At this juncture, without a more detailed discussion of the CMMI, suffice it to say that organizations that achieve Maturity Level 3 have done so by institutionalizing practices that have worked well on previous projects. The determination that the practices work well is generally established qualitatively. It's not until an organization reaches Maturity Level 4 that it begins to have a quantitative knowledge of the efficacy of its processes. Even then, there are limitations on the ability to quantitatively relate process maturity to delivered product quality.

There has been anecdotal evidence of the relationship, but no conclusive demonstration of a definite link between the processes selected for the system's development and the resultant quality of the systems themselves [12]. Establishment of such links has been more heuristic or intuitive rather than analytical. This uncertainty was proposed by Jeffrey Voas in an article in *IEEE Software* [20]. His article deals with the relationship between process and software quality. Voas points out that, intuitively, using a structured process will produce better-quality software than using ad hoc approaches; however, a formal quantitative relationship between process maturity (i.e., a CMMI maturity level rating) and quality of delivered systems or software has never been established. Voas likens the situation to a belief that clean pipes will always deliver clean water. However, this observation does not take into account that dirty water can still be injected into the water supply, despite the fact that the pipes are clean. Furthermore, the article points out that there is no guarantee that even the most superb development processes will produce a product that meets customer needs. Gibson, Goldenson, and Kost [12] provide data relating process and product quality, from case studies in organizations that used different approaches to analyze their results. However, the data is anecdotal in nature and only relates individual development organization results to maturity level. Industrywide ranges of results are provided but not as a function of maturity level. Comparison is difficult for the various measures because they were measured differently from organization to organization.

An effective process has a much better chance of producing quality systems. To prove this quantitatively is, however, nontrivial. To achieve this we need quantitative knowledge of the effectiveness of the subprocesses that comprise an organization's system development process. An effective measurement program that includes statistical knowledge of the performance of critical subprocesses is integral to obtaining that knowledge. We discuss this point in more detail later in this chapter and in subsequent chapters.

1.3.4 Interactions among System Quality Program Elements

The process of defining and establishing the requirements and controlling changes to them involves interfaces with the other two elements of the SQP: (1) establishment of methodologies and (2) quality evaluation. Two kinds of interfaces with the establishment of methodologies exist. One is the specification of the preferred methodology or methodologies for performing requirements analysis. To ensure consistency in the quality of the requirements and uniformity in the way that they are documented, the methodologies must be institutionalized. That is, there must be a common way of performing requirements analysis within each project—one that is well understood and practiced by all the affected developers. For example, a project's implementation of use cases to define requirements results from the establishment of it as the preferred methodology for performing requirements analysis.

The second interface has to do with baselining requirements and controlling changes to them. This process, or methodology, is known as a "configuration management process." This is a management methodology implemented to:

■ Prevent uncontrolled changes to baselined items
■ Improve the likelihood that the development effort will result in quality systems (i.e., systems that meet the end users' needs) and that system maintenance will not degrade it

The discipline of configuration management assists project management in ensuring that the requirements will be managed and controlled.

The interface between the requirements-focused element of the SQP and the systems quality evaluation element concerns evaluation of the requirements. They must be correct, complete, understandable, testable, adequate, consistent, and feasible (among other features). They must also correctly capture the user's needs. Total compliance with requirements does not guarantee quality software. If the user's needs have not been properly captured, errors exist in the requirements. Compliance with requirements will then produce a system that does not satisfy the intended end use. Clearly, requirements must be evaluated for adequacy with respect to all aspects of use, performance, and functionality while they evolve and develop.

Similarly, interfaces exist between the second element of the framework (i.e., methodology establishment and implementation) and the other two elements. The processes and methodologies that are established, in addition to focusing on correctly capturing requirements, must also focus on establishing and implementing processes and methodologies that translate those requirements into a finished product that meets the user's needs. This element must also focus on establishing and implementing methodologies to suitably perform the quality evaluation function. The interface with the quality evaluation element is that the methodologies that are being implemented must be evaluated on two scores: (1) Is the organization implementing the established processes and methodologies, and (2) are the methodologies the appropriate ones for their intended use? An additional interface with the quality evaluation element exists in that the interim and end products resulting from the implementation of these processes and methodologies must be evaluated to determine if the required quality attributes have been achieved.

The interfaces between the quality evaluation element and the other two elements have already been described in the discussions of the interfaces that exist with the other two elements.

1.4 Deployment of Strategic Quality Goals

How does an organization determine if it is on the road to achieving its strategic objectives? The organization needs a measurement program to understand past performance and current status, and to predict the ability to achieve these objectives. In this section we discuss the relationship of business goals to measures and metrics, and also a methodology for defining the essential measures. In later chapters we discuss the details of implementing the measurement program that will provide the measures and metrics to determine if the strategic business, quality, and informational needs of the organization are being met.

The GQM (Goal/Question/Metric) method is a goal-oriented approach that can be used to manage the whole measurement process. Its common application is in the field of systems process and product measurement (see [6, 18] and http://ivs.cs.uni-magdeburg.de/sw-eng/us/java/GQM/). GQM aims at evaluating the achievement of goals by making them measurable. Therefore, metrics are needed that become evident through asking the questions that are necessary to verify the goal. You start off by making up a list of goals that should be evaluated and ask the relevant questions. Then, collect metrics that have been identified as answering the questions used for evaluating the achievement of the goal. The GQM approach can be used as a systematic way to tailor and integrate process measurement's objectives into measurement goals and refine them into measurable values (see Figure 1.7).

Carried out through observation, interviews with users, and/or workshops, this process is iterative, systematic, and has been proven able to give rapid identification

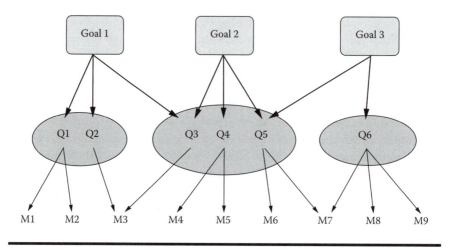

Figure 1.7   The Goal/Question/Metric paradigm.

of the structure for establishing a measurement program supportive of determining the satisfaction of the enterprise's business goals, as well as supportive of its process improvement effort. It creates a foundation of repeatable procedures for single projects or even for an entire organization. The stages of the GQM process are goal identification, measurement planning, measurement performance and validation, and analysis and interpretation of the results. It derives the metrics to be used from a thorough analysis of goals and associated questions to be answered quantitatively. Following GQM guidelines, it is possible to:

- *Establish* the goals of the measurement process.
- *Find* a proper set of questions to be able to reach the goals.
- *Choose* the right set of metrics in order to answer the questions.
- *Plan* and *execute* the data collection phase.
- *Evaluate* the results.

The meaning of the GQM main components can be described as follows:

- The *Goal* describes the measurement's purpose; stating explicit goals gives the measurement program a precise and clear context.
- The set of *Questions* refines the goal and highlights the quality focus. What should I know to be able to reach a certain goal?
- The set of *Metrics* is used to answer each question. Metric data may result from objective or subjective measurement.

Von Solingen [21] provides a detailed example of how the methodology is applied, utilizing an organization's desire to achieve a specific level of reliability in the delivered software it is producing. The methodology was originally developed for use by the software industry, but clearly can also be applied to hardware, systems, and services. Applying this methodology for all the business goals, organizational informational needs, project informational needs, customer needs, and process improvement goals can greatly facilitate the development of a measurement program that can satisfy the needs of the organization, projects, and customers.

It is also important to recognize that the capability to apply the GQM approach effectively can be negatively impacted by a number of factors. One factor is organizational process maturity as determined from an appraisal against the CMMI. Low-maturity organizations are not likely to have collected measures and metrics robust enough to do analyses that are relatively sophisticated. Another factor is ambiguity in the definitions of the measures. Poorly defined measures and metrics will lead to misunderstandings of the data being collected. These and other factors are discussed by Baker and Hantos [3] in their refinement of the GQM approach, called GQM-Rx.

To illustrate how the methodology is used, let us consider a situation common in the software industry. Assume that a software development organization wants to improve the quality of the software it produces, as measured at delivery, by 5%. Several questions related to that goal are as follows:

1. What is the quality currently at the conclusion of the acceptance test?
2. What are the contributions to quality degradation introduced in the development process?
3. What are the contributions to loss of quality introduced by supporting processes, such as QA and configuration management?
4. Which factor contributes most to the loss of quality during development?

Table 1.1 illustrates how we go from goal to questions to metrics.

We return to these measures in Chapter 4 when we discuss process performance models.

1.5 Strategic Maps and Balanced Scorecards

Measurement of organizational performance relates to the impact of products and/or services produced by an organization on the environment in which the organization operates. The environment provides the organization with essential inputs that allow its functioning and absorbs its outputs. The extent to which an organization's inputs, processes, and outputs are in harmony with the environment determines its sustainability. For long-term sustainability, an organization needs to achieve two simultaneous outcomes: satisfied stakeholders (owners, employees, customers, and community) on one hand, and its own economic health on the other hand.

Organizational performance can be measured at three levels: strategic, tactical, and operational. In the framework of this chapter, we take a detailed look at each of them.

Table 1.1   Goal/Question/Metric Method Application Example

Goal	Question	Metric
Improve the quality of the software at delivery by 5%	What is the quality currently at the conclusion of acceptance test?	Average defects per thousand lines of code (KSLOC) at the conclusion of acceptance testing
		Variance of the defects per KSLOC at the conclusion of acceptance testing
		Estimated residual defects per KSLOC
	What are the contributions to loss of quality introduced by the development process?	Average requirements defects per estimated KSLOC introduced during requirements analysis
		Variance of the requirements defects per estimated KSLOC introduced during requirements analysis
		Average design defects per estimated KSLOC introduced during design
		Variance of the design defects per estimated KSLOC introduced during design
		Average defects in the code per KSLOC introduced during coding (as determined from peer reviews and unit testing)

Table 1.1 Goal/Question/Metric Method Application Example (*Continued*)

Goal	Question	Metric
		Variance of the defects in the code per KSLOC introduced during coding (as determined from peer reviews and unit testing)
		[Note: Similar measures would exist for the integration and system testing activities]
	What are the contributions to loss of quality introduced by supporting processes, such as QA and configuration management?	Average of the noncompliances detected per hour of QA audit
		Variance of the noncompliances detected per hour of QA audit
		Average of the noncompliances detected per baseline configuration audit
		Variance of the noncompliances detected per baseline configuration audit
	Which factor contributes the most to the loss of quality during development?	Regression analysis of all the factors associated with questions 2 and 3

To manage organizational performance successfully, managers should have at least elementary knowledge of the following four fields: systems and optimization theory, statistical theory, managerial techniques and procedures, and psychology, which is in line with Deming's *Concept of Profound Knowledge* [9]. However, given the specific nature of their work and the level at which organizational performance is measured, managers should strive to deepen their understanding of individual fields in order to improve their own performance and the performance of their employees.

Three different types of measures are needed to devise an approach to the measurement of organizational performance that can be used at strategic, tactical, and operational levels within an organization:

1. Indicators of past performance (lag indicators)
2. Indicators of current performance (real-time indicators)
3. Indicators of future performance (lead indicators)

Typical *lag indicators* of business success are profitability, sales, shareholder value, customer satisfaction, product portfolio, product and/or service quality, brand associations, employee performance, etc. The most frequently used lag indicators are traditional accounting indicators. Unfortunately, they do not allow a confident prediction of future success, but tend to be suitable only for control after the event (troubleshooting instead of managing improvement). Using the quarterly financial statement indicators to manage improvement is, to Dransfield et al. [10], like steering a car along the road by looking in the rear-view mirror.

Typical *current indicators* are helpful in determining the current status of a project. They include measures such as the earned value measures. Cost performance index (CPI), for example, indicates the current status of funds expended. It measures budgeted cost of work performed against actual cost of work performed. A ratio less than one indicates that the project may be overrunning its budget. The schedule performance index (SPI), as another example, indicates the current schedule status. It compares budgeted cost of work performed to budgeted cost of work scheduled. A ratio less than one indicates that the work may be behind schedule.

Significant *lead indicators* are the result of:

■ Customer analysis (segments, motivations, unmet needs)
■ Competitor analysis (identity, strategic groups, performance, image, objectives, strategies, culture, cost, structure, strengths, and weaknesses)
■ Market analysis (size, projected growth, entry barriers, cost structure, distribution systems, trends, key success factors)
■ Environmental analysis (technological, governmental, economic, cultural, demographic, technological, etc.)

As pointed out by Aaker [1], these lead indicators are all elements external to the organization. Their analysis should be purposeful, focusing on the identification of opportunities, threats, trends, uncertainties, and choices. The danger of becoming excessively descriptive should be recognized and avoided. Readily available industry measures (such as production, sales and stock indices, as well as employment indicators) regularly published by national statistical offices and other government bodies should also be included and interpreted in this framework.

Monitoring of lag and lead indicators is crucial at the strategic and tactical level while *real-time (process) indicators* are important at the operational level.

To translate vision and strategy into objectives and measurable goals, Kaplan and Norton [15] created the balanced scorecard. It is meant to help managers keep their finger on the pulse of the business. Each organization will emphasize different measures depending on their strategy. Management is, in effect, translating its strategy into objectives that can be measured.

The balanced scorecard helps organizations identify and track a number of financial and nonfinancial measures to provide a broader view of the business. Data analysis is not limited to accounting data. A company may decide to use indicators of process efficiency, safety, customer satisfaction, or employee morale. These may capture information about current performance and indicate future success. The aim is to produce a set of measures matched to the business so that performance can be monitored and evaluated.

Leading variables are future performance indicators, and lagging variables are historic results. We already know that financial measurements are typically lagging variables, telling managers how well they have done. On the other hand, an example of a leading indicator is training cost, which influences customer satisfaction and repeat business. Some variables exhibit both lagging and leading characteristics, such as on-time deliveries, which is a lagging measure of operational performance and a leading indicator of customer satisfaction.

The balanced scorecard usually has four broad categories, such as financial performance, customers, internal processes, and learning and growth. Typically, each category will include two to five measures. If the business strategy is to increase market share and reduce operating costs, the measures may include market share and cost per unit.

Another business may choose financial indicators that focus on price and margin, willingly foregoing market share for a higher-priced niche product. These measures should be set after the strategy is in place. A list of objectives, targets, measurements, and initiatives comes with each variable. The saying "we manage what we measure" holds true. One reason the balanced scorecard works is that it raises awareness.

Although the balanced scorecard is typically used in the private sector, examples of its application in the public sector are also known. For example, the public-sector management of Charlotte (North Carolina) has been using this framework to manage the city. Their scorecard includes performance indicators that track improvements in safety of community, quality of transportation, economic development, and effectiveness of government restructuring. Other more local measures of fire prevention, garbage collection, and sidewalk maintenance are used for individual departmental scorecards. This illustrates the flexibility of the scorecard and reemphasizes the point that the scorecard measures can be determined only after the overall strategy is accepted and understood. The balanced scorecard can also be used very effectively in conjunction with GQM. Baker and Hantos illustrated the method in a tutorial at the *2003 Software Technology Conference* [4]. For each of the four categories, a goal was established. For example, for the category of internal processes, the stated goal was, "Projects achieve predictable results for schedule within the known range of the organization's defined process capability." The strategic measure core outcomes were stated to be:

- Products' overall development cycle times are on par with or better than industry segment's benchmark.
- Delivery slip by teams in the phase-gates is less than or equal to the corporate benchmark.

The questions spawned by these outcomes were the following:

- What activities are included in the cycle time?
- What constitutes slip?
- How do we aggregate team/project data into program data?
- How do we measure slip in the phase-gates and how do we aggregate it to the full cycle?

The presentation then went on to show the cause-and-effect relationships between the four categories, showing how satisfying the strategic measurement core outcomes for learning and growth impacts the internal processes, which in turn impacts the customer, which affects the financials.

The balanced scorecard, if properly implemented, is an excellent management framework that can help managers track different factors that influence performance. Linking various components of a balanced scorecard, with cause-and-effect relationships, produces a *strategic map* (Figure 1.8).

1.6 Summary

The driving force behind any program or initiative implemented by an organization is the strategic objectives of that organization. Obviously, the organization must first clearly articulate the objectives. A number of models are available to facilitate doing this, as we have discussed in this chapter. An effective quality program goes hand in hand with defining, deploying, and implementing the strategic objectives of the organization. Without such a program, the quality of the products produced or the services provided will be unpredictable, making it extremely difficult to achieve the strategic objectives of the organization. A measurement program is essential for assessing past performance, current status, and making predictions critical for the future business of the organization and the determination of quality when the products or services are delivered. In this chapter we discussed two techniques—GQM and the balanced scorecard—that facilitate defining the necessary measures to determine the achievement of goals and objectives, and ensuring that they measure critical organizational performance.

Clearly, the measurement of development and maintenance processes proposed and in use by the organization are crucial elements of any measurement program.

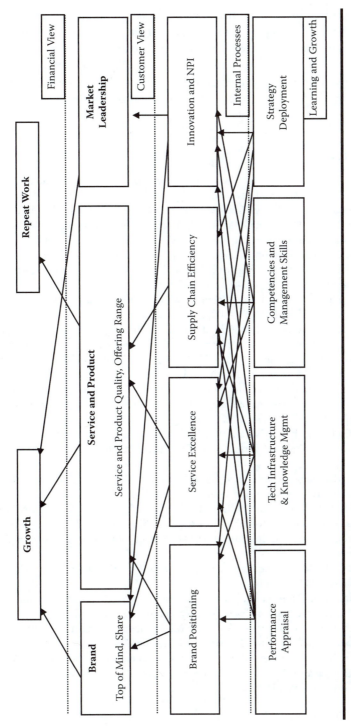

Figure 1.8 A sample strategic map.

In this chapter we discussed elements of a measurement program and measurement methods that can be employed to determine if the organization's business goals are being met. The development and maintenance processes implemented by the organization must be closely tied to the achievement of those goals and objectives; consequently, the measurement program must look at these processes to determine how well they are working in the service of business goal achievement.

References

1. Aaker D.A., *Strategic Market Management*. New York: John Wiley & Sons, 2001.
2. Baker, E. R. and Fisher, M.J., A Software Quality Framework, in *Concepts—The Journal of Defense Systems Acquisition Management*, Robert Wayne Moore, Ed.; 5(4), Autumn 1982.
3. Baker, E.R. and Hantos, P. GQM-R$_X$—A Prescription to Prevent Metrics Headaches, *Fifth Multi-Conference on Systemics, Cybernetics and Informatics,* Orlando, FL, July 22–25, 2001.
4. Baker, E.R. and Hantos, P. Implementing a Successful Metrics Program: Using the GQM-RX Concept to Navigate the Metrics Minefield, *Software Technology Conference,* Salt Lake City, UT, 28 April–1 May 2003.
5. Baker, E.R., Fisher, M. J., and Goethert, W., *Basic Principles and Concepts for Achieving Quality*, CMU/SEI-2007-TN-002, December 2007.
6. Basili, V. and Weiss, D., A Methodology for Collecting Valid Software Engineering Data, *IEEE Transactions on Software Engineering*, November 1984, pp. 728–738.
7. Burgelman, R., Maidique, M., and Wheelwright, S., *Strategic Management of Technology and Innovation, 2nd edition,* New York: Irwin, 1998.
8. Chrissis, M., Konrad, M., and Shrum, S., *CMMI Second Edition: Guidelines for Process Integration and Product Improvement*, Upper Saddle River, NJ; Addison-Wesley, 2007.
9. Deming, W.E., *Out of the Crisis*, Cambridge, MA: MIT Press, 1983.
10. Dranfield, S.B., Fisher, N., and Vogel, N., Using Statistics and Statistical Thinking to Improve Organisational Performance, *International Statistical Review*, 67(2), 99–150, 1999.
11. Galvin, R., *Encyclopedia of Statistics in Quality and Reliability*, F. Ruggeri, R. Kenett, and F. Faltin, Editors in Chief, New York: John Wiley & Sons, 2007.
12. Gibson, D.L., Goldenson, D.R., and Kost, K., Performance Results of CMMI-Based Process Improvement, Technical Report, CMU/SEI-2006-TR-004, 18 August 2006.
13. Godfrey, A.B., Buried Treasures and Other Benefits of Quality, *The Juran Report*, No. 9, Summer 1988.
14. Juran, J.M., *Juran on Leadership for Quality: An Executive Handbook,* New York: The Free Press, 1989.
15. Kaplan, R. and Norton, D., Using the Balanced Scorecard as a Strategic Management System, *Harvard Business Review*, January–February 1996, p. 75–85
16. Kenett, R.S. and Koenig S., A Process Management Approach to Software Quality Assurance, *Quality Progress*, November 1988, p. 66–70.
17. Kenett, R.S., Managing a Continuous Improvement of the Software Development Process, *Proceedings of the Annual Conference on Quality Improvement, IMPRO 89.* Juran Institute, Inc., 1989.

18. Kenett, R.S. and Baker, E.R., *Software Process Quality: Management and Control*, New York: Marcel Dekker, Inc., 1999.
19. Segev, E., *Corporate Strategy Portfolio Models*, London: Thomson Publishing, 1995
20. Voas, J., Can Clean Pipes Produce Dirty Water?, *IEEE Software*, July/August 1997, p. 93–95.
21. Von Solingen, R., The Goal Question Metric Approach, in the *Encyclopaedia of Software Engineering, 2nd ed.*, Marciniak, John J., Ed., New York: John Wiley & Sons, 2002.

Chapter 2

Basic Elements of Continuous Process Improvement

Synopsis

This chapter is a general introduction to continuous process improvement. It discusses the organizational infrastructures required to sustain process improvement, dwells on definitions of measures and basic elements of measurement programs, and covers modern Six Sigma and Enterprise Knowledge Development (EKD) methodologies. Later chapters focus on specific applications to system and software development, including high maturity practices of the Capability Maturity Model Integration (CMMI). Chapter 10 concludes with a case study integrating several tools and techniques presented in the book, and demonstrates how process improvement is achieved in a Scrum agile development environment using EKD organizational patterns for knowledge management and Six Sigma for process improvement.

2.1 Continuous Improvement Infrastructures

Modern industrial organizations, in manufacturing and services, are subjected to increased competitive pressures and rising customer expectations. Management teams, on all five continents, are striving to satisfy and delight their customers while

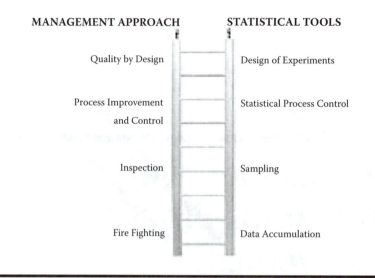

Figure 2.1   The Quality Ladder.

simultaneously improving efficiencies and cutting costs. In tackling this complex management challenge, an increasing number of organizations have proved that the apparent conflict between high productivity and high quality can be resolved through improvements in work processes and quality of designs. This achievement is provided by increased maturity in management style. Different approaches to the management of organizations have been summarized and classified using a four-step Quality Ladder [11]. The four management approaches are (1) Fire Fighting, (2) Inspection, (3) Process Improvement and Control, and (4) Quality by Design. Parallel to the management approach, the Quality Ladder lists quantitative techniques that match the management sophistication level (Figure 2.1).

Managers applying reactive firefighting can gain from basic statistical thinking. The challenge is to get these managers to see the value of evolving their organization from a state of data accumulation to data analysis and proactive actions, turning numbers into information and knowledge (see [7]). Managers who attempt to contain quality and inefficiency problems through inspection and 100% control can increase efficiency using sampling techniques. Their approach is more proactive than firefighting but the focus on end products, post factum, can be very expensive. Sampling inspection can reduce these costs, provided proper statistical analysis is used, in order to establish sample sizes for inspections. The decision of what to test, when and where, should be assessed statistically so that the performance of the approach is known and adequate. More proactive managers, who invest in process control and process improvement, can take full advantage of control charts and process control procedures. Process improvements affect how things are done, thereby affecting both cost and quality in a positive way. At the top of the Quality Ladder is the Quality by Design approach where up-front investments are secured to run experiments designed to optimize product and process performance. At that level of management sophistication, robust experimental designs are run, for instance, on simulation platforms, reliability engineering is performed routinely, and reliability estimates are compared with field returns data to monitor the actual performance of products and improve the organization's predictive capability [11–13]. Efficient implementation of statistical methods requires a proper match between the management approach and implemented statistical tools. More generally, the statistical efficiency conjecture discussed by Kenett et al. [13] states that organizations increasing the management sophistication of their management system, moving from Fire Fighting to Quality by Design, enjoy increased benefits and significant improvements with higher returns on investments.

The goal of moving up the Quality Ladder is pursued by management teams in different industries, including electronic systems design, mechanical parts manufacturing, system assembly, software-based services, and chemical processes. A particular industry where such initiatives are driven by regulators and industrial best practices is the pharmaceutical industry. In August 2002, the Food and Drug Administration (FDA) launched the pharmaceutical current Good Manufacturing Practices (cGMP) for the twenty-first century initiative. In that announcement, the FDA explained the agency's intent to integrate quality systems and risk management approaches into existing quality programs with the goal of encouraging industry to adopt modern and innovative manufacturing technologies. The cGMP initiative was spurred by the fact that since 1978, when the last major revision of the cGMP regulations was published, there have been many advances in design and manufacturing technologies and in the understanding of quality systems. This initiative created several international guidance documents that operationalized this new vision of ensuring product quality through "a harmonized pharmaceutical quality system applicable across the life cycle of the product emphasizing an integrated approach to quality risk management and science." This new approach is encouraging the implementation of Quality by Design and hence, de facto, encouraging the pharmaceutical industry to move up the Quality Ladder [15]. The Capability Maturity Model Integration (CMMI) presented in Chapter 3 invokes the same principles for the software and systems development industry.

Quality, as practiced by most system and software organizations, focuses on problem detection and correction. Problem correction, while essential, only serves to remove the defects that have been embedded in the product by development and production processes. When properly organized for continuous improvement and quality management, organizations focus on problem prevention in order to improve the quality of the product and improve their competitive position. Continuous improvement, as a companywide management strategy, was introduced only in the 1950s. For thousands of years, improvement and innovation were slow. New scientific breakthroughs often occurred by chance, with an "intelligent observer" being in the right place at the right time. As significant events occurred, someone asked why and, after some experimentation, began to understand the cause and effect uncovered by that observation. The discovery of x-rays by

W. C. Roentgen is a classic example. Champagne was also discovered that way. By the late 1800s, an approach to centrally plan innovation and improvement began to appear. Thomas Edison built his laboratory in 1887 and conducted thousands of experiments. Throughout the twentieth century, several industrial, academic, and government laboratories, often employing tens of thousands of researchers, were established. The art and science of experimental design became widely used to drive improvements in products and processes, and in developing entirely new products and services. In the latter half of the twentieth century, another phenomenon took place, first in Japan and then quickly in other parts of the world. Large numbers of employees in organizations were taught the basics of the scientific method and were given a set of tools to make improvements in their part of the company. They were empowered to introduce changes in processes and products in order to achieve improved production and product performance. Six Sigma and Lean Sigma are examples of this approach. Similarly, as we will see in Chapter 4, the use of statistical and probabilistic methods, such as those associated with the CMMI high maturity practices, enable organizations to quantitatively assess and predict the benefits of process improvements and process and technological innovations. They also permit quantitative assessment of actions proposed for the elimination of recurring problems. These practices are particularly successful when coupled with Six Sigma methods. Continuous improvement in software and systems development builds on this experience.

Problem prevention has two aspects: (1) preventing the recurrence of existing problems and (2) preventing the introduction of new problems. That is, problem prevention results in quality improvement. Quality improvement can be considered to be of two types: (1) reactive (driven by problems) and (2) proactive (driven by the desire to improve quality and efficiencies). Reactive quality improvement is the process of understanding a specific quality defect, fixing the product, and identifying and eliminating the root cause to prevent recurrence. Proactive quality improvement is the continual cycle of identifying opportunities and implementing changes throughout the product realization process, which results in fundamental improvements in the level of process efficiency and product quality. It is a continual process sometimes primed by problem correction initiatives. A reactive causal analysis of a quality defect will sometimes trigger a change in a process, resulting in a proactive quality improvement that reduces defect levels. In Chapter 4 we discuss this approach as part of the CMMI high maturity practices.

Systematic and large-scale companywide quality improvement initiatives necessitate competencies at the organizational, team, and personal levels. The first organizational competency is the implementation of a Quality Council. Other names used for such forums are Continuous Improvement Committee, Quality Board, or Steering Committee. Sometimes, the role of the Quality Council is split between top-level management and middle management. In such cases, top management assumes the role of a strategic steering committee, whereas middle management is responsible for overseeing the details of the process improvement activities.

Specifically, in system and software development and maintenance organizations, it is common to find System or Software Engineering Process Groups (SEPGs) or Software Engineering Committees that operate in the capacity of a Quality Council. The management steering committee sets process improvement goals, provides budgets and other resources, and is the final approval authority for the proposed process improvements. In many cases the SEPG or Software Engineering Committee acts as an administrative arm of process improvement. Overall, the Quality Council is responsible for:

- Defining and helping implement the improvement process
- Initiating improvements
- Supporting the improvement process
- Keeping the process going and tracking progress
- Disseminating information
- Gathering and analyzing data for project prioritization

Specifically, the Quality Council actions consist of:

- Establishing a project system by choosing quality improvement projects, appointing project teams, and soliciting and screening nominations for quality improvement projects
- Setting responsibilities for carrying out quality improvement projects by defining team charters and appointing facilitators, team members, and team leaders
- Identifying training needs, planning the training, and identifying the trainees
- Establishing support for quality improvement project teams
- Providing for coordination of process improvement efforts
- Establishing measures for progress on improvement
- Designing a plan for publicity and recognition of achievements

One might consider all the above activities "strategic." At the "tactical level," the organizational entities that drive the improvements are the Quality Improvement Teams or Process Action Teams. These are ad hoc teams that are commissioned by and are responsible to the Quality Council for the implementation of specific process improvement initiatives. The teams are responsible for:

- Initiating improvements
- Diagnosing problems
- Determining causes and solutions
- Estimating the benefits for specific process improvements and securing support
- Addressing people issues
- Implementing the change
- Establishing methods to maintain the level of improvement
- Tracking progress and evaluating results

Organizational structures and entities in and of themselves are not sufficient. Other factors must be considered in order to accomplish successful process improvement. Chief among these are the commitment and involvement of management. Without that, virtually nothing is possible. Lip service on the part of management will be readily recognized; consequently, there will be no motivation for the organization to implement any changes in the way they do business. "Talking the talk" without "walking the walk" will be recognized as empty words. Management should exhibit the following behavioral characteristics to ensure that the organization understands that management is sincerely committed to process improvement. In particular:

- Commitment must be genuine and self-evident.
- Guidance, support, and training through close involvement must be provided.
- Teams must be kept focused on business objectives.
- A relationship of trust and respect between all members must be established.
- All recommendations and contributions must be acknowledged.

Another element involved in structuring process improvement is the role of the champion. Typically there are no formal organizational structures defining the role of the champion, but it is an important role. The champion is the person who promotes a change in the process. In organizations that are just starting the journey of process improvement, he/she may be the person out beating the drums to implement a more formal process discipline in the organization. In organizations where process improvement is successfully implemented, everyone in the organization is empowered to be a champion; that is, anyone in the organization can propose a process improvement initiative, and everyone is encouraged to do so. The proposed changes are submitted to the SEPG or Quality Council for consideration and action.

Joseph M. Juran is probably the individual who has contributed most to establishing proper structures for quality improvement (see [6–10]). Juran's Quality Trilogy of Plan, Control, and Improve is possibly the most simple, complete, and pure representation of managing for quality ever devised. Juran found it useful to teach about managing quality using an analogy that all managers easily understand: managing budgets. Financial management is carried out by three managerial processes: budget planning, financial control, and cost reduction. Managing quality makes use of the same three fundamental processes of planning, control, and improvement. The trilogy exemplifies the essence of Quality with a capital Q:

- *Quality Planning*: The process for designing products, services, and processes to meet new breakthrough goals.
- *Quality Control*: The process for meeting goals during operations.
- *Quality Improvement:* The process for creating breakthroughs to unprecedented levels of performance (see Figure 2.2).

While Juran's Trilogy is simple and accurate, it is important to understand that the elements represented are infinite and layered. To properly implement the trilogy

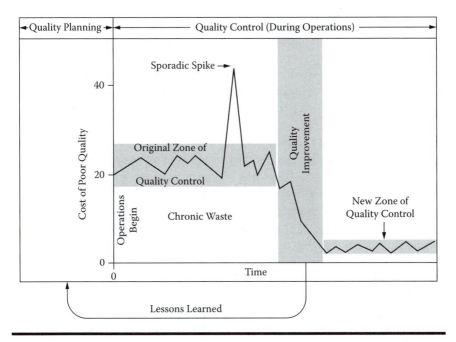

Figure 2.2 The Juran Trilogy. Adapted from The Quality Trilogy: A Universal Approach to Managing for Quality. Juran, J. M. *Proceedings of the ASQC 40th Annual Quality Congress,* **Anaheim, California, May 20, 1986.**

in products, services, and processes, one must understand that the trilogy is three dimensional and limitless. For example, the Quality Planning of a new market offering will consist of Quality Planning for products, services, processes, suppliers, distribution partners, etc., with impact on the delivery of the new offering. On another layer, this phase must also plan for the support and maintainability of the offering. On yet another layer, the planning phase must account for design and integration of data collection, control, improvement processes, people, and technologies. Finally, the planning phase must design the evaluation of the planning phase itself. In this simple example, we see four layers of Quality Planning. This phase will typically be iterated and improved with every cycle and within each cycle. The same is true of all other phases [21].

2.2 Process Improvement Projects and Measurement Programs

As described in the previous section, since the 1960s, improvement of quality and efficiency within many organizations has been tackled by focused improvement projects. A project-by-project improvement strategy relies on successful employee participation in quality improvement project identification, analysis, and implementation. Many factors influence a project's success; they often are overlooked and, in many cases, may be unknown. A project-by-project quality improvement program must be supported by quality principles, analysis techniques, effective leaders and facilitators, and extensive training. While quality improvement teams can vary by size (even during the duration of the project), a typical team size is five to seven people [2, 6, 8, 9].

The proper selection and formulation of a problem are critical for effective quality improvement. The link between problem/project selection and team member selection must be direct and relevant. The regular management team, a Quality Council, or the team itself can select the problem or project. Typically, management will identify the more strategic problems with clear business justification. Bottom-up approaches to project selection are very adequate when handling ground fruit or low-hanging fruit. Whatever the origin of the project proposal, an organization set up to handle improvement projects systematically gains competitive advantage.

While the end result of a quality improvement project remains the primary concern from a management perspective, time is critical for the success and support of quality improvement teams. The typical time frame for project completion is 3 to 5 months, which is about 32 man-hours per team member, 2 hours per week for 16 weeks (see [2, 6, 8, 9]). Team factors are also critical to success. Managers should be involved in the project selection process and the outcome of the project. The team size and the length of the project are both important to ensure that group dynamics and motivation are effectively employed. A structured improvement plan is essential, including a list of steps to be followed, with clear definitions, completion marks, and the tools to be used, many of them statistical.

Typical areas of improvement include defect reduction, performance to standard, cost reduction, and customer satisfaction improvement. Some teams develop quality measures by using basic flowcharting, work simplification, and data collection tools. Teams should be able to compare data collected before and after they have solved the problem and implemented their solution. This enables teams to track improvements and demonstrate the return on investment (ROI) of the improvement project.

Surveys are effective in tracking customer satisfaction before and after project solution implementation. With surveys, teams can assess the impact of improvements using customer satisfaction ratings. For more information on customer surveys, see [14]. If you include survey data, there is practically no aspect of the enterprise where measurements cannot be made available. Lord Kelvin stated that "when you can measure what you are speaking about and express it in numbers, you know something about it; but when you cannot measure it, when you cannot express it in numbers, your knowledge is of a meager and unsatisfactory kind: it may be the beginnings of knowledge but you have scarcely in your thoughts advanced to the stage of Science" [19]. This was greatly simplified by Tom DeMarco as, "you cannot control what you cannot measure" [5].

Why do development projects fail? A common reason is that the development organization has not established a proper process infrastructure to ensure that the process will endure. When there are no consistent, orderly processes for the development and maintenance efforts, there are also no meaningful, consistent measures of the process. Consequently, *the processes cannot be managed in an effective manner, and the quality of the products produced or developed becomes unpredictable.*

Measurement is an integral part of a process management strategy. Chapter 3 discusses how the SEI's Capability Maturity Model Integration (CMMI) and other process models such as the ISO standards emphasize the role of process as determinants of product quality. In this chapter we focus on the measurement of system and software development and maintenance processes. Without such a measurement program, our knowledge of the process is superficial at best, and so is our ability to know how well our processes are performing with regard to producing quality products and systems. Process measurement is introduced below in the context of organizational objectives and management strategies. One or more management strategies can be used in an organization. However, for process measurement to be a relevant activity, the net result should be a focus on increasing the quality and responsiveness of the organizational process.

Measurement ties several disciplines together, creating an environment where inputs, processes, and outputs are controlled, and improved. In fact, the need for process measurement was made an integral part of the Capability Maturity Model (CMM) for software [16] when this process improvement model was first introduced. Measurement and Analysis was a practice included in every key process area (KPA) in the CMM. It was "elevated" from a practice that was placed in every KPA in the CMM to a process area that is put into effect at Maturity Level 2 in the staged representation of the CMMI [3]. In the CMMI, it is considered a basic foundational building block for achieving process improvement.

During the planning phase of a new project, measurement information from prior projects may be used to estimate the parameters of the new project. During the execution of the project, process and resource measures can be collected and analyzed against such estimates.

Quantitative and statistical methods can be used to control critical processes when corrective action needs to be applied. As subsystems and systems are created, product measures are collected and analyzed, yielding product metrics. The number of open problems, per subsystem, is an example of a product metric.

The answer to the question *"Why measure?"* involves several other questions, such as:

■ Are the processes more productive now than they were during the last reporting period? Less productive? Why?

- Are the processes delivering higher-quality products now than they were during the last reporting period? Lower quality? Why?
- Are our customers more satisfied with our products and services now than they were during the last reporting period? Less satisfied? Why?
- Are the processes more reliable now than they were during the last reporting period? Less reliable? Why?
- Do the processes cost more now than they did during the last reporting period? Cost less? Why?

Note that if one inserts any other business process or service, the questions would still apply to organizations in general. System and software processes should be treated like any other business processes. Authors such as W. Edwards Deming, J. M. Juran, and Phil Crosby have made process modeling, measurements, and the subsequent improvements, key items in improving business performance. When we apply the same continuous improvement methodology to our system and software processes, the need to *measure* those processes becomes critical.

The definitions for "measure" and "metric" are not consistent throughout literature and industry publications. Often the terms "measure," "indicator," and "metric" are used synonymously. The CMMI [3] defines two other terms: "base measure" and "derived measure," where a base measure is essentially an item of raw data that is an attribute of a process or a product, and a derived measure is the result of mathematically manipulating two or more base measures. In this section we define the terms for the mathematical "yardsticks" of process and product quality evaluation that we will use in this book. Although in Chapter 3 we focus on the CMMI as a process improvement model, we define and use the terms "measure" and "metric" that are of common usage in industry as our "yardsticks."

- *Measure:* As a noun, a measure is defined as a number that assigns values on a scale. In terms of an organization's measurement program such as we describe in this chapter, it is an attribute of a product, process, or service. Examples may include number of errors, lines of code, or work effort. It is essentially the same thing as the base measure described in the CMMI. As a verb, measure means to ascertain or appraise by comparing to a standard.
- *Metric:* There is no single universal definition of metrics. In the context of this book, a metric is a combination of two or more measures or attributes. Examples include (1) fault density for evaluating software components, (2) Flesch readability index for evaluating written documents, and (3) man-months for tracking project resources. It is essentially the same thing as the derived measure described in the CMMI [3].

In evaluating a metric, we can consider several characteristics. In particular, we will want the metric to be:

■ *Understandable:* It must relate to some quality or characteristics that the practitioner and stakeholder find meaningful and simple to understand. It must be clearly defined so that there are no ambiguities with respect to its meaning.

  – *Field tested:* It must have a proven record by other industrial organizations or proven on a local pilot effort.

  – *Economical:* It should be easily extracted from existing products and processes.

  – *High leverage:* It should help identify alternatives that have high impact on cost, schedule, and quality.

  – *Timely:* It should be available within a suitable time frame to meet the objectives of the measurement. Monthly labor expenditure reports will have little benefit for a project that has a development cycle of 3 months.

  – *Operational definition:* The measure should be defined in such a way that it is clear to the users what is being measured, what values are included and excluded, and what the units of measure are; the measurement is repeatable, given the same definition, and the sources of the data are clearly identified in terms of time, activity, product, reporting mechanism, status, and measurement tools used. The importance of this was demonstrated by the failure of the Mars Polar Orbiter in 1999 [20]. One team calculated everything in English units but the team receiving the data expected everything in metric units. Clear, unambiguous operational definitions could have prevented this from happening.

■ *Measurement:* Measurement is defined as the activity of computing and reporting metrics values. Comprehensive measurement activities lead to operational control that include:

  – Setting of standards or goals for relevant metrics
  – Computation of relevant metrics in a timely manner
  – Comparison of reported metrics value with appropriate standards or goals
  – Evaluation of differences to determine required action

■ Periodic review of standards, metrics definition, list of required actions

2.3 Six Sigma (DMAIC and DFSS Methodology)

The Six Sigma business improvement strategy was introduced by Motorola in the 1980s and adopted successfully by General Electric and other large corporations in the 1990s. The key focus of all Six Sigma programs is to optimize overall business results by balancing cost, quality, features, and availability considerations for products and their production into a best business strategy. Six Sigma programs combine the application of statistical and nonstatistical methods to achieve overall business improvements. In that sense, Six Sigma is a more strategic and more aggressive initiative than simple improvement projects. One particularly eloquent testimonial of the impact of Six Sigma is provided by Robert W. Galvin, previous Chairman of the Executive Committee, Motorola, Inc. In the foreword to the book *Modern Industrial Statistics: Design and Control of Quality and Reliability* by Kenett and Zacks [11], Galvin states:

> At Motorola we use statistical methods daily throughout all of our disciplines to synthesize an abundance of data to derive concrete actions. ... How has the use of statistical methods within Motorola Six Sigma initiative, across disciplines, contributed to our growth? Over the past decade we have reduced in-process defects by over 300 fold, which has resulted in a cumulative manufacturing cost savings of over 11 billion dollars.

An employee of an organization who participates in a Six Sigma team is referred to as a Green Belt. A Green Belt will have sufficient knowledge to support and champion Six Sigma implementation and to participate in Six Sigma projects as team leader or team member. A managerial level or technical specialist assigned full responsibility to implement Six Sigma throughout the business unit is referred to as a Black Belt. This term, as well as Six Sigma, Green Belts, etc., were coined by Motorola and adopted by other organizations as a de facto industry standard. A Black Belt is a Six Sigma implementation expert. Each project is expected to have at least one Black Belt as a team member. Many companies have experienced the benefits of Six Sigma with, on average, savings between $ 50,000 and $250,000 per project. Black Belts, with 100% of their time allocated to projects, can execute five or six projects during a 12-month period, potentially adding over $1 million to annual profits.

Six Sigma improvement projects are orchestrated from the executive-level Quality Council and consist of a Define–Measure–Analyze–Improve–Control (DMAIC) roadmap. DMAIC can be likened to a life cycle where in the Define phase, a particular process problem is identified and the solution is planned, Measure is where the relevant data is collected and the extent of the problem is determined, Analyze is where the cause-and-effect relationships are determined, Improve is where the plan to improve the process is put into effect, and Control is where the change is piloted to determine if organizational business goals are being achieved and the change is monitored after deployment to ensure that process improvement goals are being achieved. Typically, executives support Six Sigma as a business strategy when there are demonstrated bottom-line benefits.

The term "Measure" as a phase in the Six Sigma DMAIC strategy can be misleading. Within Six Sigma training, the Measure phase encompasses more than just measurements. It typically includes the tracking of key process output variables (KPOVs) over time, quantifying the process capability of these variables, gaining insight into improvement opportunities through the application of cause-and-effect diagrams and failure mode and effects analysis (FMEA), and quantifying the effectiveness of current measurement systems. These activities help define how a KPOV is performing, and how it could relate to its primary upstream causes, the key process input variables (KPIVs). The Measure step is repeated at successive business, operations, and process levels of the organization, measuring baseline information for KPOVs in order to provide an accurate picture of the overall variation that the customer sees.

Measurements provide estimates of the process capability relative to specifications and drive investigations of all significant sources of variation, including measurement system analyses, which provide us with insight into the accuracy of measurements. This information is then used throughout the Analyze, Improve, and Control phases to help make overall business improvements.

Several unique Six Sigma metrics, such as sigma quality level, yield, and cost of poor quality (COPQ), have been developed to help quantify and reduce the "hidden factory," that is, hidden, non-added-value production costs. Six Sigma relies heavily on the ability to access information. When the most important information is hard to find, access, or understand, the result is extra effort that increases both the hidden factory and the COPQ [10].

Many organizations set a dollar threshold as they begin to prioritize Six Sigma projects. Successful projects thereby provide returns that will pay for up-front investments in Six Sigma training and full-time Black Belt team leaders. The definition of what types of savings are considered is also critical and drives specific behaviors to the identification of the low-hanging fruit first. In many organizations, soft or indirect benefits fall into categories such as cost avoidance related to regulatory or legal compliance, or benefits related to improving employee morale or efficiency. Such benefits cannot be tied directly to operating margins or incremental revenues through a specific measurement.

Why is it called Six Sigma? *Sigma* is a letter in the Greek alphabet used to denote the standard deviation of a process. A standard deviation measures the variation or amount of spread about the process mean. A process with "Six Sigma" capability exhibits twelve standard deviations between the upper and lower specification limits. Essentially, under Six Sigma, process variation is reduced so that no more than 3.4 parts per million fall outside the specification limits. The higher the sigma level, the more assurance we have in the process producing defect-free outcomes.

Six Sigma roadmaps typically consist of a 6-month intensive program combining training and the implementation of projects (http://www.kpa.co.il/english/six_sigma/index.php). The training of Black Belts typically consists of 16 full-time equivalent training days with projects that usually span 6 months. Following completion of the projects and the training program, the project leaders get certified as Black Belts. A typical small- or medium-sized development organization will identify two or three Black Belt candidates and launch two or three Six Sigma projects defined after an initial one-day training of the management team. Projects are conducted in teams led by a Black Belt. A Master Black Belt mentors the various projects and provides the necessary training.

To graduate as a certified Black Belt, the candidates must demonstrate knowledge in the tools and methodologies of the Six Sigma Body of Knowledge, in addition to completing a Six Sigma project. As part of Six Sigma Black Belt training, the Master Black Belt reviews, critiques, and advises the Black Belt candidates throughout their training. A typical Six Sigma roadmap covers the following topics in a mixture of frontal and hands-on training and implementation steps as outlined below:

Month 1, DEFINE:	Six Sigma Basics
	Elicitation of potential Six Sigma projects
	Screening of Six Sigma projects and Black Belt candidates
	Defining and launching of Six Sigma project
Month 2, MEASURE:	Process baselining
	VOC and QFD
	Seven basic tools
	Process flow mapping
	Data collection and analysis
	Defect metrics
	Cycle-time assessment
	Benchmarking
Month 3, ANALYZE:	Project reviews
	Process capability analysis
	Measurement system evaluation
	FMEA
	Systems thinking
	Statistical thinking
	Control charts
	10X metrics/capability
	Statistical inference
	Decision and risk analysis
	Project management
	Financial impact assessment

Month 4, IMPROVE:	Project reviews
	Regression modeling
	Design of experiments
	Tolerancing
	Variance components
Month 5, IMPROVE:	Project reviews
	Robust design
	Pugh concepts
	DFA/DFM
	Lean manufacturing
Month 6, CONTROL:	Project reviews
	Evaluation of results
	Post project assessment
	Lessons learned
	Where else?

Design for Six Sigma (DFSS) is a Six Sigma quality process focused on the Quality Planning of the Juran Trilogy. The DMAIC methodology requires that a process be in place and functioning. In contrast, DFSS has the objective of determining the needs of customers and the business, and driving those needs into the product solution created. Unlike the DMAIC methodology aimed at improving an existing product or process, the DFSS approach is used during development of new products or services. The DFSS model is based on the five-step DMADV cycle: Define, Measure, Analyze, Design, and Verify. DMADV is applied to new processes or products whose goal is to achieve Six Sigma quality levels. We now review these basic steps:

- Define: Defining project goals and customer demands (both external and internal).
- Measure: Measuring and quantifying the expectations and requirements of customers.
- Analyze: Analyzing the options of the processes required to meet the needs of the customer.
- Design: Detailed designing of processes to meet the needs of the customers.
- Verify: Verification of the designed performance and the ability to meet the needs of the customer.

DFSS is especially relevant to integrated systems with significant system development. DFSS seeks to avoid manufacturing/service process problems by using systems engineering techniques of problem avoidance and robust designs at the outset (i.e., fire prevention). These techniques include tools and processes to predict, model, and simulate the product delivery system as well as the analysis of the system life cycle itself to ensure customer satisfaction from the proposed system design solution. In this way, DFSS is closely related to systems engineering, operations research, systems architecting, and concurrent engineering. DFSS is largely a design activity requiring specialized tools, including quality function deployment (QFD), axiomatic design, TRIZ, Design for X, Design of Experiments (DoE), Taguchi methods, tolerance design, Robustification, and response surface methodology (see [11]). While these tools are sometimes used in the classic DMAIC Six Sigma process, they are uniquely used by DFSS to analyze new and unprecedented systems/products. DMAIC improvements and DMADV designs result in changes in processes, activities, definition of roles and responsibilities, and interactions between different parts of the organization. The next section presents a comprehensive methodology for mapping processes and managing interventions in how things are done.

2.4 Enterprise Knowledge Development (EKD)

As described above, improvement requires change. At the organizational level, change is taking an organization from point A to point B along a specific path with various levels of planning and control. To plan that path, or change scenario, a mapping of the organization has to occur to represent the present, the "as-is" model, and the future, the "should-be" model. In mapping an organization, a language and tools are required. In this section we present such a tool, labeled EKD. Enterprise Knowledge Development (EKD) was developed in the FP4 ELEKTRA project supported by the European Commission as a business process management methodology for large organizations [22]. It provides a controlled way of analyzing and documenting an enterprise, its objectives and support systems. The approach represents an integrated set of techniques and associated tools for the purpose of dealing with situations of mapping business processes and re-engineering of information systems. At the center of the approach is the development of enterprise knowledge models pertaining to the situations being examined [17, 18]. The definition of such models is carried out using appropriate modeling techniques and tools. The process followed in developing these models, and subsequently using them, complements Six Sigma, Lean Sigma, and Design For Six Sigma initiatives. During the process of developing these models, the participant parties engage in tasks that involve deliberation and reasoning. The objective is to provide a clear, unambiguous picture of how enterprise processes function currently in an "as-is" model or in a modified "should-be" model.

Change is derived by the implementation of five key activities:

1. *Introduce:* What are the new actions that need to be added?
2. *Improve:* What are the actions that already exist that need to be improved?
3. *Extend:* What are the actions that already exist and that need to be extended?
4. *Cease:* What are the actions that currently exist and that need to be ceased?
5. *Maintain:* What are the actions that currently exist and that need to be maintained?

The EKD approach considers the concept of an "enterprise process" as a composite of four key enterprise components: (1) the roles played by enterprise actors in order to meet the process goals; (2) the activities involved in each role; (3) the objects involved, together with their evolution from creation to extinction (within the context of the enterprise process); and (4) the rules that determine the process components. In other words, an enterprise process may transcend any functional divisions and presents a dynamic view of an enterprise system. Systems are composed of interfacing or interdependent parts that work together to perform a useful function. System parts can be any combination of things, including people, information, software, equipment, products, or raw materials. In effect, operational models should describe what a system does, what controls it, what it works on, what it uses to perform its functions, and what it produces.

Actor–role modeling is about representing the organizational and behavioral aspects of an enterprise. This aspect of modeling is concerned with the way that a business process is performed through the involvement of enterprise actors in discharging their responsibilities via their role in a process and the interaction of their role with other roles that collectively bring about the realization of the business processes. Through enterprise actor–role modeling, EKD encourages identification of key operational components that can be measured (activity duration, actor skills, resource costing, etc.). Such measurable components can then be subjected to "what-if" scenarios in order to evaluate alternative designs for the operation of an enterprise.

A high-level view of the association between actors and their different roles is supported through the *actor–role diagram* (ARD). The actor–role diagram depicts the actors of the enterprise and the roles they play. For each role involved in the model, information is given about the responsibilities assigned to the role in terms of the activities that the role carries out and the enterprise resources that the role requires. The diagram also presents the dependencies that exist between the roles. An additional element represented in this view is the goal (or goals) that the role must satisfy. The ARD can be used to get a "first-cut" view of the organizational aspects regarding the responsibilities of individuals or groups in their involvement in the operation of a business process.

Figure 2.3   The role box and its goals.

A detailed view of the activities in which a role is engaged is supported by the *role–activity diagram* (RAD). This diagram describes in detail how the role performs each of its responsibilities, in terms of activities undertaken, and is based on the notation of the RAD approach. An important point to note is the distinction between the *actor* (i.e., the physical enterprise entity) and the *role*, a notion that expresses the responsibility of performing various activities within the enterprise. Roles are assigned to actors and summarize a set of skills or capabilities necessary to fulfill a task or activity. A role can be acted by a person or a group. A role can be acted by person X on one day and person Y on another day. The role is separate from the actors that play the role. For example, a managing director may play multiple roles, such as "setting the budget," "approving expenses," etc. A role is a collection of components of either operational or structural nature expressing responsibilities. Operational components represent the activities that the role must perform. Structural components represent resource objects that are required by one or more activities being carried out by the role. These may be physical or informational. The role can thus comprise behavioral aspects of the organizational life as well as hierarchical and social aspects. The notation for a role and its goals is shown in Figure 2.3.

2.4.1 EKD Actor and Role Components

2.4.1.1 The Actor and the Actor's Role

It is essential to distinguish between the role and the *carrier* of the role, that is, the actor who at a specific moment might play the role. The role exists independently of what organizational entity is chosen to play it; this particular choice can be changed over time, as particular actors may change in carrying the responsibilities of a role. A role should therefore be considered a stable and accountable concept, one that summarizes a set of responsibilities and the activities that are performed in order to fulfill these responsibilities. An actor is the physical entity that personifies an instance of the role at any given moment; an actor may play several roles at the same time. In that sense, while the selection of roles and their responsibilities can be considered a design choice, the selection of actors for playing individual roles can be considered an implementation strategy. In EKD, the notation for the actor concept is the actor box (Figure 2.4).

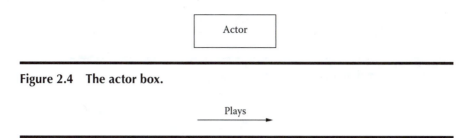

Figure 2.4   The actor box.

Figure 2.5   The actor-plays-role arrow.

2.4.1.2 The Actor–Role Relationship

The relationship between a role and the actor who, at a particular time, incarnates that role is represented by the plays relationship, which is shown as an arrow in Figure 2.5. The arrow connects an actor box with a role box and illustrates that the actor plays the particular role.

2.4.1.3 The Role–Role Relationship

Roles often depend on other roles in order to perform the duties assigned to them. There are two parties involved in the dependency: the requester role (i.e., the one that needs something in order to fulfill its responsibilities) and the provider role (i.e., the one that can provide the missing component). This relation can be of various types:

- *Authorization dependency:* This denotes hierarchical dependencies that can exist between roles; the provider role gives authorization to the requester role. For instance, in order to perform changes to a customer's accounting data, an authorization of the responsible manager must be given.
- *Goal dependency:* This relationship reflects the fact that the achievement of a goal that the role brings about depends on the achievement of a goal of another role. For instance, the goal of the customer service provision to satisfy customers promptly depends on satisfaction of the goal of immediate availability of programmers expressed by the human resource full-time equivalent (FTE) staff numbers. Generation of artifacts or deliverables is also considered a goal.
- *Coordination dependency:* This type of dependency expresses the need for one role to wait for completion of another role's responsibilities before it can complete its own. For example, in order for the material provision service to purchase material inputs appropriate for a purchase order must first be received from engineering.
- *Resource dependency:* This illustrates the need for one role to use a resource that can be provided by another role. For instance, the construction service requires material that is under the supervision of the warehousing service.

Figure 2.6 The dependency arrow.

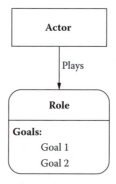

Figure 2.7   A sample actor–role relationship.

The dependency relationships are represented by the dependency arrow, as depicted in Figure 2.6. The dependency arrow is accompanied by the initial of the dependency type and superscripted by any additional details on the dependency. The initials of the dependency types are **A** for authorization dependency, **R** for resource dependency, and **C** for activity coordination dependency (the "ARC" dependency). The dependency arrow is directed from the provider role toward the requester role.

2.4.1.4 Using the Notation

The role box consists of the name of the role and its components, which are drawn within its boundary. A sample role box and the actor that plays the role are presented in Figure 2.7.

For the dependency relations, we draw the dependency arrow between the two role boxes. A sample actor–role diagram comprising two roles, their respective actors, and a coordination dependency is illustrated in Figure 2.8.

2.4.1.5 Guidelines for the Actor–Role Diagram

1. An actor cannot exist without playing any role; that is, at any moment, every actor within the organization should be assigned at least one role.
2. If more than one dependency is identified between roles, multiple arrows are drawn, each annotated by the respective dependency type and details.

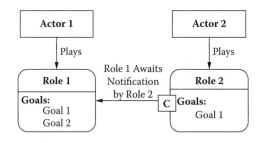

Figure 2.8   A sample actor–role diagram.

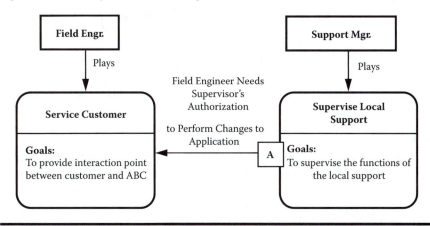

Figure 2.9   An example actor–role diagram with "authorization" dependencies.

2.4.1.6 Examples of Actor–Role Diagrams

A number of examples in this section demonstrate the concepts (and the use of the notation) for actor–role modeling. The actor–role diagram of Figure 2.9 represents a situation of two actors, each playing a different role with different goals, and the dependency of the "service customer" role on the "supervise local support" role. The former requires the latter in order to perform certain activities. This kind of dependency shows the structural relationship between the two roles as being one of authorization (indicated by the "A" on the dependency line).

2.4.1.7 EKD Role–Activity Diagram

Basic concepts and notation include:

■ *Role:* The concept of a role is also used in the role–activity diagram. In this case, however, it is used to group together the details of the activities being carried out within the enterprise, including information on their order of execution and their interdependencies.

Figure 2.10   Activity sequence components.

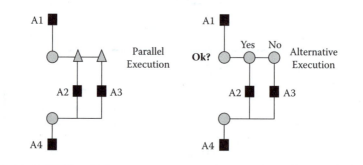

Figure 2.11   Part refinement and case refinement.

- *Activity sequence:* The activities that a role performs are represented as sequences of nodes. Nodes are drawn as black squares, which are connected with a straight, vertical line. These primitives are illustrated in Figure 2.10.
- *Activity refinement:* Apart from sequential execution of activities, we can also have parallel execution of more than one activity. Moreover, selection can be made between two or more activities, according to whether or not a condition is satisfied. These controls over activities are called part refinement and case refinement, respectively, and they are illustrated in Figure 2.11. In part refinement, two or more parallel activity sequences are initiated with triangles, while in case refinement the alternative activity sequences are initiated with circles, which also indicate the value of the criterion that determines which of the alternatives is executed. With respect to Figure 2.11: in the left part, activities A2 and A3 are executed in parallel after activity A1; in the right part, one of the two activities is selected according to the answer to the question posed.
- *Role interaction:* At some point during an activity sequence, a role may need to interact with another role. This means that both roles will be involved in activities that make their respective actors communicate in various ways. The notation for this interaction is represented as a horizontal line connecting the two activities across the boundaries of the roles, as shown in Figure 2.12.

    The level of detail of the interaction is not specified, nor is the duration of the interaction. Depending on the case, an interaction can be decomposed into smaller, more detailed activities and interactions, or it can be represented as a single interaction. This choice, however, is left to the model designer and is related to the importance of the specific interaction in the overall model.

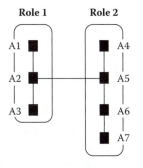

Figure 2.12   Role interaction.

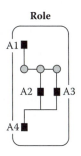

Figure 2.13   A role–activity diagram.

- *Using the notation:* The activities of which a role consists are refined in this level. The role box now contains a sequence of activities, possibly refined according to one of the refinement schemes previously presented, as illustrated in Figure 2.13. Roles therefore have goals achieved by activating specific activities.
- *Guidelines for the role–activity diagram:* Dependencies identified at the actor–role level model are represented as interactions at the role–activity level; indeed, interdependencies between roles are translated into specific activities at this level, which constitute the interface between the roles.

2.5 Organizational Patterns

The software development community is a useful source of examples of pattern use, in particular by those advocating and practicing object-oriented approaches and reuse. What these efforts have in common lies in their attempt to exploit knowledge about best practice in some domain. Best-practice knowledge is constructed in "patterns" that are subsequently used as the starting point in the programming, design, or analysis endeavors. Patterns, therefore, are not invented but rather they are discovered within a particular domain with the purpose of being useful in many similar situations. A pattern is useful if it can be used in different contexts.

2.5.1 The Notion of Pattern

Alexander defines a pattern as describing "a problem which occurs over and over again in our environment and then describes the core of the solution to that problem, in such a way that you can use this solution a million times over, without ever doing it the same way twice"[1]. Here, the emphasis is put on the fact that a pattern describes a recurrent problem and it is defined with its associate core solution. According to Alexander, what repeats itself is a fabric of relationships. For example, when a statistical consultant is first approached by a customer and gets a first look at some data, or a detailed description of it, the statistical consultant is initiating a basic investigation to understand the context of the problem. This example represents a structural pattern that is repeatable in many different settings—for example, in a troubleshooting assignment in manufacturing, or a market research study. Aligned with this structural pattern, there is a pattern of events that is also repeatable; in our example, the basic investigation preceding the statistical analysis takes place time and time again within a company. It is important to note that a pattern relates a problem to a solution.

2.5.2 Pattern Template

A pattern is more than just a description of some thing in the world. A pattern should also be a "rule" about when and how to create that thing. Therefore, a pattern should be made explicit and precise so that it can be used time and time again. A pattern is explicit and precise if:

- It defines the problem (e.g., "We want to improve yield in a manufacturing process."), together with the forces that influence the problem and that must be resolved (e.g., "Managers have no sense for data variability," "Collaboration of production personnel must be achieved," etc.). Forces refer to any goals and constraints (synergistic or conflicting) that characterize the problem.
- It defines a concrete solution (e.g., "How basic problem investigations should be done"). The solution represents a resolution of all the forces characterizing the problem.
- It defines its context (e.g., "The pattern makes sense in a situation that involves the initial interaction between the statistical consultant and his customer."). A context refers to a recurring set of situations in which the pattern applies.

A pattern should be visualizable and should be identifiable, so that it can be interpreted equally well by all who might share the pattern. In this sense, "visualization" may take the form of "statements in natural language," "drawings," "conceptual models," and so on. In the literature there are many different proposals for the description of those desirable properties [1, 4]. One general example is the pattern template presented below:

Name:	Should be short and as descriptive as possible.
Examples:	One or several diagrams/drawings illustrating the use of the pattern.
Context:	Focuses on the situation where the pattern is applicable.
Problem:	A description of the major forces/benefits of the pattern as well as its applicability constraints.
Solution:	Details the way to solve the problem; it is composed of static relationships as well as dynamic ones describing how to construct an artifact according to the pattern. Variants are often proposed along with specific guidelines for adjusting the solution with regard to special circumstances. Sometimes, the solution requires the use of other patterns.

The structure of the pattern includes information describing the pattern, examples of use along with information describing the relationships that the pattern has with other patterns, and so on.

There exists a growing collection of documented patterns, even if these as yet mainly consist of software patterns. Sites available on the Internet with example software development patterns include:

- http://hillside.net/patterns/
- http://www.cmcrossroads.com/bradapp/docs/patterns-intro.html
- http://c2.com/cgi-bin/wiki?JimCoplien

The Process Patterns Resource Page is:

- http://www.ambysoft.com/processPatternsPage.html

For Organizational Patterns, see:

- http://www.bell-labs.com/cgi-user/OrgPatterns/OrgPatterns?Organizational Patterns
- http://people.dsv.su.se/~js/ekp/ekp.html
- http://crinfo.univ-paris1.fr/EKD-CMMRoadMap/index.html

For the Antipatterns Repository, see:

- http://c2.com/cgi/wiki?AntiPatterns/

Anti-patterns are defined as telling you how to go from a problem to a bad solution, telling you how to go from a bad solution to a good solution (or something that looks like a good idea), but which backfires badly when applied. Recognizing "bad" business practice may provide knowledge or impetus for identifying and describing the relevant good practice.

To make the EKD knowledge easy to organize and access for the benefit of the organization, one needs to systematize and structure the knowledge and experience gained in different parts of the company. The knowledge engineer's task is to provide a framework where potential patterns can be compared in a systematic way according to commonly accepted criteria, so as to enable a satisfactorily informed decision.

A commonly recurring problem is the trade-off that is often made between having "easily organized knowledge," which is inherently incomplete so as to be structured, as compared to having knowledge that reflects reality, which is often not easily structured or even understood. The choice then becomes where on a knowledge continuum we wish to place ourselves. Between highly structured and unrealistic knowledge at one end and unstructured but realistic knowledge at the other, we tend to be somewhere in the half of the continuum toward the unstructured. In practical terms this means that when defining patterns, it is more important that these reflect real problems and solutions rather than flashy and technically brilliant presentations. Knowledge, expressed by generic patterns, should facilitate the creativity process by reducing the need to "reinvent the wheel" when facing new problems and situations. The essence of the use of patterns is that they are applied to recurring problems. A pattern is of no use if it aims to solve a problem that is extremely unlikely to occur within the foreseeable future for those businesses that are envisaged to have access to the patterns.

To enhance reuse, patterns need to be evaluated and assessed periodically. Table 2.1 presents a set of criteria that can be used to classify a pattern on a High–Medium–Low scale. The criteria focus on usefulness, quality, and cost. Obviously each organization should develop its own criteria, in line with its strategy and organizational culture.

In using patterns, we advocate an approach to describing repeatable solutions to recognizable problems. In this context, both the problem and the solution must be uniquely identifiable and accessible. The pattern usage framework must therefore make the distinction between product or artifact patterns and process patterns, and includes an indexing schema for accessing them. The pattern's typology aims to distinguish between the way to solve a problem and the elements that will be used for the solution, while the indexing hierarchy characterizes each pattern by the problem that it addresses through the usage perspective and the knowledge perspective. The template presented above represents the usage perspective, and the EKD modeling presents the knowledge perspective. To enable reuse and pattern retrieved, a signature is required in either a formal or informal format. An example of an electronic patterns repository based on EKD is available at http://crinfo.univ-paris1.fr/EKD-CMMRoadMap/index.html.

Table 2.1 Pattern Classification Criteria

Criteria	Sub-criteria	High Value	Medium Value	Low Value
Usefulness	**Degree of triviality** The degree to which the pattern addresses a problem which is of little importance because the problem or solution is obvious.	The pattern is concerned with issues that are or most likely will be of concern to other parts of the company.	While the pattern deals with a pertinent problem, the solution is already well known.	The pattern is concerned with a problem which does warrant the creation of a pattern since it is so trivial with the proposed solution being obvious to domain experts.
	Grade of implementability Extent that pattern is thought to be practical and implementable. Is change compatible with business strategy? Have trade-offs been taken into account?	The pattern is useful in that it prescribes practical, easy to understand and implement solutions.	The pattern may be of some use despite some practical problems in implementation and some difficulty in understanding the solution.	The pattern is not usable. The solution is impractical and difficult to understand. The pattern only proposes "paper-based" change rather than real change
	Degree of confidentiality	The pattern does not disclose any confidential business information.	Some information may be able to be used by other projects.	The pattern discloses sensitive project information.

Continued

Table 2.1 Pattern Classification Criteria (Continued)

Criteria	Sub-criteria	High Value	Medium Value	Low Value
Quality	**Degree of complexity** The number of factors and the their relationships.	The pattern addresses only a few manageable main concepts and ideas.	The pattern is complex but may still be useful in that the complexity is needed.	The large number of factors that affect the implementation of the solution minimizes the chances that the solution can be implemented.
	Addition of Value The local and global benefits accruing to the business with the implementation.	The consequences of a successful implementation are great value to the project directly affected as well as other projects	The local and global benefits are unclear, difficult to determine or marginal.	There are no local or global benefits or there is a conflict between these so that in total no value is added
	Level of genericity Abstraction level of the problem that the pattern addresses.	The pattern addresses a problem that is general enough for all the company.	The pattern addresses a problem that applies only to part of the company.	The pattern addresses a problem that is only relevant to the project in which it was discovered.
	Grade of understandability Visualizable and identifiable stakeholders.	The pattern is easy for decision makers, domain experts and those to be affected, to comprehend.	The pattern is only partially understandable to decision makers, domain experts and those to be affected.	The pattern is incomprehensible to stakeholders.

External compatibility The extent to which the pattern could be used by other companies.	The pattern has taken into account differences in national, and organizational cultures and ways of working amongst identified future external users of the pattern.	The pattern has partially takes into account differences in national, and organizational cultures and ways of working amongst identified future external users of the pattern.	The pattern does not take into account differences in national, and organizational cultures and ways of working amongst identified future external users of the pattern.
Cost **Level of experience in their use.**	The pattern has been implemented within the company.	The pattern has been partially or sporadically used.	The pattern has never been implemented.
Economic feasibility of the proposed solutions.	Proposed solution is relatively easy to implement. Organizational support exists in terms of sufficient resources as well as managerial support. The solution is politically and socially acceptable.	Proposed solution is difficult but feasible to implement. Organizational support is lukewarm. Resources are available but may not be sufficient. There may exist political and social difficulties in making the pattern feasible.	Proposed solution is not feasible. Organizational support will be difficult to obtain. The resources will not be made available. The existing difficult social and/or political climate would make an implementation impossible.

Chapter 9 of this book applies EKD to describe how an organization implements and improves by implementing agile development methods. The application of EKD is, however, generic and relevant to any process improvement activity.

2.6 Summary

In this chapter we introduced concepts related to establishing an infrastructure for continuous process improvement. We discussed organizational infrastructures required to sustain process improvement, such as establishing an SEPG and a management steering committee, dwelled on definitions of measures and basic elements of measurement programs, and covered modern Six Sigma and Enterprise Knowledge Development (EKD) methodologies. We discussed measurement programs and their importance with respect to the implementation of process improvement. In general, we presented generic methods for implementing process improvement. In later chapters we will discuss these concepts in more detail, showing how these concepts are actually implemented in facilitating process improvement.

References

1. Alexander, C., *The Timeless Way of Building*, New York: Oxford University Press, 1979.
2. Aubrey, C.A. and Felkins, P.K., *Teamwork: Involving People in Quality and Productivity Improvement*, Milwaukee, WI: Quality Press, 1988.
3. CMMI Product Team. CMMI for Development, Version 1.2 (CMU/SEI-2006-TR-008), Pittsburgh, PA: Software Engineering Institute, Carnegie Mellon University, 2006.
4. Coplien, J., Borland Software Craftsmanship: A New Look at Process, Quality and Productivity, *Proceedings of the 5th Annual Borland International Conference,* Orlando, FL, 1994.
5. DeMarco, Tom, *Controlling Software Projects,* New York: Yourdon Press, 1982.
6. Godfrey, A.B. and Kenett, R.S., Joseph M. Juran, A Perspective on Past Contributions and Future Impact, *European Network for Business and Industrial Statistics (ENBIS) Sixth Annual Conference on Business and Industrial Statistics*, Wroclaw, Poland, 2006.
7. Hoerl, R. and Snee, R.D., *Statistical Thinking—Improving Business Performance*, Pacific Grove, CA: Duxbury, 2002.
8. Juran, J.M., The Quality Trilogy: A Universal Approach to Managing for Quality, *Proceedings of the ASQC 40th Annual Quality Congress*, Anaheim, California, May 20, 1986.
9. Juran, J.M. and Godfrey, A.B., *Juran's Quality Control Handbook*, New York: McGraw-Hill Book Co., 1999.
10. Juran, J.M., *Juran on Leadership for Quality—An Executive Handbook,* New York: Free Press, 1989.

11. Kenett, R.S. and Zacks, S., *Modern Industrial Statistics: Design and Control of Quality and Reliability*, San Francisco, CA: Duxbury Press, 1998. (2nd edition, 2003; Chinese edition, 2004.)

12. Kenett, R., Practical Statistical Efficiency, in *Encyclopedia of Statistics in Quality and Reliability*, Ruggeri, F., Kenett, R.S., and Faltin, F., Editors in Chief, New York: Wiley, 2007.

13. Kenett, R.S., De Frenne, A., Tort-Martorell, X., and McCollin, C., The Statistical Efficiency Conjecture, in *Applying Statistical Methods in Business and Industry—The State of the Art*, Coleman, S., Greenfield, T., Stewardson, D. and Montgomery, D., Editors, New York: Wiley, 2008.

14. Kenett, R.S., On the planning and Design of Sample Surveys, *Journal of Applied Statistics*, 33(4), 405–415, 2006.

15. Nasr, M., Quality by Design (QbD)—A Modern System Approach to Pharmaceutical Development and Manufacturing–FDA Perspective, *FDA Quality Initiatives Workshop*, February 28, 2007, North Bethesda, MD, 2007.

16. Paulk, M.C., Weber, C.V., Curtis, B., and Chrissis, M.B., *The Capability Maturity Model: Guidelines for Improving the Software Process*, Reading, MA: Addison-Wesley, 1995.

17. Rolland, C., Nurcan, S., and Grosz, G., 1998. A Unified Framework for Modelling Co-Operative Design Processes and Co-Operative Business Processes, in *Proceedings of the 31st Annual International Conference on System Sciences*, Big Island, HI.

18. Rolland, C., Nurcan, S., and Grosz G., Enterprise Knowledge Development: The Process View, *Information and Management Journal*, 36(3), 165–184, 1999.

19. Thomson, Sir William (Lord Kelvin), *Popular Lectures and Addresses*, London: McMillan and Co., 1894.

20. Perlman, David, Simple Error Doomed Mars Polar Orbiter, *San Francisco Chronicle*, October 1, 1999.

21. Wikipedia, http://en.wikipedia.org/wiki/Juran%27s_Trilogy, (accessed March 2007).

22. http://crinfo.univ-paris1.fr/PROJETS/elektra.html, (accessed 30 October 2008).

ASSESSMENT AND MEASUREMENT IN SYSTEMS AND SOFTWARE DEVELOPMENT

II

Chapter 3

CMMI, People CMM, Personal and Team Software Process, eSCM, ITIL, and ISO Standards

Synopsis

George Box [2], the statistician and scientist, is famous for stating repeatedly that "all models are wrong, but some are useful." The quotation refers to statistically based models; however, the Software Engineering Institute (SEI) at Carnegie Mellon University (CMU) adopted it to characterize its approach to process improvement models. Successful process improvement is best achieved when the underpinnings for it are based on a model. But as stated by Box and pointed out by the SEI, there is no one model that is best suited for every organization in every industry. Different models tend to have a different focus. In most cases, they are developed for different business domains, although some overlap does exist. For example, project management practices are fairly consistent from model to model. Every model has strengths and elements that fit well in some environments, but not in others. An organization interested in process improvement must look at available models and determine which one fits best for the business needs.

In this chapter we provide an overview of different process models, including:

- The Capability Maturity Model Integration (CMMI), including:
  - CMMI for Development (CMMI-DEV)
  - CMMI for Acquisition (CMMI-ACQ)
  - CMMI for Services (CMMI-SVC)
- People Capability Maturity Model (P-CMM)
- Personal and Team Software Process
- Information Technology Infrastructure Library (ITIL)
- Information Technology Service Standard (ISO 20000, which incorporates the service elements of ITIL)
- CMU's eSourcing Capability Models
- ISO/IEC quality management Standard 9000
- ISO software life cycle Standard 12207
- ISO software processes assessment framework Standard 15504
- ISO system engineering—system life-cycle processes Standard 15288.

This chapter provides a broad context regarding the various models' suitability for an organization's process improvement initiatives, and helps in tailoring the model that will be most useful for a specific organization. The purpose is not to compare and contrast the models or to make a recommendation as to which is best to use. It is only to describe what models are available and in what domains they are used. We will, however, emphasize the CMMI because it is a process improvement model in wide use throughout the world in software and systems development and maintenance. The SEI's September 2008 report of appraisal results indicates that 3,553 Version 1.1 and Version 1.2 Class A SCAMPI appraisals based on Version 1.1 and Version 1.2 of the CMMI had been performed and reported in 2,168 participating companies in 63 countries through June 2008 [1]. Clearly, the data indicates widespread acceptance of the model for process improvement.

3.1 Capability Maturity Model Integration (CMMI)

3.1.1 History

The CMMI is the latest in a series of process improvement models developed by the Software Engineering Institute (SEI) at Carnegie Mellon University. The first of these was developed under contract to the U.S. Department of Defense (DoD) as a means of discriminating between capable contractors and those that weren't as competent. The original model, first reported in 1987 [22], was called the Process Maturity Model. It defined five levels of maturity and described rather general characteristics pertaining to each level of maturity. The maturity levels were organized somewhat similar to the maturity levels we see in the CMMI: Initial, Repeatable,

Defined, Managed, and Optimizing. An assessment methodology was also developed, and the first set of results from that assessment methodology was reported by the SEI in 1987 [23]. The assessment methodology was commercialized in October 1990 when the SEI conducted the first training class for lead assessors. Since that time, most assessments (now known as appraisals) have been conducted by what are now referred to as SEI Partners.

In 1993, the SEI released the Capability Maturity Model for Software (commonly referred to as the CMM or SW-CMM) [36], which was effectively an update to the Process Maturity Model. This model was a vast improvement over the previous incarnation in that there was now a structure of building blocks, called key process areas, which denoted what kinds of practices should be in place in order to achieve each level of maturity. These practices did not specify "how" the organization should implement it, but rather "what" the organization should do. For example, Activity 2 in the key process area (KPA) "Software Configuration Management" stated, "A documented and approved configuration management plan is used as the basis for performing SCM activities." The practice did not specify the detailed content of the plan, but merely indicated that a plan should exist covering the activities to be performed, the schedule for them, assigned responsibilities, and the resources required.

Each KPA was organized into a set of common features that structured the KPA to provide practices describing the actual activities to be performed in implementing that KPA, as well as supporting infrastructure practices to ensure that the KPA would be institutionalized, that is, would be performed consistently and in the same way all the time. Each KPA had the same common features. The implementing practices were listed under the common feature simply called "Activities Performed." The institutionalization common features were called "Commitment to Perform," "Ability to Perform," "Measurement and Analysis," and "Verifying Implementation." The institutionalization common features followed a logical train of thought. Commitment to Perform was management's statement to the organization that the organization was committed to performing these activities, and this was denoted in the form of policy statements and assignment of leadership, as appropriate. Ability to Perform meant that the organization had to provide the necessary resources to implement the KPA. This included training in the activities of the KPA, tools, and adequate funding and personnel resources. The philosophy was that it was not sufficient for management to state a policy that they wanted the KPA performed, but that they had to demonstrate their intentions by providing the necessary resources. Measurement and Analysis was intended as a monitoring and control function. Measures were required, usually in the form of data that showed trends of actuals versus plan, so that corrective action could be applied if the data indicated that the activities of the KPA were going off the tracks. For example, for the KPA of Software Quality Assurance (SQA), a measure could be the plot of audits actually performed versus audits planned. If the trend showed an increasing lag between audits conducted and audits planned, this would have been a signal that some causal analysis needed to be performed and corrective action possibly implemented, such as providing additional SQA resources. Finally, Verifying Implementation was intended to ensure that the activities of the KPA were being conducted in accordance with the established process. This involved audits to determine if the process was being followed, as well as reports to higher levels of management of the progress on the KPA's implementation.

An assessment methodology was developed along with the model as a means of evaluating an organization's progress toward process improvement. This method was called the CMM-Based Assessment for Internal Process Improvement (CBA-IPI) [16]. The methodology assembled a team of practitioners from within the organization and an SEI-authorized lead assessor to combine reviews of documentation with interviews of project leads and technical practitioners as a means of collecting data to assess the organization's progress in implementing the CMM. The extent to which the goals of each KPA and each KPA itself were satisfied was rated by the team, and if so desired by the organization's sponsor for the appraisal, a maturity level rating was also assigned. To the knowledge of the authors, no organization chose to enter into a CBA-IPI assessment without electing to obtain a maturity level rating. A detailed discussion of the methodology can be found in an earlier book by the authors of this book [31].

Due to the success of the Software CMM®, a number of other CMMs began to emerge, together with companion assessment methodologies. These addressed issues such as human capital management, software acquisition, security, systems engineering, and even the effectiveness of a lead assessor's capabilities. Some, such as the People CMM (P-CMM®) [10–12] and the Software Acquisition CMM (SA-CMM®) [8, 9], became established as released models. Responsibility for the Systems Engineering CMM (SE-CMM) was transferred by the SEI to the Electronics Industry Alliance (EIA), which later released it as the interim standard EIA/IS 731 [17]. It became a source document for the CMMI®. No other CMMs were ever formally established by the SEI.

3.1.2 CMMI

3.1.2.1 Motivating Factors

As the effort to develop other CMMs to characterize how other efforts were performed by industry (and in particular the defense industry), there was a growing concern that industry would be inundated with assessments to characterize how well all these activities were being performed. In particular, in the defense industry, multiple assessments could become part of the source selection process. For example, a defense contractor bidding on an intelligence system could potentially undergo separate systems engineering, software engineering, and security assessments as part of the source selection process. Supporting these assessments resulted in considerable cost to the company in terms of time and money in preparation for and participation in them. Because each procuring agency often required its own separate set of assessments, a company could potentially face multiple sets of assessments over a single year. Consequently, industry went to the DoD and asked them to come up with a single integrated model that would cover all the disciplines that were of interest to the acquisition community. The intent was to have a basic model that covered processes that applied to all disciplines (such as project management) and to which add-ons could be applied to cover other disciplines of interest. This would permit, if it was so desired, to have security added as an additional process area to look at as part of an assessment process. As a consequence, the CMMI was born. It had a number of similarities to the CMM, but in some ways might be thought of as an advanced version of the CMM. In the first integrated model released, the activities of project management, process management, software and systems engineering, and support (e.g., quality assurance, configuration management) were integrated into the model.

3.1.2.2 Version 1.1

Although Version 1.1 has been formally retired, we dwell a fair amount on it because Version 1.1 represented a major change in structure and content from the CMM. To elaborate on what the changes were, that can best be accomplished by discussing Version 1.1. Version 1.2 also implemented some significant changes in content, but the magnitude of the change from version to version was less than the change to the CMMI from the CMM and can be fairly easily described as "deltas" to the first released version.

Version 1.1, the first formally released version of the CMMI, came out in March 2002 [7].* It covered the disciplines of Systems Engineering, Software Engineering, Integrated Product and Process Development (IPPD), and Supplier Sourcing (SS). Systems Engineering and Software Engineering constituted the basic model, with IPPD and SS as add-ons, in that process areas additional to those included in the basic model were needed to cover those disciplines. To help in applying the practices of the model for the various disciplines, amplifying information was provided where some interpretation may be needed specific to the discipline. The IPPD discipline was intended to address an environment where concurrent engineering is utilized. Concurrent engineering focuses on having a project team that includes representatives from all phases of a system's life cycle: early planners, system engineers, developers, manufacturing personnel, logisticians, maintenance personnel, and support personnel. It can also include the customer. The idea is to ensure that the concerns of all those involved in all phases of the life cycle are properly addressed and that they participate in the planning and execution of the project.

---

* Version 1.1 has been retired and replaced by Version 1.2. A discussion of Version 1.2 follows this discussion.

Supplier Sourcing was intended to address the acquisition issues for projects where acquisition was the primary mode for system development. In this case, the development organization tended to acquire most or all the software and/or hardware components, and integrated them into a deliverable system. The development organization did some development work to produce the hardware or software that integrated (or "glued") the components together.

An assessment methodology, now referred to as an appraisal, was also developed. It is referred to as the SCAMPI methodology. SCAMPI stands for Standard CMMI Appraisal Method for Process Improvement [34]. Section 3.1.4 discusses the SCAMPI method.

The basic CMMI for System Engineering/Software Engineering model was comprised of 25 major components known as Process Areas (PAs). These were ostensibly the same kind of components that had existed in the CMM, known there as Key Process Areas. Each PA was comprised of both Specific and Generic Goals. The Specific Goals were comprised of two or more Specific Practices that addressed what kinds of practices (not "how") were expected to be in place in order to implement the intent of the goal. For example, Specific Goal 2 in Configuration Management addresses the practices an organization is expected to perform to track and control configuration changes. The Generic Goals were comprised of two or more Generic Practices that addressed the kinds of practices an organization was expected to perform to provide an infrastructure for ensuring that the Specific Practices of that PA would be performed consistently from project to project, that is, that the practices were institutionalized.* They were called "generic" because the names of the Generic Goals and Practices were the same for each PA. Only the content differed. For example, for Generic Goal 2 for each PA, there was a Generic Practice 2.7, "Identify and Involve Relevant Stakeholders" [4]. What changed from PA to PA were the stakeholders identified and how they were involved in that PA.

The model had two representations: staged and continuous. The staged representation provided a path for an organization to improve its overall capability to develop systems, using a structure of building blocks, added in incremental stages analogous to that of building a block wall fence, with one layer built upon the previous one. These layers were called Maturity Levels, and are defined in Table 3.1. Note the similarities between the maturity levels of the staged representation and the "rungs" on the Quality Ladder described in Chapter 2. The staged representation and appraisal methodology yielded maturity level ratings that rated the organization's process performance capability as a whole, and enabled comparisons across organizations. A Maturity Level 3 rating meant the same thing for each organization that had the same rating. It provided an easy migration path to the CMMI for those organizations that had previously implemented the CMM because of the similarity in structure.

---

* As used in the CMMI, institutionalization means that regular performance of the practices has become a way of life.

Table 3.1   Maturity Levels and Their Descriptions

Maturity Level	Name	Description
1	Initial	Process implementation is ad hoc and chaotic; meeting schedule and cost targets and quality requirements are unpredictable; the organization is highly dependent on the heroics of a few competent people for any project's successes
2	Managed	A project management discipline is in place; an infrastructure exists to ensure that projects perform effective project management practices in a consistent manner from project to project
3	Defined	A standard process exists in the organization for performing project and process management and engineering and support activities; tailoring guidelines exist for adapting the standard process to meet the unique needs of individual projects; project management is proactive, rather than reactive; projects contribute work products, measures, etc., to the organizational process assets
4	Quantitatively Managed	Quantitative management of the organization's and project's processes are implemented; quantitative process performance and quality goals and objectives are established; projects quantitatively manage their projects to meet these goals; quality and process performance is understood statistically in order to address special causes of variation
5	Optimizing	The organization is focused on reducing the common cause of variation by continuously improving its processes; process and innovative improvements are proposed, evaluated in terms of effect of process performance baselines and capability of meeting quality and process performance goals and objectives, and deployed to the organization in a planned and managed fashion; the most frequent causes of defects and problems in project processes are analyzed and corrected, and elevated to the organizational level, as appropriate

For the Systems Engineering/Software Engineering Disciplines, the staged representation had seven process areas at Maturity Level 2, eleven at Maturity Level 3, two at Maturity Level 4, and two at Maturity Level 5. Table 3.2 identifies the PAs at all five maturity levels. To accommodate the other disciplines (for example, IPPD), additional PAs were added at Maturity Level 3: one for Supplier Sourcing (Integrated Supplier Management) and two for IPPD (the Organizational Environment for Integration and Integrated Teaming PAs). Two additional goals were added to the Integrated Project Management PA to address additional aspects of IPPD.

The continuous representation allowed an organization to selectively choose PAs for improvement. Within a PA, *capability levels* were established that provided benchmarks for the organization to measure the extent to which the organization had improved in its capability to perform the practices contained therein. The capability levels are defined in Table 3.3. The continuous representation "Afford[ed] an easy comparison of process improvement to International Organization for Standardization and International Electrotechnical Commission (ISO/IEC) 15504, because the organization of process areas is similar to ISO/IEC 15504" [4]. It also afforded an easy migration path to the CMMI for those organizations heavily engaged in systems engineering that had adopted the EIA/IS 731 standard on system engineering [17] because of both the structure of the continuous representation and the fact that the EIA/IS 731 (a continuous representation model) was a source document for the CMMI. What the continuous representation could not do was provide an easy comparison among organizations. An organization saying that it is at Capability Level 4 has no meaning. To compare between organizations, an organization has to specify what the capability level rating is for each PA that was formally rated. Two organizations can only compare capability level ratings PA by PA.

In general, the PAs were grouped (and are in Version 1.2, as well) into four categories: Project Management, Process Management, Engineering, and Support. The Project Management PAs covered all aspects of managing a project, performing risk management, managing suppliers, and using quantitative and statistical techniques to manage projects. These included the following:

- Project Planning (PP)
- Project Monitoring and Control (PMC)
- Supplier Agreement Management (SAM)
- Integrated Project Management (IPM)*
- Risk Management (RSKM)

---

* There were four Specific Goals (SGs) in IPM in Version 1.1. SG 1 and SG 2 applied to all disciplines, but SG 3 and SG 4 applied only if IPPD is being implemented; consequently, if an organization was not implementing IPPD, it did not have to be concerned with SG 3 and SG 4.

Table 3.2   Version 1.1 CMMI Process Areas by Maturity Level

Maturity Level	Name	Focus	Process Areas	Applicable Disciplines
1	Initial		None	None
2	Managed	Basic Project Management	Requirements Management	All
			Project Planning	All
			Project Monitoring and Control	All
			Supplier Agreement Management	All
			Measurement and Analysis	All
			Process and Product Quality Assurance	All
			Configuration Management	All
3	Defined	Process Standardization	Requirements Development	All
			Technical Solution	All
			Product Integration	All
			Verification	All
			Validation	All
			Organizational Process Focus	All
			Organizational Process Definition	All
			Organizational Training	All
			Integrated Project Management, Specific Goals 1 and 2	All

Continued

Table 3.2   Version 1.1 CMMI Process Areas by Maturity Level (*Continued*)

Maturity Level	Name	Focus	Process Areas	Applicable Disciplines
			Integrated Project Management, Specific Goals 3 and 4	IPPD
			Risk Management	All
			Decision Analysis and Resolution	All
			Organizational Environment for Integration	IPPD
			Integrated Teaming	IPPD
			Integrated Supplier Management	SS
4	Quantitatively Managed	Quantitative Management	Organizational Process Performance	All
			Quantitative Project Management	All
5	Optimizing	Continuous Process Improvement	Organizational Innovation and Deployment	All
			Causal Analysis and Resolution	All

- Quantitative Project Management (QPM)
- Integrated Teaming (IT), if IPPD was included in the scope of the effort
- Integrated Supplier Management (ISM), if Supplier Sourcing was included in the scope of the effort

The Process Management PAs covered establishing qualitative and quantitative process management requirements, establishing the processes, providing training, and quantitatively evaluating and implementing organizational process changes. These five PAs included the following:

- Organizational Process Focus (OPP)
- Organizational Process Definition (OPD)

Table 3.3   Capability Levels and Their Descriptions

Capability Level	Name	Description
0	Incomplete	Process is not performed or only partially performed
1	Performed	The specific practices of the process are performed, but no infrastructure exists to ensure that the process will be performed the same way from project to project
2	Managed	The specific practices of the process are performed and an infrastructure exists to support performing the process consistently; organizational policies exist that define what activities the projects have to do, but each project is free to perform these activities differently
3	Defined	The process is a Managed process that can be tailored (using the organizational tailoring guidelines) from the organizational standard for the process to accommodate the unique needs of the project; projects contribute work products, measures, etc., to the organizational process assets
4	Quantitatively Managed	The process is a Defined process that is controlled to reduce special causes of variation using statistical techniques and other quantitative measures, based on process performance goals and objectives established for the process
5	Optimizing	The process is a Quantitatively Managed process that is improved to reduce the common causes of variation in performing the process, using causal analysis of the more frequent sources of defects and problems and addressing process changes and innovative changes; effects on process performance baselines are evaluated in changes to the process when proposed and implemented

- Organizational Training (OT)
- Organizational Process Performance (OPP)
- Organizational Innovation and Deployment (OID)

The Engineering PAs covered all engineering aspects of software and system development from requirements definition to delivery of the product. Included in this group of PAs are the following:

- Requirements Management (REQM)
- Requirements Development (RD)
- Technical Solution (TS)
- Product Integration (PI)
- Verification (VER)
- Validation (VAL)

The Support PAs cover all aspects of support services for development, such as configuration management, quality assurance, and the measurement and analysis program. Included in this group of PAs are the following:

- Configuration Management (CM)
- Process and Product Quality Assurance (PPQA)
- Measurement and Analysis (MA)
- Organizational Environment for Integration (OEI), if IPPD was included in the scope of the effort
- Decision Analysis and Resolution (DAR)
- Causal Analysis and Resolution (CAR)

It is beyond the scope of this book to go into detail about the content of each of the process areas. For those who are interested, the details can be found in SEI Technical Report CMU/SEI-2002--TR-012 [4].

3.1.2.3 Version 1.2

Most of the basic concepts underlying the CMMI did not change in Version 1.2 [5]. The concepts of the two representations and the constituent structural elements remained unchanged. Many of the PAs remained unchanged. Some of the text was tightened up, and some practices were added and deleted here and there. Nonetheless, there were some significant changes in content and in one particular basic concept. Disciplines were eliminated as a structural organizing element for the model, and what were considered disciplines in Version 1.1 became additions to other existing PAs. Constellations became a major organizing structural element instead, as discussed in the next paragraph. What were referred to as the IPPD and Supplier Sourcing disciplines in Version 1.1 were significantly restructured as a consequence. The Integrated

Supplier Management PA (the PA that covered the Supplier Sourcing discipline) was removed, and elements of it were added to Supplier Agreement Management. The Integrated Teaming and Organizational Environment for Integration PAs (the PAs that covered the IPPD discipline) were removed and additional IPPD-related goals were added to the Integrated Project Management and Organizational Process Definition PAs to address these concerns. The common features were dropped as a method for organizing the Generic Practices for Generic Goals 2 and 3.

A new concept was added: constellations. In this concept, there are core PAs having invariant content that will serve all constellations, and clusters of additional new or modified PAs will be added to constitute the constellation. As with Version 1.1, where amplifying information was provided to assist in applying the practices to the various disciplines, amplifying information is provided for the core PAs to help in applying them to the constellations. The first constellation that was created was the CMMI for development, or CMMI-DEV [5] (which is Version 1.2), and one was released for organizations that primarily do acquisition, CMMI-ACQ [7] (it is also called Version 1.2). An additional constellation for services (for organizations that primarily perform services, such as logistics, help desk, computer operations, etc.) has also been released. The concept of constellations permits extending the basic model for other uses, provided the core part of the model remains intact unaltered (except for amplifications).

3.1.2.4 Version 1.3

A new version, Version 1.3, is currently under development. The focus of this new version is perfective, in that a number of clarifications are needed throughout the model based on experience in using the model for process improvement and applying it in appraisals. In particular, considerable clarification will be provided with respect to the high maturity Process Areas and Practices. Another consideration is to provide a lower maturity version of the Causal Analysis and Resolution PA, in addition to the high maturity version, because some of the concepts of this PA can be applied very effectively at lower maturity levels.

Some other considerations for Version 1.3 include:

- Providing amplifications to illustrate how some of the lower maturity level PAs evolve to the higher maturity levels
- Simplifying the Generic Practices and improving their clarity
- Improving the efficiency of the SCAMPI method
- Providing consolidation across the CMMI constellations

3.1.3 CMMI for Acquisition (CMMI-ACQ)

The CMMI model that focuses on organizations that primarily perform acquisition was released in November 2007, and is referred to as the CMMI-ACQ, Version 1.2.

It is the model that addresses the Acquisition constellation and is intended as an extension to the CMMI for Development (CMMI-DEV). The SEI technical report that documents the model [7] is a stand-alone document; it covers the PAs associated with acquisition as well as the other PAs referred to as the Core Process Areas. The original intent was to have a set of Core Process Areas that would apply to all constellations, with amplifying material that would explain how affected practices would be used differently for each constellation. Content applicable to the other constellations would be covered in the texts for those constellations. The text of the CMMI-ACQ technical report, however, is oriented toward how all the PAs are used when doing acquisition. The applicability of this constellation is for organizations that are primarily in the business of acquiring systems or services, as opposed to developing the systems or performing the services. For example, a number of organizations no longer maintain a development organization of any substantial size. They outsource to outside organizations the design, development, and lower-level test of the systems they acquire. These may be other divisions or subsidiaries within their own corporate structure, or they may be external vendors. Often, the acquiring organization will develop a set of requirements for the system, system components, or services they want to acquire and contract out the development. The acquiring organization will then receive and perform acceptance testing on the system or system components when the vendor has completed their work. The acquiring organization will also perform any remaining system integration that needs to be performed, and then do final system testing against the requirements for the system to determine if the system will perform as intended.

Because the acquiring organization does little or no development or performs the services required, to account for that fact, the Engineering PAs in the CMMI-DEV are replaced by a set of PAs that address product and service acquisition. Accordingly, the category of Engineering PAs in the CMMI-ACQ is replaced by the category of Acquisition PAs. There are six such PAs:

- Acquisition Requirements Development (ARD)
- Solicitation and Supplier Agreement Development (SSAD)
- Agreement Management (AM)
- Acquisition Technical Management (ATM)
- Acquisition Verification (AVER)
- Acquisition Validation (AVAL)

In addition, there are Practices that are added to four of the Core Process Areas to cover Acquisition. Project Planning has a Practice that addresses developing an acquisition strategy. Project Planning also has a Practice that addresses developing a plan to transition the system or services to an operational state. Project Monitoring and Control has a Practice that monitors the implementation of the transition plan. Integrated teaming is covered by single Practices in Integrated Project Management (IPM) and Organizational Process Definition (OPD). The concept of integrated teams in this context is very similar to the IPPD concept discussed in the CMMI-DEV (above). The intent is to ensure that all parties associated with the acquisition of the system or services, from basic concept to operations and maintenance, are involved in the management of the acquisition. In the CMMI-DEV, IPPD is an add-on, whereas in the CMMI-ACQ, integrated teaming is not. OPD has a Practice that addresses defining the ground rules for establishing and operating integrated teams, and IPM has a Practice that addresses forming and operating the integrated teams in accordance with these ground rules.

The six PAs that address Product and Service Acquisition appear as Maturity Level 2 and 3 PAs in the staged representation. Requirements Management is grouped under the Project Management PAs in the CMMI-ACQ (it is grouped under the Engineering PAs in the CMMI-DEV). The CMMI-ACQ technical report [7] has an overview of the Generic Practices and Goals in one section of the text only, and does not include them in the descriptions of the individual PAs, unlike the CMMI-DEV.

3.1.4 CMMI for Services (CMMI-SVC)

The CMMI model that focuses on organizations that primarily perform services was released in February of 2009, and is referred to as the CMMI-SVC, Version 1.2 [49]. It is the model that addresses the Services constellation and is intended as an extension to the CMMI for Development (CMMI-DEV). Like the technical report for the CMMI-ACQ, it is a stand-alone document; it covers the PAs associated with services as well as the other process areas referred to as the Core Process Areas. The text of the CMMI-SVC Technical Report, however, is oriented toward how all the PAs are used when performing services. The applicability of this constellation is for organizations that are primarily in the business of performing any kind of services, such as help desk, computer operations, building maintenance services, and even pest control, as opposed to developing systems or performing acquisition.

The Engineering PAs in the CMMI-DEV are replaced by a set of PAs that address the provision of services, as well as two that are oriented toward project management. There are seven such PAs:

- Capacity and Availability management (CAM)
- Incident Resolution and Prevention (IRP)
- Service Continuity (SCON)
- Service Delivery (SD)
- Service System Development (SSD)*
- Service System Transition (SST)
- Strategic Service Management (STSM)

---

* This process is referred to as ìan additionî in that if a service organization does not need to develop a new service or make major changes to an existing one, SSD is optional.

There is one process area, Supplier Agreement Management (SAM) that is shared with the CMMI-DEV. It has two Specific Practices that are modified from what they are in the CMMI-DEV. In addition, Project Planning, Organizational Process Definition, and Integrated Project Management each have a practice unique to services.

The category of Engineering PAs that appear in the CMMI-DEV is replaced by the category of Service Establishment and Delivery. IRP, SD, SSD, SST, and STSM are the PAs that comprise the Service Establishment and Delivery category. CAM and SCON are Project Management PAs.

CMMI-SVC also has the staged and continuous representations. Unlike the CMMI-DEV and CMMI-ACQ, the constellation-specific PAs (in this case, the Service Establishment and Delivery process areas) are not all at Maturity Level 3 in the staged representation. Service Delivery is a Maturity Level 2 PA, while the rest are at Maturity Level 3. The CAM and SCON PAs are Maturity Level 3 Project Management PAs.

The concept of a project is different in the CMMI-SVC from that in the CMMI-DEV or CMMI-ACQ. In the CMMI-SVC, a project is interpreted as a management set of interrelated resources that delivers one or more products or services to a customer or end user [49]. For example, if a pest control company had separate agreements separate customer arrangements for termite eradication for restaurants as opposed to residences as opposed to office buildings, the resources necessary to provide these services to each type of client would constitute a project. An individual service agreement with a homeowner for termite eradication would not typically be a project in this context.

3.1.5 SCAMPI Class A Appraisal Methodology

A companion assessment methodology (now referred to as an appraisal methodology) called SCAMPI* (Standard CMMI Appraisal Method for Process Improvement) was developed along with the model from the very beginning. The "pre-release" versions of the CMMI (versions prior to Version 1.1) used a methodology that was very much like the CBA-IPI assessment methodology in use for the CMM, although it was referred to as a SCAMPI methodology. When CMMI Version 1.1 (the formal release) came out, a more extensive SCAMPI methodology was released with it. The methodology was codified in the Method Definition Document, or MDD [34], and updated for Version 1.2 [42]. Like the CBA-IPI, the appraisal methodology uses a combination of document reviews and interviews. The SCAMPI method requires a team comprised of a minimum of four people and a maximum of nine, and also requires an SEI-certified Lead Appraiser. The major differences between the CBA-IPI and SCAMPI methodologies are that the document reviews are more

---

* A story, perhaps apocryphal, was circulating around the time of release that the name SCAMPI came about because the person responsible for naming the methodology loved shrimp, cooked Italian style.

extensive in a SCAMPI appraisal, the rules for data corroboration are more extensive, and the determination of goal satisfaction within a process area is different.

Three classes of SCAMPI appraisals were specified as appraisal methodologies [3, 6]. The Class A SCAMPI is the most stringent of the appraisal methodologies and is the only one that can result in either a maturity level or capability level rating. It can only be led by a lead appraiser certified by the SEI to lead a Class A SCAMPI. Class B and C SCAMPIs are more informal and cannot be used for either a maturity or a capability level rating. Class B SCAMPIs, however, must be led by an SEI-authorized Class B Lead Appraiser. Class B and C appraisals are discussed more fully in the next section.

The SCAMPI methodology requires that a direct artifact exists for each Specific and Generic Practice in scope for the appraisal for each focus and non-focus project in scope for the appraisal, that is, a project that has been designated to provide work products. The rules for artifacts for non-focus projects are somewhat different from that for a focus project, but discussion of the difference is beyond the scope of this book. A direct artifact is a work product that is produced as a direct outcome of performing that practice. The direct artifact must be corroborated by an indirect artifact or an affirmation. An indirect artifact is a work product that is produced as a result of some other activity (often, another practice) that relates to the activity performed as a consequence of implementing the practice in question. An affirmation is a verbal statement confirming (or denying) that the practice is being performed. Affirmations are garnered through interviews with project leads or technical practitioners. Coverage rules exist that establish the minimum number of affirmations that must be obtained.

The data collected for each practice is evaluated by the team, and by using consensus (not a majority vote), the team determines if a practice has been fully, largely, partially, or not implemented. For a goal to be satisfied, each practice in the goal must be rated either largely or fully implemented. For a practice to be fully implemented, there must not be any weaknesses observed in its implementation. If it is largely implemented, the artifacts/affirmations must be present and adequate but one or more weaknesses can exist. If one or more practices within a goal are largely implemented, the weaknesses observed must not adversely impact the ability to satisfy the goal. If all the goals within a PA are satisfied, likewise the PA is satisfied, but to the extent that the Generic Goals within that PA apply to either a maturity level or capability level. For example, for an appraisal performed in accordance with the staged representation of the CMMI, for the PAs associated with Maturity Level 2, all the practices within Generic Goals 2 and 3 must be characterized as either fully implemented or largely implemented for the PA to be rated as satisfied at Maturity Level 3. If Generic Goal 3 is not satisfied, the PA can only be rated as satisfied at Maturity Level 2. If Generic Goal 2 is not satisfied, the PA is not satisfied, no matter what the rating of Generic Goal 3 is.

The continuous representation, on the other hand, looks internally within a PA and rates PAs individually. The capability level rating depends on the extent to which the six Generic Goals (Generic Goals 0 through 5) for that PA have been satisfied. The capability level rating is one level lower than the lowest level at which a

Generic Goal is not satisfied. Thus, for example, within a given PA, if Generic Goals 0 through 3 and 5 were satisfied, but Generic Goal 4 was not, the highest capability level rating that could be achieved is Capability Level 3. On the other hand, if all six Generic Goals were satisfied, the rating that is achieved is Capability Level 5. For the staged representation, if all the PAs within a maturity level are satisfied, that maturity level is satisfied as long as all the PAs at lower maturity levels are also satisfied. There is a "weakest link" effect here, as alluded to above. The maturity level rating awarded to an organization is one level lower than the corresponding level at which a PA is rated as "not satisfied." Maturity level satisfaction depends on successful implementation of each lower maturity level. For example, for a Maturity Level 3 rating, all the PAs associated with Maturity Levels 2 and 3 must be satisfied [34, 42].

The Version 1.2 SCAMPI method release resulted in a number of relatively minor procedural changes in reporting, the definition of the organizational unit being appraised, the permissible time interval between the start and completion of an appraisal, the period of validity for a maturity level rating, project sampling, and several other matters. The most significant change was the requirement for a lead appraiser to be certified by the SEI as a high maturity lead appraiser in order to lead an appraisal targeted for Maturity Level 4 or 5.

3.1.6 SCAMPI B and C Appraisals

When the Version 1.1 SCAMPI methodology was released, an overarching document was also produced, called the Appraisal Requirements for the CMMI® (ARC) document [3]. It set the requirements for conducting three levels of appraisals using the CMMI: Class A, B, and C. It established *what* the appraisal methodologies should accomplish, not *how* they should accomplish it. In it, it defined that only a Class A appraisal could result in a maturity level or a capability level rating. By default, that appraisal had to be a SCAMPI appraisal because only the SEI-authorized Lead Appraisers using the SCAMPI method could submit maturity or capability level ratings to the SEI, and only maturity level ratings posted at the SEI website were recognized in the community. Class B and C appraisals were intended to be less formal appraisals, where a Class B required a moderate amount of objective evidence, and could be used as a baseline appraisal or an interim one in preparation for a Class A appraisal. The Class B appraisal could be likened to something like a CBA-IPI appraisal, and is often used in preparation for a Class A SCAMPI appraisal to obtain a sense of how likely the organization is to achieve the target maturity level rating. The Class C appraisal required less objective evidence (only one artifact or affirmation per practice), but was nonetheless performed as an appraisal to produce a gap analysis by many lead appraisers.

The SEI did not initially define any methodologies for performing Class B and C appraisals, and lead appraisers generally developed their own approach, consistent with the requirements in the ARC. In 2005, at about the time Version 1.2 of the CMMI and SCAMPI method was nearing release, the SEI released a method description document for performing SCAMPI Class B and C appraisals [41]. The methodologies were more formalized than what the lead appraisers had developed, required planning and reporting submittals to the SEI by the lead appraisers, and required that SCAMPI B and C appraisals be led by SEI-authorized SCAMPI B and C Lead Appraisers. This change did not negate the ability of lead appraisers to conduct their own form of ARC-compliant Class B and C appraisals, but they could not call them SCAMPI appraisals. When the Version 1.2 SCAMPI Method Definition Document [42] was released, the ARC document was updated [6], but the method for performing a Class B and C SCAMPI did not change.

3.2 People CMM

The People Capability Maturity Model, called the People CMM, or P-CMM for short, relates to developing the workforce capability of an organization. Whereas the models in the CMMI constellations relate to an organization's capability to perform development, acquisition, or service functions in the organization, the P-CMM is complementary to it in that it relates to the capability of the workforce in an organization to perform these functions. It contains practices for "attracting, developing, organizing, motivating, and retaining [a] workforce [12]." It examines a number of dimensions, such as the readiness of the workforce to perform the functions critical to the organization's success, what the results are likely to be, and the likelihood that the organization can benefit from investments in process improvement or new technology. Like the staged representation of the CMMI, the P-CMM has five levels of maturity (it has no continuous representation). The titles of the maturity levels are slightly different from that of the CMMI, but have similar characteristics, more or less, but a different effect. The maturity level titles, from Level 1 to 5, are Initial, Managed, Defined, Predictable, and Optimizing. Their defining characteristics and effects are shown in Table 3.4.

Each maturity level, from Level 2 to 5, is comprised of three to seven PAs. The PAs are not organized as they are in the CMMI into categories such as Project Management or Process Management. Similar to the CMMI, each PA is comprised of Goals, which are comprised of Practices. Unlike the CMMI, there are no Specific and Generic Goals and Practices. Instead, each PA has a Goal that addresses institutionalization of the Practices, where institutionalization means establishing an infrastructure to ensure that the PA is performed in a consistent manner within the organization. The institutionalization Practices are organized into four categories, known as the Common Features:* Commitment to Perform, Ability to Perform, Measurement and Analysis, and Verifying Implementation.

---

* The practices of Generic Goals 2 and 3 in the CMMI were organized along the lines of the common features in the staged representation of CMMI, Version 1.1, but were dropped as an organizing structure in Version 1.2.

Table 3.4   People CMM Maturity Levels

Maturity Level	Name	Description	Effect
1	Initial	Workforce practices are ad hoc and inconsistent; responsibilities rarely clarified	Difficulty in retaining talent; lack of training to perform whatever practices exist
2	Managed	Focus on workforce practices is to establish basic practices at the unit level, addressing immediate problems; policies and accountability established; how the practices are performed may vary from unit to unit	Organization not ready to implement organizationwide work practices at this level; units are stable environments; employee turnover reduced
3	Defined	Unit-level workforce practices standardized across the organization; infrastructure established to tie practices to business objectives; workforce competencies are developed and strategic workforce plans are developed to ensure that the level of talent needed for the workforce competencies are acquired or developed	Workforce competencies are strategic; a common organizational culture begins to develop; career opportunities become more clarified, centered around increasing levels of capability in workforce competencies
4	Predictable	Quantitative management of workforce capability and performance; process capability baselines are established for critical competency-based processes, corrective action is applied, as appropriate	Unit and organizational workforce processes become more predictable; more accurate predictions about future performance; management has better input for strategic decisions

Table 3.4   People CMM Maturity Levels (*Continued*)

Maturity Level	Name	Description	Effect
5	Optimizing	Improvements are evaluated quantitatively to ensure alignment with business objectives and improvement in process capability baselines	Entire organization is focused on continual improvement

- *Commitment to Perform* is an indicator of management's intentions to have the practices of that PA implemented, and generally takes the form of policies, management sponsorship, and organizationwide roles.
- *Ability to Perform* addresses the provisioning of resources to enable the practices of the PA to be implemented. It is not enough for management to state its desire to have the practices performed, but if truly committed to having the PA performed, management must provide resources such as funding, personnel, tools, training to perform the roles of the PA, and assigned responsibility for the roles.
- *Measurement and Analysis* addresses establishing measures to manage and control the activities of the PA.
- *Verifying Implementation* addresses the reviews and audits that are performed to verify that the PA is being implemented in accordance with its process description.

The other goals address the activities that are performed to implement the PA.

Figure 3.1 shows the mapping of the PAs to the maturity levels.

As with the CMM and CMMI, an assessment methodology was also developed [19]. This methodology enables an organization to achieve a maturity level rating relative to its implementation of the P-CMM. The assessment methodology used to assess an organization's implementation of the P-CMM is different from the CBA-IPI and SCAMPI methodologies.* With the release of CMMI 1.2 and its associated SCAMPI assessment method, the SEI has transitioned to the use of SCAMPI with the P-CMM.

As noted earlier, the P-CMM is complementary to the CMMI. The SEI Technical Report [13] notes that at Maturity Level 3 on the P-CMM:

> By defining process abilities as a component of a workforce competency, the People CMM becomes linked with the process frameworks established in other CMMs and with other process-based methods, such as

---

* See Reference 9 for a description of the CBA-API assessment methodology, and Reference 18 for the SCAMPI methodology.

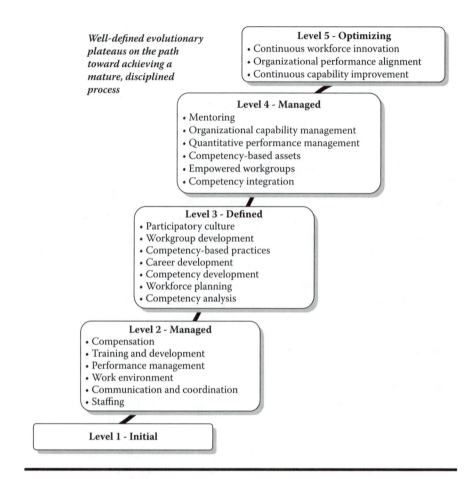

Well-defined evolutionary plateaus on the path toward achieving a mature, disciplined process

Level 5 - Optimizing
- Continuous workforce innovation
- Organizational performance alignment
- Continuous capability improvement

Level 4 - Managed
- Mentoring
- Organizational capability management
- Quantitative performance management
- Competency-based assets
- Empowered workgroups
- Competency integration

Level 3 - Defined
- Participatory culture
- Workgroup development
- Competency-based practices
- Career development
- Competency development
- Workforce planning
- Competency analysis

Level 2 - Managed
- Compensation
- Training and development
- Performance management
- Work environment
- Communication and coordination
- Staffing

Level 1 - Initial

Figure 3.1   The mapping of process areas to maturity levels.

business process reengineering. A process ability is demonstrated by performing the competency-based processes appropriate for someone at an individual's level of development in the workforce competency. To define the process abilities incorporated in each workforce competency, the organization defines the competency-based processes that an individual in each workforce competency would be expected to perform in accomplishing their committed work. Within a workforce competency, a competency-based process defines how individuals apply their knowledge, perform their skills, and apply their process abilities within the context of the organization's defined work processes.

Clearly, an assessment of the organization's capabilities on the P-CMM could be an indicator of an organization's ability to achieve higher levels of maturity on the CMMI. In 2004, a pilot enterprisewide appraisal of Tata Consultancy Services

(TCS) was conducted using Version 1.1 of the SCAMPI methodology against Version 1.1 of the CMMI and Version 2 of the P-CMM, combined [38]. TCS had developed what it believed to be a model-independent quality system manual in order to be compliant with a number of different process improvement models. The results showed that the SCAMPI methodology could be used very effectively as the appraisal methodology for both models. The focus of the appraisal was not on correlations between the two models, but on the applicability of the SCAMPI appraisal methodology to both CMMI and P-CMM-based appraisals. Clearly, some correlations between the two models did exist, and some artifacts could be used as evidence of satisfaction of practices in both models. Other investigators have shown that mappings and synergies between the two models do exist [20, 35, 40].

3.3 Personal and Team Software Process

The CMMI constellations and the P-CMM address issues related to how organizations develop or acquire systems and services, and how the human resource aspects at the organizational level interact with the ability to perform development, acquisition, or services. The P-CMM addresses developing teams and workgroups, and empowering them to make improvements, based on analyses of their performance. The Team Software Process (TSP) and the Personal Software Process (PSP) address issues regarding development at the next lower levels of organizations: TSP at the team level and PSP at the personal level. Although one may look at the organization of models as a top-down architecture, in reality, the TSP extends and builds upon the concepts contained in the PSP. These techniques provide detailed guidance for how to implement individual- and team-based processes to quantitatively manage and improve software development.

3.3.1 PSP

Change comes about slowly and in small steps. Those who attempt to implement massive, sweeping changes—whether we are talking about a person's way of interacting with the world, changing social structures, modifying governmental institutions, or anything else—are doomed to failure. We find that changes in small increments work best. So, too, it is with implementing process change with software developers. Software engineers are used to working with methods that on the surface of things seemed to have worked best in the past, whether they are truly beneficial or not for producing a quality product, and are reluctant to change. As stated by Watts Humphrey, "Engineers are understandably skeptical about changes to their work habits; although they may be willing to make a few minor changes, they will generally stick fairly closely to what has worked for them in the past until they are convinced a new method will be more effective. This, however, is a chicken-and-egg problem: engineers only believe new methods work after they use them and see the results, but they will not use the methods until they believe they work" [24]. The only way to bring about this change was through the use of a radical intervention, and this was to remove them from their work environment and expose them to a highly intensive training course running 125 to 150 hours over 1 to 3 months.

As described by Hayes and Over [18]:

> The PSP course incorporates what has been called a "self-convincing" learning strategy that uses data from the engineer's own performance to improve learning and motivate use. The course introduces the PSP practices in steps corresponding to seven PSP process levels. Each level builds on the capabilities developed and historical data gathered in the previous level. Engineers learn to use the PSP by writing ten programs, one or two at each of the seven levels, and preparing five written reports. Engineers may use any design method or programming language in which they are fluent. The programs are typically around one hundred lines of code (LOC) and require a few hours on average to complete. While writing the programs, engineers gather process data that are summarized and analyzed during a postmortem phase. With such a short feedback loop, engineers can quickly see the effect of PSP on their own performance. They convince themselves that the PSP can help them to improve their performance; therefore, they are motivated to begin using the PSP after the course.

The PSP is based on the practices of the CMM [26],* but scaled down to make it consistent with how they may be applied by individuals. The PSP methods are organized into seven process versions and address different aspects of how PSP is applied [26]. The process versions are labeled PSP0, PSP0.1, PSP1, PSP1.1, PSP2, PSP2.1, and PSP3. Each process version contains a similar set of scripts, logs, forms, and standards. PSP0 and PSP0.1 formulate a baseline understanding of PSP, focusing on developing and understanding the need for establishing an initial base of historical size, time, and defect data. These process versions address issues such as understanding the current process in use, the need for basic measures, the importance of measures of size (which relate to productivity), etc. In PSP training courses, in learning this aspect of PSP, students develop three small programs, using their own methods but following a rigorous process. They are required to estimate, plan, perform, measure, compare actuals to estimates, and conduct a post-mortem when finished.

PSP1 and PSP1.1 focus on personal project management techniques, introducing size and effort estimating, schedule planning, and schedule tracking methods. PSP2 and PSP2.1 fold in methods for managing quality. The notion of personal

---

* Although the PSP was originally developed based on the practices of the CMM, it can still be readily applied in organizations that are implementing the CMMI.

reviews for design and code are introduced. The use of design notation and templates for design are also introduced. Importantly, the need for measures to manage process and product quality is also addressed. As noted by Hayes and Over [18]:

> The goal of quality management in the PSP is to find and remove all defects before the first compile. The measure associated with this goal is yield. Yield is defined as the percent of defects injected before compile that were removed before compile. A yield of 100% occurs when all the defects injected before compile are removed before compile.

The importance of the use of historical data is emphasized for PSP2. Hayes and Over [18] point out that:

> Starting with PSP2, engineers also begin using the historical data to plan for quality and control quality during development. Their goal is to remove all the defects they inject before the first compile. During planning, they estimate the number of defects that they will inject and remove in each phase. Then they use the historical correlation between review rates and yield to plan effective and efficient reviews. During development, they control quality by monitoring the actual defects injected and removed versus planned, and by comparing actual review rates to established limits (e.g., less than 200 lines of code reviewed per hour). With sufficient data and practice, engineers are capable of eliminating 60% to 70% of the defects they inject before their first compile.

Clearly, an implication of this is that each developer develops and maintains a database of planned and actual measurements. These include software size, time expended on performing the tasks, and defects. Furthermore, if the database is to be useful, operational definitions for each of these measures must exist in order to even build up a meaningful organizational database. (The importance of this will be seen in the next section in the discussion of the Team Software Process.) There should never be any attempt to use the data to critique a developer's performance. The intent is to foster a self-critique.

PSP3, Cyclic Personal Process, addresses scalability. It is important for students who are learning PSP to be able to scale upward efficiently from what they have learned in the PSP training course to larger, real-life projects, without sacrificing either quality or productivity. PSP3 addresses scalability by decomposing large programs into smaller elements, developing them, and then integrating them. This reduces the development problem to something that is analogous to developing software using methods described in PSP2.1.

Data collected from approximately 300 engineers who had taken the PSP training course indicate dramatic improvement in a number of areas [24]. The ability to estimate software size improved by a third. Schedule estimation improved by a factor of 2. Compile and test defects dropped from 110 per 1000 lines of code (KLOC) to 20 per KLOC. Most importantly, the engineers reported that they were now spending a greater proportion of their time on design, rather than on fixing defects.

One important aspect of the PSP is that implementing process improvement models, whether it's the CMMI, ISO 12207, or any other model, enables the developers within the organization to accept the disciplined methodologies associated with these standards. Once a developer understands the value in his/her own use of disciplined practices as a result of using PSP, he/she will be more willing to accept the disciplined practices associated with implementing a process improvement model.

3.3.2 TSP

As noted previously, the Team Software Process [15, 25, 33] builds upon and extends the skills and disciplines learned in the Personal Software Process. The idea is to motivate the developer to learn to work as a member of a team that is self-directed, and to make team members feel personally responsible for the quality and the timeliness of the products they produce. It can support multidisciplinary teams that range in size from two engineers to more than a hundred engineers. TSP involves both developers and managers, in that both have a role in implementing it. Developers and managers build self-directed teams that plan and track their own work, establish goals related to their own piece of the total development effort, and own their processes and plans. Managers coach and motivate their teams and help them achieve and perform at peak levels. The organization finds it easier to accelerate process improvement by making CMMI high maturity behavior a way of life in the organization.

TSP complements CMMI implementation by helping project and line managers implement the project management practices contained in the Project Planning, Project Monitoring and Control, Integrated Project Management, and Risk Management Process Areas. It facilitates the team (or teams) being able to hit the ground running when the project starts. The TSP process recognizes that up-front project planning is performed on the basis of a lot of unknowns; consequently, the process addresses planning at four points during the development cycle: at project start, at the start of architectural design, at the start of implementation (e.g., coding), and at the start of integration and test. The initial point is referred to as the launch, and the other three points are referred to as relaunches. As with the PSP, there are a number of scripts, forms, and standards that are a part of the TSP process.

The launch, which is the team-building component of TSP, consists of nine meetings spread over 4 days. It focuses on planning the team's work and covers the following topics: establishing product and business goals, assigning roles and establishing team goals, laying out the development strategy, building top-down and next-phase plans, developing the quality plan, building bottom-up and consolidated plans, conducting the risk assessment, preparing the briefing to management and the launch report, holding a management review, and conducting the launch post-mortem. Some of the topics for these meetings involve team contribution; others do not. The team does not establish product and business goals; that is laid out for them by other elements of the organization (e.g., management and marketing). To build bottom-up and consolidated plans, the team members build their own plans (as they would do under PSP), and then check them for consistency and balance and consolidate them.

An identical set of activities is performed for each relaunch.

The measurement database that each developer maintains is consolidated into a team database. The team database consists of individual and team planned and actual measures of size, time, and defects, and adds to that team task completion dates. Weekly, the team conducts meetings to evaluate progress and discuss issues. The team also prepares reports to management on progress on a regular schedule. Clearly, comparisons of planned versus actual measures constitute part of these reports. TSP might be considered from some perspectives as a micro-implementation of the CMMI. In fact, a study conducted of the mappings of the CMMI to the TSP showed that TSP instantiates "a majority of the project-oriented specific practices of CMMI" [33].

3.4 eSourcing Capability Model (eCSM)

With an increasing amount of IT work being performed as projects—or what might be considered subprojects by outside providers or internal providers having a long-lived relationship with the client organization, such as shared service centers or central IT departments—industry perceived a need for best practice guidance that addressed the nature of the relationship, the multiple ongoing projects or activities that may be performed, and covering the full life cycle, not just development and maintenance. This need led Carnegie Mellon University to establish the IT Services Qualification Center (ITSqc) in 2000. The ITSqc at Carnegie Mellon University develops capability models to improve the relationship between service providers and their clients. Two such models, covering both sides of outsourcing agreements, have been developed: (1) The eSourcing Capability Model for Service Providers (eSCM-SP) Version 2, which was released in April 2004; and (2) the eSourcing Capability Model for Client Organizations (eSCM-CL), which was released in September 2006.

3.4.1 Barriers to Successful Sourcing Relationships

Managing and meeting client expectations is a major challenge in these business relationships, and examples of failure abound. The eSourcing relationships between clients and their service providers must overcome many challenges to be successful. The eSourcing relationship challenges include:

- Clients often have little experience in outsourcing and have no standard criteria for selecting a provider.
- Success criteria for the relationship are not well understood or agreed upon from inception by both parties.
- Clients' expectations often change as the nature of the services change, due to rapid shifts in technology and tools, and providers are not always able to keep up with those changes.
- The necessary trade-offs between the service's quality, speed, and cost are not always articulated and understood.
- The transfer of personnel, equipment, and knowledge between the client and service provider is often problematic.
- Service providers often have trouble analyzing and reporting their progress in terms that are meaningful for clients.

To help organizations succeed in the face of these challenges, the ITSqc created "best practices" capability models for both sides of the eSourcing relationship. eSourcing refers to IT-enabled sourcing, that is, service delivery that uses information technology as a key component or as an enabler. Its services are commonly provided remotely, using telecommunication or data networks, and they range from routine, noncritical tasks that are resource intensive and operational in nature to strategic business processes that directly impact revenues. Service providers use information technology as a key component of, or as an enabler for, delivering their services. These services include desktop maintenance, data-center support, and applications management, as well as human resources, customer care, engineering services, and finance and accounting.

3.4.2 eSCM for Service Providers

The eSourcing Capability Model for Service Providers (eSCM-SP) provides IT-enabled sourcing service providers a framework to improve their capability to deliver consistently high quality services and aids them in establishing, managing, and continually improving relationships with clients. The intent of the eSCM is to present service providers with a set of best practices that helps them effectively manage sourcing relationships, and it presents clients with a way to evaluate and compare service provider capabilities.

The eSCM-SP helps sourcing organizations manage and reduce their risks and improve their capabilities across the entire sourcing life cycle. The eSCM-SP was developed specifically to address the difficulties in providing eSourcing services. The model's Practices can be thought of as the best practices associated with successful sourcing relationships. It addresses the critical issues related to IT-enabled sourcing (eSourcing).

The eSCM-SP was developed for three purposes. It helps IT-enabled service providers appraise and improve their ability to provide high-quality sourcing services, and it gives them a way to differentiate themselves from the competition.

Service providers use the eSCM-SP and its accompanying capability determination methods to evaluate their eSourcing capabilities, and to become eSCM-SP-certified. This status provides an advantage over their competitors.

Prospective clients can evaluate service providers based on their eSCM-SP level of certification and Practice Satisfaction Profile.

Each of the Model's 84 Practices is distributed along three dimensions: Sourcing Life-cycle, Capability Areas, and Capability Levels.

Most quality models focus only on delivery capabilities, but the eSCM-SP's Sourcing Life-cycle includes not only delivery, but also initiation and completion of contracts where many commonly encountered problems arise. The Capability Areas addressed by the eSCM-SP are:

- Knowledge Management
- People Management
- Performance Management
- Relationship Management
- Technology Management
- Threat Management
- Contracting
- Service Design and Deployment
- Service Transfer
- Service Delivery

The eSCM-SP offers a five-level improvement path that service providers can travel to enhance value and sustain excellence over time. Achieving a higher level of capability brings new levels of client trust as it improves the provider's organizational efficiency. By grouping its Practices into increasing levels of capability, the eSCM-SP describes an improvement path for a service provider. Providers may advance from a minimal level of delivering services to the highest level where they are proactively enhancing value for clients, regardless of the requirements or scope of sourcing efforts.

The eSCM-SP was designed to complement existing quality models so that service providers can capitalize on their previous improvement efforts. The model's structure complements most existing quality models, such as ISO-9000, ISO 20000, ISO 17799, the CMMIs®, CobiT®,* and COPC-2000®,† so it can be implemented in parallel with these other frameworks. A series of documents comparing

---

* CobiT is an IT governance standard that "integrates and institutionalizes good practices to ensure that the enterprise's IT supports the business objectives" [30].
† COPC-2000 "is a Performance Management Framework designed to deliver results in Customer Service Provider (CSP) contact center environments including Call Centers, E–Commerce Centers and Transaction Processing Operations" [14].

the eSCM-SP with other models and standards is available from the ITSqc website (http://itsqc.cmu.edu/downloads).

3.4.3 eSCM for Client Organizations

Over the past several years, many organizations, from manufacturing firms to banks to hospitals, have been delegating computer-intensive activities to internal or external service providers because they are focusing on "core competencies" or lack their own in-house capabilities. In many cases, they have not been satisfied with the results.

The actions of the client organization and those of the service provider in these sourcing relationships are both critical for success. To address both aspects of the eSourcing relationship, the eSCM for Client Organizations (eSCM-CL) was developed, which addresses the challenges of sourcing relationships from the client's perspective [21]. Existing frameworks do not comprehensively address the best practices needed by client organizations to successfully source and manage IT-enabled services. ITSqc's investigations show that most current quality models do not address all phases of the client's sourcing process. ITSqc has developed a best practices model that allows client organizations to continuously evolve and improve their capabilities, and innovate new ones to develop stronger, longer-term, and more trusting relationships with their service providers.

The eSCM-CL enables client organizations to appraise and improve their capability to foster the development of more effective relationships, better manage these relationships, and experience fewer failures in their client–service provider relationship.

The eSCM-CL addresses a full range of client organization tasks—developing the organization's sourcing strategy, planning for sourcing and service provider selection, initiating the agreement, managing service delivery, and completing the service. The eSCM-CL is based on the eSCM-SP; it contains client-focused counterparts to more than half of the eSCM-SP Practices. Each Practice in the eSCM-CL is distributed along the same three dimensions as the eSCM-SP, although the Sourcing Life-cycle is expanded to cover activities that the client must perform, such as establishing a sourcing strategy, identifying potential service providers, developing a sourcing approach and activities later in the sourcing relationship dealing with assuring alignment and value from sourcing activities.

The eSCM-CL extends the eSCM-SP sourcing life-cycle phases of Initiation, Delivery, and Completion by adding a new Analysis phase at the beginning of the life cycle. During this phase, client organizations analyze their operations and identify sourcing actions to implement. The Capability Areas addressed by the eSCM-CL, by phase, are:

- Ongoing:
  - Sourcing Strategy Management

- – Governance Management
- – Relationship Management
- – Value Management
- – Organizational Change Management
- – People Management
- – Knowledge Management
- – Technology Management
- – Threat Management
- ■ Analysis:
  - – Sourcing Opportunity Analysis
  - – Sourcing Approach
- ■ Initiation:
  - – Sourcing Planning
  - – Service Provider Evaluation
  - – Sourcing Agreements
  - – Service Transfer
- ■ Delivery:
  - – Sourced Services Management
- ■ Completion:
  - – Sourcing Completion

The eSCM-CL enables client organizations to appraise and improve their capability to develop more effective relationships, to better manage these relationships, and to experience fewer failures in their eSourcing relationships. The eSCM-CL capability determination methods indicate the client organization's current capabilities and can highlight areas where internal improvement efforts can be focused. A client organization can also pursue certification by the ITSqc in the eSCM-CL to demonstrate their capability level.

3.4.4 eSCM Capability Determinations

A set of Capability Determination methods have been defined to be used with the eSCM models. There are five different versions of the eSCM Capability Determination method that can be used to systematically analyze evidence of the organization's implementation of the eSCM Practices. They are used to objectively determine current capabilities relative to an eSCM model and to identify targets for future improvement. The methods also provide a consistent way for senior management and stakeholders to evaluate their organization's current capability. The five Capability Determination methods are (1) Full Self-appraisal, (2) Mini Self-appraisal, (3) Full Evaluation, (4) Mini Evaluation, and (5) Full Evaluation for Certification. The determination process for these five methods is very similar and uses the same kinds of evidence, but each is tailored to meet different needs. The Full Evaluation for Certification is an independent, third-party external evaluation of an organization's capability. It is the only Capability Determination method that can lead to certification by the ITSqc, and should be used when an organization wants a public record of its capabilities. It is based on evidence of the client organization's implementation of all of the Practices in an eSCM model. The evaluation data is rigorously reviewed by a certification board at the ITSqc and, when warranted, results in certification by the ITSqc of the organization's capability. The certificate is issued with a Capability Level and a Capability Profile, as well as a final report, with ratings and observations about each Practice and Capability Area, indicating strengths and areas for improvement. Unless otherwise specified by the sponsor, the certification level, summary of the coverage of the organization, and any rating exceptions of the evaluation are published on the ITSqc website (http://itsqc.cmu.edu).

3.5 Information Technology Infrastructure Library® (ITIL®)

3.5.1 Description

ITIL is a collection of best practices that are geared toward facilitating, maintaining, and improving the delivery of high-quality information technology (IT) services. The focus is on achieving high financial quality and value in IT deployment, operations, and support. ITIL groups all IT activities into five broad areas, with two broad areas being the most widely used: Service Support and Service Delivery. Extensive procedures exist to help an organization implement the ITIL functions, and include organizational structure and skill set considerations. ITIL has been codified in ISO standard 20000 and will be further developed to be an ISO/IEC 15504 (discussed later) conformant process reference model.

ITIL first appeared in the 1980s in the United Kingdom. Although developed in the 1980s, it wasn't until well into the 1990s that ITIL began to catch on. In the 1980s, the U.K. Government's Central Computer and Telecommunications Agency (CCTA) commissioned a project to address the development of an IT infrastructure management approach. The first released version of it (now called Version 1) comprised a series of publications, covering procedures in a number of areas related to IT operations and support. As interest in ITIL began to emerge from different parts of the world, the intent became to make it guidance, rather than a formal methodology.

Some aspects of ITIL, such as Service Management, have been codified in international standards, for example, ISO/IEC 20000. "ITIL is often considered alongside other best practice frameworks such as the Information Services Procurement Library (ISPL), the Application Services Library (ASL), Dynamic Systems Development Method (DSDM), the Capability Maturity Model [Integration] (CMMI), and is often linked with IT governance through Control OBjectives for Information and related Technology (COBIT) [45]." Some organizations have integrated both ITIL and the CMMI into a complementary process management framework [37].

In Version 2, seven books were created (information merged, supplemented, edited, and updated) to cover the broad spectrum of IT management, deployment, maintenance, and support "to link the *technical* implementation, operations, guidelines and requirements with the strategic management, operations management, and financial management of a modern business" [45]. These seven books addressed two areas, as follows:

- IT Service Management Set:
  - Service Delivery
  - Service Support
- Other Operational Guidance:
  - Information and Communication Technology (ICT) Infrastructure Management
  - Security Management
  - The Business Perspective
  - Application Management
  - Software Asset Management

In actuality, two additional books were published. One was intended to provide implementation guidance, primarily for Service Management. The other was intended to help smaller organizations implement ITIL.

In its current incarnation (Version 3), it has expanded to 31 volumes. Version 3 has five core texts, covering the following areas, and the following topics within each area:

1. *Service Strategy:* This volume is the heart of the ITIL Version 3 core. It is a view of ITIL that aligns information technology with business objectives. Specifically, it covers:
   - Strategy and value planning
   - Roles and responsibilities
   - Planning and implementing service strategies
   - Business planning and IT strategy linkage
   - Challenges, risks, and critical success factors
   The focus here is on translating the business strategy into an IT strategy, and selecting best practices for the specific domain.
2. *Service Design:* This volume provides guidance for creating and maintaining IT policies and architectures for the design of IT service solutions. Specifically, it covers:
   - The service life cycle
   - Roles and responsibilities

- Service design objectives and elements
- Selecting the appropriate model
- Cost model
- Benefit and risk analysis
- Implementation
- Measurement and control
- Critical Success Factors (CSFs) and risks

This volume also considers the impact of outsourcing, insourcing, and co-sourcing.

3. *Service Transition:* This volume provides guidance for transitioning IT services into the business environment. It covers change management and release practices over the long term. Specifically, it covers:
   - Managing change (organizational and cultural)
   - Knowledge management
   - Risk analysis
   - The principles of service transition
   - Life-cycle stages
   - Methods, practices and tools
   - Measurement and control
   - Other best practices

   While this does not look at deployment per se, what it does look at is transitioning service design to the production environment, which is a significant consideration for deployment.

4. *Service Operation:* This volume covers delivery and control processes. It is concerned with ensuring service stability:
   - Principles and life-cycle stages
   - Process fundamentals
   - Application management
   - Infrastructure management
   - Operations management
   - CSFs and risks
   - Control processes and functions

   The concern here is managing services in the production environment, including daily operational issues and fire fighting. Clearly, with effective service operation, "fire fighting" can be minimized.

5. *Continual Service Improvement (CSI):* This volume covers the processes involved in improving the business's service management, and service retirement:
   - The drivers for improvement
   - The principles of CSI
   - Roles and responsibilities
   - The benefits
   - Implementation

- Methods, practices, and tools
- Other best practices

Essentially, it describes how to improve services, post deployment.

3.5.2 Certification

There are three levels of certification under ITIL. They are the Foundation Certificate, Practitioners Certificate, and Managers Certificate. The Foundation Certificate, as its name implies, is the lowest level of certification within the ITIL structure, and is a prerequisite for certification at higher levels. It is the entry-level professional qualification in IT Service Management. It is obtained by taking and satisfactorily completing a multiple-choice test on topics covered in an ITIL Foundation's course.

The ITIL Practitioner certification, the next level of certification, is geared for those who are responsible within their organization for designing specific processes within the IT Service Management discipline and performing the activities that belong to those processes. It focuses on ensuring that there is a depth of understanding and application of those processes, treating each process as a specialization. An example would be change management. Included among the specializations are Service Desk and Incident Management, Change Management, Problem Management, Configuration Management, Release Management, Service Level Management, and Planning to Implement IT Service Management. There are courses that focus on developing the depth of understanding necessary to be a competent practitioner, and certification is awarded upon completion of practical assignments in each discipline, which prove the ability to apply the necessary skills in practice.

The Manager's Certificate in IT Service Management encompasses IT Service Management Service Support (based on ITIL) and IT Service Management Service Delivery (based on ITIL). It requires that the applicant be certificated at the Foundation level. It is targeted for managers and consultants in IT service management who are involved in implementing or consulting in ITIL. The prerequisites for certification at this level, in addition to the Foundation Certificate are:

■ Higher educational level or level obtained through practical experience or self-study
■ Good spoken and written language skills
■ Speaking skills, presentation skills, empathy, meeting skills, teamwork skills
■ At least two years of professional experience as a manager or consultant in the field of IT management

A written exam consisting of essay questions is administered to the applicants as a means of testing their knowledge of the practices and their ability to apply them, given a scenario.

There is no external audit scheme to rate an organization's implementation of the ITIL practices, as of the time of this writing; however, there is an ISO work item proposed to develop an additional part of the ISO/IEC standard 15504 to include an exemplar process assessment model for ISO/IEC 20000 (ITIL) (see Section 3.6 in this chapter). Organizations can be rated as compliant with some aspects of ITIL under ISO/IEC 2000, but no formal "ITIL-compliant" rating is currently available.

3.5.3 Connections between ITIL and eSCM

The ITIL focus is on service management and the eSCM models are concerned with sourcing capabilities. ITIL Version 3 provides some guidance on sourcing options and supplier management from a service management perspective. The eSCM models provide comprehensive guidance on the development and improvement of sourcing capabilities throughout the entire sourcing life cycle. For example, eSCMs focus on enterprise-level sourcing practices, such as contracting and pricing. ITIL Version 3 addresses service transition, design, and operation.

Supplier management is one aspect of a complete set of sourcing capabilities. ITIL Version 3 has a different structure than the eSCMs without defined capability levels or the means for evaluating organizational capabilities, in that the eSCMs focus on organizational capabilities, and ITIL focuses on individual certification.

ITIL Version 3 and the eSCM models were developed to be complementary and there are no known incompatibilities. In fact, ITIL Version 3 recommends the use of eSCMs for a more diligent approach to sourcing, and eSCMs reference ITIL for guidance about IT Services management.

3.6 ISO Standards and Models

The ISO (the International Organization for Standardization) and the IEC (the International Electrotechnical Commission) have developed standards for the international community developing systems. There are a number of ISO and ISO/IEC Standards that are applicable or have been applied to systems and software development, acquisition, maintenance, and quality. Among them are ISO 9001, ISO/IEC 12207, ISO/IEC 15504, and ISO/IEC 15288 standards. Many organizations implement these standards in order to be certified with ISO as being compliant with one or more of the various ISO standards. There are good business reasons for doing so. In many parts of the world, certification as being compliant with one or a number of ISO standards is essential in order to conduct business in these countries. Duly-authorized certification bodies conduct ISO audits of organizations to determine their compliance.

3.6.1 ISO 9001

ISO 9001 is standard under the ISO 9000 family of standards. ISO 9000 is a family of standards for quality management systems. Earlier versions of the ISO 9000 family included ISO 9001, ISO 9002, ISO 9003, and ISO 9000-3. ISO 9001 was a quality assurance standard for the design, development, production, installation, and servicing of systems, typically under a two-party agreement. ISO 9002 focused on quality assurance for the production, installation, and servicing of systems, and did not include the development of new products. An example application of ISO 9002 was for businesses that manufactured parts for clients that provided them with the drawings and specifications for the parts. ISO 9003 was a quality assurance standard for final inspection and test, covering only the final inspection and test of finished parts. Because the requirements of these earlier versions of the ISO 9000 family were targeted for manufacturing systems, they were often considered somewhat difficult to apply to software development. Consequently, the need arose to develop a guide on the application of these requirements to software. The guide was ISO 9000-3, which provided interpretations of the ISO 9001 requirements for software development and maintenance. Nonetheless, some organizations utilized ISO 9001 as the basis for developing their software quality system, rather than ISO 9000-3. ISO 9000-3 was considered a guidebook, and not a software quality system standard.

The ISO 9000:1994 family of standards was updated in 2000. The language was modified to address "products." This terminology was defined such that the standard could be better applied to systems, software, or services. The different variants of the ISO 9000 family (e.g., ISO 9002) were eliminated through this change in terminology, and now one standard applies to all possible applications of it; however, a guidebook for software life-cycle processes (ISO 90003) has been published and one for system life-cycle processes (ISO 90005:2008) was published in 2008. ISO 9001:2000 now specifies requirements for design, development, manufacture, installation, and/or product service for any product or that provides any form of service for a client (e.g., outsourced help desk services). It contains requirements that an organization needs to implement to achieve customer satisfaction through consistent products and services to meet customer needs and expectations.

For the most part, ISO 9001 specifies high-level requirements for the elements covered in the standard, and does not prescribe what practices should be implemented to achieve these requirements. Effectively, it requires that the organization have a Quality Manual and a Quality Policy to address these elements, and the organization specifies what practices it will apply in implementing these elements. Wikipedia [46] summarizes the requirements of ISO 9001:2000 as follows:

> The quality policy is a formal statement from management, closely linked to the business and marketing plan and to customer needs. The ... policy is understood and followed at all levels and by all employees. Each employee [is evaluated against] measurable objectives ....
>
> Decisions about the quality system are made based on recorded data .... The system is regularly audited and evaluated for conformance and effectiveness.

... a documented procedure [is needed] to control quality documents in [the] company. Everyone must have ... up-to-date documents and [know] how to use them.

To maintain the quality system and produce conforming product[s], [an organization] need[s] to provide suitable infrastructure, resources, information, equipment, measuring and monitoring devices, and environmental conditions.

[The organization must] map out all key processes in [the] company; control them by monitoring, measurement and analysis; and ensure that product quality objectives are met. If [a process] ... can't [be] monitor[ed] by measurement, [the organization should] make sure [that] the process is well enough defined [so] that [adjustments] ... can be [made] if the product does not meet user needs.

For each product ... [the company] need[s] to establish quality objectives; plan processes; and [the company needs to] document and measure results to use as a tool for improvement. For each process, [the company needs to] determine what kind of procedural documentation is required. (Note: a 'product' is hardware, software, services, processed materials, or a combination of these.)

[The company ...] ... determine key points where each process requires monitoring and measurement, and ensure that all monitoring and measuring devices are properly maintained and calibrated.

[The company must] ... have clear requirements for purchased product. Select suppliers appropriately and check that incoming product meets requirements.

[The company] need[s] to determine the skills required for each job in [the] company, suitably train employees and evaluate the effectiveness of the training.

[The company] need[s] to determine customer requirements and create systems for communicating with customers about product information, inquiries, contracts, orders, feedback and complaints.

For new products, [the company must] plan the stages of development, with appropriate testing at each stage testing and documenting whether the product meets design requirements, regulatory requirements and user needs.

[The company ...] ... regularly review performance through internal audits and meetings. [The company must] determine whether the quality system is working and what improvements can be made. [The company must] deal with past ... and potential problems. Keep records of these activities and ... resulting decisions, and monitor their effectiveness. (Note: [the company] need[s] a documented procedure for internal audits.)

[The company] need[s] documented procedures for dealing with actual and potential nonconformances (problems involving suppliers or customers, or internal problems). Ensure no one uses bad product, determine what to do with bad product, deal with the root cause of the problem and keep records ... to improve the system.

A detailed reading of both ISO 9001:2000 and the CMMI will reveal synergies between the standard and the model. ISO 9001 specifies requirements for which the CMMI identifies practices that organizations can implement to meet these requirements.

Organizations can be certified as having satisfactorily implemented the requirements of ISO 9001. There is a requirement that an organization be audited by a duly accredited certification body as part of the certification process. The applying organization is audited based on an extensive sample of its sites, functions, products, services, and processes. Any problems or noncompliances observed during the audit are reported to management. A determination is made of the severity of the problems, and if they are not major, the certification body will issue an ISO 9001 certificate for each geographical site it has visited; however, the organization must develop a satisfactory improvement plan showing how any reported problems will be resolved, and this must be delivered to the certification body before the certificate is issued. Periodic follow-up audits are conducted, typically every 6 months subsequent to the initial registration in order to maintain the registration, and a renewal audit is conducted generally every 3 years. Registered organizations must, therefore, demonstrate continuous compliance with their quality system documentation.

In essence, establishing conformance to the ISO 9001 standard plays a role in strategic quality planning. Certification is a lengthy process, involving the establishment of quality plans, procedures, etc. Preliminary audits may be used as a means of determining the readiness for an external third-party audit. As a result, the process of certification can take from 1 to 2 years from the date of inception to accomplish.

3.6.2 SPICE (ISO 15504)

Another ISO standard, ISO/IEC 15504, often referred to as SPICE (Software Process Improvement Capability dEtermination), is a framework for assessing process capability. It has gone through several evolutions, from originally being a technical report focused primarily on software with a normative component that described a set of 40 software life-cycle processes in terms of a purpose and outcomes, together with a measurement framework of capability levels and process attributes. The current version of ISO/IEC 15504 has broadened its scope beyond software and, in Part 2, purely defines the requirements for conducting process assessments, describes the measurement framework of capability levels and process attributes, and describes the requirements for process reference models to be used as a basis for a conformant process assessment model [28]. It has some resemblance to the continuous representation of the CMMI in that it measures the capability levels of PAs, and allows the organization being assessed to select from the model the processes to be measured. The continuous representation of the CMMI could be considered an ISO/IEC 15504-conformant process assessment model, and work is being undertaken to determine this.

In its present form, it is considerably reduced in size and complexity from the technical report and contains seven parts [28]. They are:

Part 1: Concepts and Vocabulary
Part 2: Performing an Assessment
Part 3: Guidance on Performing an Assessment
Part 4: Guidance on Use for Process Improvement and Process Capability Determination
Part 5: An Exemplar Process Assessment Model (PAM) for Software Life Cycle Processes
Part 6: An Exemplar Process Assessment Model (PAM) for Systems Life Cycle Processes
Part 7: Assessment of Organizational Maturity (2008)

Part 5 is an exemplar PAM based on ISO/IEC 12207, "Software Development Life Cycle Processes" (ISO 12207 is discussed in Section 3.6.3). Part 6 is another exemplar PAM based on ISO/IEC 15288, "Systems Development Life Cycle Processes," which is discussed in Section 3.6.4. Part 7 is an exemplar for a staged maturity model similar to the staged representation of the CMMI. There are plans for an additional part for a PAM for ITIL (ISO/IEC 20000).

Two different classes of models are specified in the framework of ISO 15504: process reference models and process assessment models. The purpose of the process reference models is to provide a description of the processes that are to be measured. The standard contains requirements for process reference models "that can be met by current and emerging standards" [39]. The major benefit of this is that:

> [I]t expands the available scope of the Standard. Instead of being constrained to the processes of the software life cycle, it will now define a common assessment process that can be applied to any processes defined in the required manner. Processes defined through an appropriate Process Reference Model are assessed in relation to the measurement scale for capability set out in the Standard. The scale is based upon the existing Capability Dimension of ISO/IEC TR 15504-2 [39].

The standard now relies on external process reference models for process descriptions, instead of prescribing them. This permits the application of the standard to a variety of process reference models, such as ISO/IEC 15288.

The purpose of the process assessment models is to support conducting an assessment or an appraisal. "Process assessment models may have significant differences in structure and content, but can be referenced to a common source (a Process Reference Model), providing a mechanism for harmonization between different approaches to assessment" [39]. This calls for consistency between the process attributes in ISO/IEC 15504 Part 2, a process reference model, and a process assessment model. In other words, the process assessment model must provide a direct mapping to the process reference model, and to the process attributes described in Part 2 of ISO/IEC 15504.

Relationships and mappings can be found between ISO 15504, ISO 12207, ISO 9001, ISO 15288, and the CMMI [32, 39]. In fact, when the CMMI was developed, the intent of producing two representations (i.e., the staged and the continuous) was to have the model be conformant with ISO/IEC 15504. The purpose of the continuous representation was to have a form of the model be ISO/IEC 15504-compliant. In a pilot SCAMPI appraisal conducted for Critical Software, S.A., of Portugal [32], the results demonstrated the ability to fulfill 15504 requirements and generate a 15504 process profile relative to a separate process reference model (the CMMI). Consequently, the validity of the conformity of the CMMI with ISO 15504 was demonstrated.

3.6.3 ISO 12207

ISO 12207, first published in 1995, is a life-cycle standard that defines processes for developing and maintaining software. The standard establishes the tasks and activities that can occur during the life cycle of software, including the processes and activities that can be implemented during the acquisition, fielding, support, operations, and maintenance of a software system. It is a modular standard in that it can be adapted to the needs of the using organization. The standard does not have to be adopted in its entirety, in that the organization can choose the processes that are of the most concern to them. The definition of the life-cycle processes has a structure of processes consisting of activities that are comprised of tasks and outcomes. In Annex F, Amendments 1 and 2, each process in ISO 12207 is described with a purpose statement, which is the high-level objective for performing the process, and the outcomes that are the achievable result of successful implementation of the process [44]. There are currently 49 processes defined in the standard [29].

This standard is the evolutionary result of a number of efforts to catalog and characterize the kinds of practices that organizations should implement during the development and maintenance cycles to produce quality software. This is different from models like the CMMI, which catalog processes that organizations should implement in their system development and maintenance efforts, in that ISO 12207 organizes the practices in a life-cycle structure, whereas the CMMI is life cycle phase-independent in its approach.

In 1979, the U.S. Department of Defense (DoD), recognizing that software delivered to it by its various contractors was not meeting schedule, cost, or quality objectives, convened a conference at the Naval Post-Graduate School in Monterey, California. This conference, called Monterey I, produced a series of recommendations; notably among them were recommendations to develop software development and quality policies, and software development and quality standards [43]. DoD Standards 2167, a standard for software development, and 2168, a standard for software quality, resulted from these initiatives. A problem raised by the using community with DoD-STD-2167 was that it was perceived as legislating top-down development utilizing a waterfall life-cycle model approach. The DoD responded to the objections and revised the standard. It was superseded by DoD-STD-2167A, which was still perceived as legislating a waterfall life-cycle model approach (but not top-down development), and then later by MIL-STD-498, which was felt to be more amenable to object-oriented development approaches, and accommodated more diverse life-cycle model implementations. The standard went through several additional iterations as a consequence of the DoD deciding to implement acquisition reform, an initiative that was intended for the DoD to rely more on contractor selection of commercial standards for producing hardware and software and implementing processes, rather than for the DoD to mandate government standards, some of which resulted in much higher system acquisition costs for the government than was necessary. Numerous DoD and Military standards were retired as a consequence of acquisition reform. Ultimately, ISO 12207 evolved out of all of this, incorporating some additional IEEE standards that served as transitional standards until ISO 12207 could be released. There is an IEEE version of 12207 for U.S. Government acquisitions so that the government could acquire whatever accompanying product documentation was felt necessary to support the system. Table 3.5 addresses the processes contained in the standard.

As noted in the discussion of ISO 15504, Part 5 is an exemplar Process Assessment Model based on ISO 12207. Assessments are now performed on organizations' implementation of this standard.

3.6.4 ISO 15288

ISO 15288 [27] defines life-cycle processes for systems development and maintenance, covering the system life cycle from concept to retirement. It is similar in concept to ISO 12207. Whereas ISO 12207 is more narrowly concerned with software, ISO 15288 is broader in scope and includes all aspects of a system: hardware, software, and human interfaces with a system. Also included are processes for acquiring and supplying system products and services.

The standard was first published in 2002, and is the first ISO standard to deal with system life-cycle processes. There was at one time an Electronic Industries Alliance (EIA) systems engineering maturity model (EIA/IS-731); however, this was incorporated into the CMMI, which now covers systems, software, services,

Table 3.5   ISO 12207 Processes

Focus	Category	Process
Primary Life Cycle Processes	Acquisition	Acquisition Preparation
		Supplier selection
		Supplier monitoring
		Customer acceptance
	Supply	Supplier tendering
		Contract agreement
		Software release
		Software acceptance
	Operation	Operational Use
		Customer support
	Engineering	Requirements elicitation
		System Requirements Analysis
		System Architecture Design
		Software Requirements Analysis
		Software Design
		Software Construction
		Software Integration
		Software Testing
		System Integration
		System Testing
		Software Installation
	Maintenance	Maintenance
Supporting Life Cycle Processes	Quality Assurance	Quality Assurance
		Verification
		Validation

Continued

Table 3.5   ISO 12207 Processes (*Continued*)

Focus	Category	Process
		Joint Review
		Audit
	Configuration Control	Documentation Management
		Configuration Management
		Problem Management
		Change Request Management
	Product Quality	Usability
		Product Evaluation
Organizational Life Cycle Processes	Management	Organizational Alignment
		Organization Management
		Project Management
		Quality Management
		Risk Management
		Measurement
	Improvement	Process establishment
		Process assessment
		Process improvement
	Resource and Infrastructure	Human Resource Management
		Training
		Knowledge Management
		Infrastructure
	Reuse	Asset Management
		Reuse Program Management
		Domain Engineering

and acquisition. Like ISO 12207, ISO 15288 organizes the standard into practices for different stages of a life cycle that are good practices. The stages are Concept, Development, Production, Utilization, Support, and Retirement [48]. The difference between ISO 15288 and the CMMI is that the CMMI defines good practices to implement in general, and organizes them by PA. Also like ISO 12207, the standard is modular and allows the using organization to construct a life cycle from the various components appropriate to its own needs, considering business needs and the system domains with which they are involved. The components can be used to construct a standard process for the organization or a project-specific environment. It can also be used as the contractual basis for specifying the process to implement in two-party agreements for the development of a system. Part 6 of ISO 15504 provides an Exemplar Assessment Model for assessments with ISO 15288.

The standard contains 26 processes, 123 outcomes, and 208 activities. The processes are shown in Table 3.6.

3.6.5 ISO Standard Interrelationships

There is an interrelationship between the other ISO standards described in this chapter. ISO 12207 and ISO 15288 define best practices for software and system life cycles, respectively. They are linked to ISO 9001 in that ISO 9001 defines the requirements for a quality system management designed to ensure that the outcomes of performing the practices of these standards will result in a quality product. They are also linked to ISO 15504 as a means of performing assessments against an organization's or a project's implementation of these standards to determine if the organization or the project is in compliance with the provisions of these standards. Thus, the ISO standards have evolved into a complementary set of standards that can be practically applied to define the kinds of processes that organizations should implement in their development and maintenance efforts, and an assessment methodology to determine how well they are accomplishing this.

3.7 Summary

Yogi Berra, the former Hall of Fame catcher for the New York Yankees, was noted for his many sayings. One of his most famous sayings is, "If you don't know where you are going, you will wind up somewhere else." Interestingly enough, this quotation also applies to process improvement. Many organizations set out to implement process improvement, but use nothing with which to guide the effort. They don't know where they are going and will wind up implementing a process improvement program that will produce unpredictable results.

In this chapter, we presented a number of models that can be used to guide process improvement efforts. As we have seen, some of them had a similar focus (e.g., systems development), and others had a different focus (e.g., systems

Table 3.6   ISO 15288 Processes

Category	Process
Agreement Processes	Acquisition
	Supply
Enterprise Processes	Enterprise Environment Management
	Investment Management
	System Life Cycle Processes Management
	Resource Management
	Quality Management
Project Processes	Project Planning
	Project Assessment
	Project Control
	Risk Management
	Configuration Management
	Information Management
	Decision Making
Technical Processes	Stakeholder Requirements Definition
	Requirements Analysis
	Architectural Design
	Implementation
	Integration
	Verification
	Transition
	Validation
	Operation
	Maintenance
	Disposal
Special Processes	Tailoring

development performed internally, information technology services, outsourcing, etc.). Many of these models had areas of overlap (e.g., management of the activities that were the focus of the model). Clearly, the model that an organization chooses to guide its process improvement efforts should be one that is in sync with the type of business it is in and in sync with the purpose of its process improvement effort. We must keep in mind that "[A]ll models are wrong, but some are useful" [2].

There is a corollary to this business of choosing an appropriate model. A process improvement model is essentially a roadmap. It tells you how to get from here to there with respect to process improvement. But roadmaps have limitations. Specifically, if you are lost and don't know where you are, a roadmap isn't going to help. To get from point A to point B, you have to know where point A is. That is the function of an assessment, appraisal, or an audit based on the model that an organization has chosen to use for process improvement. We discuss that aspect in Chapter 5, that is, performing an appraisal to determine what the current state of the practice is in order to determine what the plan of action should be in order to implement process improvement. In that chapter, we will use the CMMI to illustrate that point.

References

1. *Appraisal Program.* Process Maturity Profile, CMMI®, SCAMPI[SM] Class A Appraisal Results, 2008 Mid-Year Update, Pittsburgh, PA: Software Engineering Institute, Carnegie Mellon University, September 2008.
2. Box, G.E.P. and Draper, N.R., *Empirical Model-Building and Response Surfaces,* New York: John Wiley & Sons, 1987, p. 424.
3. CMMI Product Team. Appraisal Requirements for CMMI[SM], Version 1.1 (ARC, V1.1) (CMU/SEI-2001-TR-034), Pittsburgh, PA: Software Engineering Institute, Carnegie Mellon University, December 2001.
4. CMMI Product Team. CMMI for Systems Engineering, Software Engineering, Integrated Process and Product Development, and Supplier Sourcing, (CMU/SEI-2002-TR-012), Pittsburgh, PA: Software Engineering Institute, Carnegie Mellon University, 2002.
5. CMMI Product Team. CMMI for Development, Version 1.2 (CMU/SEI-2006-TR-008), Pittsburgh, PA: Software Engineering Institute, Carnegie Mellon University, 2006.
6. CMMI Upgrade Team. Appraisal Requirements for CMMI[SM], Version 1.2 (ARC, V1.2) (CMU/SEI-2006-TR-011), Pittsburgh, PA: Software Engineering Institute, Carnegie Mellon University, August 2006.
7. CMMI Product Team. CMMI for Acquisition, Version 1.2 (CMU/SEI-2007-TR-017), Pittsburgh, PA: Software Engineering Institute, Carnegie Mellon University, November 2007.

8. Cooper, J., Fisher, M., and Sherer, S.W., Editors, Software Acquisition Capability Maturity Model (SA-CMM), Version 1.02, Technical Report CMU/SEI-99-TR-002. Pittsburgh, PA: Software Engineering Institute, Carnegie Mellon University, 1999.

9. Cooper, J. and Fisher, M., Software Acquisition Capability Maturity Model (SA-CMM) Version 1.03, CMU/SEI-2002-TR-010. Pittsburgh, PA: Software Engineering Institute, Carnegie Mellon University, 2002.

10. Curtis, B., Hefley, W.E., and Miller, S., Overview of the People Capability Maturity Model, [CMU/SEI-95-MM-01 (AD-A301167S)]. Pittsburgh, PA: Software Engineering Institute, Carnegie Mellon University, September 1995.

11. Curtis, B., Hefley, W.E., and Miller, S., People Capability Maturity Model. [CMU/SEI-95-MM-02 (AD-A300822)]. Pittsburgh, PA: Software Engineering Institute, Carnegie Mellon University, September 1995.

12. Curtis, B., Hefley, W.E., and Miller, S., *The People Capability Maturity Model: Guidelines for Improving the Workforce*, (ISBN 0-201-60445-0). Reading, MA: Addison Wesley Longman, 2002.

13. Curtis, B., Hefley, W.E., and Miller, S.A., People Capability Maturity Model (P-CMM) Version 2.0, (CMU/SEI-2001-MM-01), Pittsburgh, PA: Software Engineering Institute, Carnegie Mellon University, July 2001.

14. Customer Operations Performance Center, COPC-2000® CSP Standard, http://www.copc.com/standards.aspx (accessed 26 December 2008).

15. Davis, N. and Mullaney, J., The Team Software Process[SM] (TSP[SM]) in Practice: A Summary of Recent Results, (CMU/SEI-2003-TR-014), Software Engineering Institute, Carnegie Mellon University, September, 2003.

16. Dunaway, D.K. and Masters, S., CMM[SM]-Based Appraisal for Internal Process Improvement (CBA-IPI): Method Description, Pittsburgh, PA: Software Engineering Institute, Carnegie Mellon University, April 1996.

17. Electronic Industries Alliance, EIA Interim Standard: Systems Engineering (EIA/IS-731), Washington, D.C.: Electronic Industries Alliance, 1994.

18. Hayes, W. and Over, J.W., The Personal Software Process[SM] (PSP[SM]): An Empirical Study of the Impact of PSP on Individual Engineers, (CMU/SEI-97-TR-001), Pittsburgh, PA: Software Engineering Institute, Carnegie Mellon University, December 1997.

19. Hefley, W.E. and Curtis, B., People CMM®-Based Assessment Method Description, (CMU/SEI-98-TR-012), Pittsburgh, PA: Software Engineering Institute, Carnegie Mellon University, 1998.

20. Hefley, W.E. and Miller, S.A., Software CMM® or CMMI®? The People CMM® Supports Both, Software Engineering Process Group (SEPG) Conference, Boston, MA, February 24–27, 2003.

21. Hefley, W.E. and Loesche, E.A.., The eSourcing Capability Model for Client Organizations (eSCM-CL): Part One, Model Overview, [ITsqc Working Paper Series # CMU-ITSQC-WP-06-001]. Pittsburgh, PA: IT Services Qualification Center, Carnegie Mellon University, 2006.

22. Humphrey, W.S., Characterizing the Software Process: A Maturity Framework, CMU/SEI-87-112, June 1987.

23. Humphrey, W.S. and Kitson, D.H., Preliminary Report on Conducting SEI-Assisted Assessments of Software Engineering Capability, SEI Technical Report SEI-87-TR-16, July 1987.

24. Humphrey, W.S., Three Dimensions of Process Improvement, Part II: The Personal Process, *CrossTalk*, March 1998.

25. Humphrey, W.S., Three Dimensions of Process Improvement, Part III: The Team Process, *CrossTalk*, April 1998.

26. Humphrey, W.S., The Personal Software Process[SM] (PSP[SM]), (CMU/SEI-2000-TR-022), Pittsburgh, PA: Software Engineering Institute, Carnegie Mellon University, February 2000.

27. International Organization for Standardization, ISO 15288, System Life Cycle Processes, 2002.

28. International Organization for Standardization, ISO 15504, Software Process Improvement and Capability Determination, 2007.

29. International Organization for Standardization, ISO 12207, Systems and Software Engineering—Software Life Cycle Processes, 2008.

30. IT Governance Institute, CobiT 4.1, 2007.

31. Kenett, R.S. and Baker, E.R., *Software Process Quality: Management and Control*, New York: Marcel Dekker, Inc., 1999.

32. Kitson, D., Kitson, J., Rout, R., and Sousa, P., The CMMI® Product Suite and International Standards, *National Defense Industries Association 6th Annual CMMI Technology Conference & User Group*, Denver, CO, November 11–17, 2006.

33. McHale, J. and Wall, D.S., Mapping TSP to CMMI, (CMU/SEI-2004-TR-014), Software Engineering Institute, Carnegie Mellon University, April 2005.

34. Members of the Assessment Method Integration Team, Standard CMMI[SM] Appraisal Method for Process Improvement (SCAMPI[SM]), Version 1.1: Method Definition Document, (CMU/SEI-2001-HB-001), Pittsburgh, PA: Software Engineering Institute, Carnegie Mellon University, 2001.

35. Nandyal, R., Shoehorning CMMI Initiatives with People CMM, *Software Engineering Process Group (SEPG) Conference,* Nashville, TN, March 6–9, 2006.

36. Paulk, M.C., Weber, C.V., Curtis, B., and Chrissis, M.B., *The Capability Maturity Model: Guidelines for Improving the Software Process.* Reading, MA: Addison-Wesley Publishing Co., 1995.

37. Phifer, B. and Hayes, T., New Rules for Old Applications: Introducing ITIL into Your CMMI Improvement Program, *Software Engineering Process Group (SEPG) Conference*, Austin, TX, March 26–29, 2007.

38. Radice, R., Interpreting SCAMPISM for a People CMM® Appraisal at Tata Consultancy Services, (CMU/SEI-2005-SR-001), Pittsburgh, PA: Software Engineering Institute, Carnegie Mellon University, February 2005.

39. Rout, T., ISO/IEC 15504—Evolution to an International Standard, *Software Process: Improvement and Practice*, 8(1), 27–40, January/March 2003.

40. Sanchez, K., CMMI and People CMM: Anything in Common?, *Software Engineering Process Group (SEPG) Conference*, Nashville, TN, March 6–9, 2006.

41. SCAMPI B and C Project, Handbook for Conducting Standard CMMI Appraisal Method for Process Improvement (SCAMPI) B and C Appraisals, Version 1.1, (CMU/SEI—2005-HB-005), Pittsburgh, PA: Software Engineering Institute, Carnegie Mellon University, December 2005.

42. SCAMPI Upgrade Team. Standard CMMI® Appraisal Method for Process Improvement (SCAMPI[SM]) A, Version 1.2: Method Definition Document (CMU/SEI-2006-HB-002), Pittsburgh, PA: Software Engineering Institute, Carnegie Mellon University, 2006.

43. Software Engineering Institute, Services Supplement for CMMI, Version 1.2, training materials, October 2008.
44. U.S. Joint Logistics Commanders, *Proceedings of the Joint Logistics Commanders Joint Policy Coordinating Group on Computer Resource Management Computer Software Management Subgroup Software Workshop*, Monterey, CA, August 1979.
45. Uthayanaka Tanin, *Implementing ISO/IEC 12207 One Process at a Time,* Singapore: SPIN, August 7, 2006.
46. Wikipedia, http://en.wikipedia.org/wiki/ITIL, (accessed 29 December 2008).
47. Wikipedia, http://en.wikipedia.org/wiki/ISO_9001, (accessed 29 December 2008).
48. Wikipedia, http://en.wikipedia.org/wiki/ISO_15288, (accessed 29 December 2008).
49. CMII Product Team. *CMII for Services, Version 1.2* (SMU/SEI-2009-TR-001), Pittsburgh, PA: Software Engineering Institute, Carnegie Mellon University, February 2009.

Chapter 4

System and Software Measurement Programs

Synopsis

In Chapter 2 we discussed measurement programs in relation to process improvement initiatives. In this chapter we dive deeper into measurement programs in system and software development organizations and discuss their implementation and the benefits to be accrued from such implementations. We begin with a description of measures used successfully by organizations to satisfy their information needs. We show how an organization developing systems and software selects measures, expanding our discussion of GQM in Chapter 1. We follow with two case studies and conclude with a discussion of how advanced quantitative techniques can be applied in managing and controlling development and maintenance efforts. This discussion demonstrates how statistical and probabilistic techniques described by the CMMI's high maturity practices can be used in managing such efforts.

4.1 Defining Measures for Process Control and Improvement

A measurement program serves many purposes. It provides an organization with the ability to manage and control projects, understand the health and well-being of the company's business, and improve how the organization does business. It is also an important element with respect to process improvement. In our discussion of the CMMI in Chapter 3, we showed how the understanding of process effectiveness moves from the qualitative at lower maturity levels to the quantitative at higher maturity levels. An effective measurement program is essential for evaluating the effectiveness of improvements made to a process, as well as the return on investment. In the final analysis, if a process improvement is effective, but saves very little money, it is hardly worthwhile implementing.

As already stated, in managing processes we measure to understand, control, and improve processes. There are several ways to determine what to measure in an organization. The more successful ones tie the measurement program to the organizational business goals, objectives, and information needs. Establishing such a link is the starting point for designing an effective and efficient measurement program. Measurements derived from the business goals and objectives, and accumulated at the organizational level, need to provide proper coverage of these goals and objectives. Operational or tactical measures are typically collected on projects or organizational elements at lower levels in order to fulfill informational needs at those levels. For example, the requirements of a specific customer or a specific type of product may drive additional measures that a development project will collect. In another example, flight safety considerations may drive the need for critical measurements made on avionics systems, and these measures may be unique to the product line. Nonetheless, a standard set of measures and metrics that are driven from the organization's business and informational needs will always exist.

Chapter 1 introduced Basili's Goal/Question/Metrics (GQM) paradigm [3]. In this chapter we show how it is applied to system and software development. GQM can be used as an effective methodology to guide the development of a measurement process from planning to execution and evaluation. Without the use of a structured methodology for determining what measures to collect, organizations will often essentially select measures arbitrarily, resulting in measurements that no one uses, and at great cost to the organization. For illustration purposes, we provide four real-life GQM examples.

Example 1:

Goal:	Manage and measure the progress of a project
Subgoal:	Manage and measure the progress of tasks and the overall project
Question:	*How good is my ability to plan?*
	What are the actual and planned completion dates?
	Is there a milestone slip?
	Do I have problems with resource allocation?

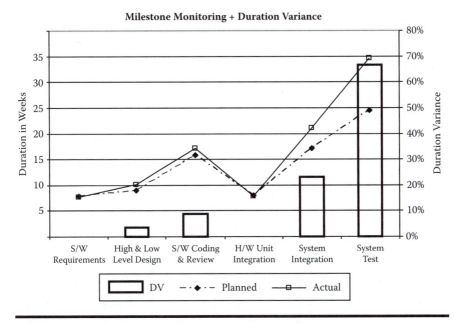

Figure 4.1   Milestone monitoring example.

Metric:	Planning Duration Variance
Metric:	Milestone Monitoring

The supplier is required to provide (every month) the planned completion date and the actual completion date for each of the following life-cycle phases: Requirements, Design, Code, Integration Test, and System Test. Figure 4.1 shows no duration variance at Hardware Unit Integration where duration variance is the absolute relative difference between planned duration and actual duration of a development phase. Somehow, it seems that this step was poorly carried out, given the large duration deviation at System Test.

Example 2:

Goal:	Manage and measure the progress of the project
Subgoal:	Manage and measure the progress of the integration tests
Question:	*When will the release be ready for deployment?*
	What is the quality of the tested software?
Metric:	Fault/Fix History

In this example, the supplier (a software development organization) is required to provide a weekly plot of cumulative faults (Figure 4.2) and open, unfixed faults classified by severity (Figure 4.3). The relevant period is Integration Test and System Test.

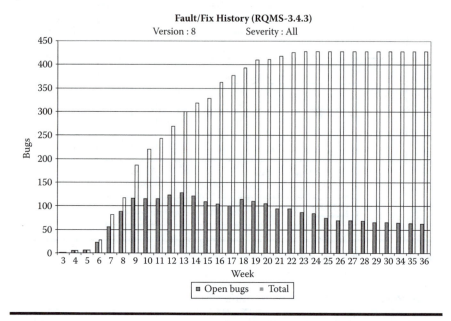

Figure 4.2 Accumulated total problem reports and weekly number of open bugs during integration and system testing.

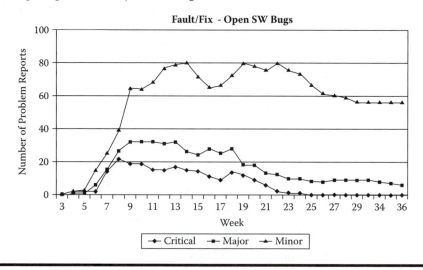

Figure 4.3 Open bugs during integration and system testing.

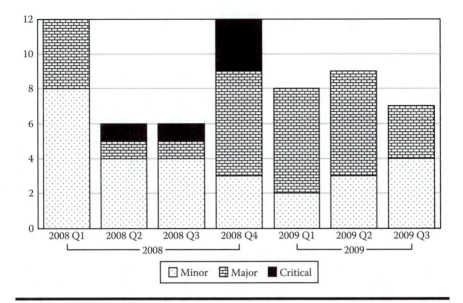

Figure 4.4   Problem reports by severity.

Example 3:

Goal:	Improved released product quality
Question:	What area of the product has the highest concentration of defects?
	How many failures were found by severity?
Metric:	Number of Problem Reports

Figure 4.4 shows the number of cases reported from the field by severity, and groups (Failures, Change requests, etc.) per month.

Example 4:

Goal:	Increase customer satisfaction — confidence
Question:	How satisfied and confident our customers with the quality of our products and services?
	How satisfied and confident our customer with our response to customers requests?

	How satisfied are our customers with the ability to meet commitments?
Metric:	Customer Service Response Time

Figure 4.5 shows the length of time from receipt of any form of communication to the appropriate supplier organization that records a documented problem until the customers receive the disposition of the problem.

When an organization's goals and objectives are stated at a very high level, formulating questions that can be converted to metrics might be difficult. For example, if an organization's goal is to increase market share by 5% each year in a specific market segment, it may be difficult to determine, on that basis, specific metrics that can be applied to development or manufacturing processes. An intermediate step, using Critical Success Factors, can help in formulating questions and defining the corresponding metrics. A Critical Success Factor (CSF) can be characterized as "the few key areas where things must go right for the business to flourish" and "areas of activity that should receive constant and careful attention from management" [17]. The CSFs indicate what has to go right in order to achieve the high-level goals. Questions can then be formulated to let management know if the CSFs are being achieved. Metrics are then derived by analyzing the questions to determine what quantitative means exist to answer them. Time to market, for example, is the elapsed time, or duration, to bring a new product through the design, development, and deployment steps. Time-to-market analysis often must include management processes, which may also include an analysis of performance and quality.

Once the metrics are defined, they must be prioritized. Then, the underlying measures must be selected and collected. Measurement programs for systems and

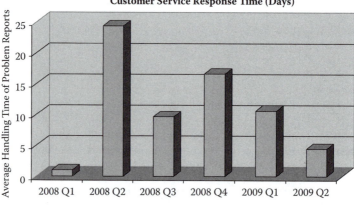

Figure 4.5 Customer service response delay.

software processes should be tailored to meet organizational needs. A number of sources are available in the literature to help organizations plan and implement measurement programs. For example, *Practical Software and Systems Measurement* (PSM) [14] is a publication that provides guidance on establishing measurement programs and provides example measures and metrics. Another example is ISO/IEC 15939 [8], an international standard that defines a measurement process for software development and systems engineering. It describes a process for establishing measurement programs, but does not provide the detailed guidance that the PSM does. The CMMI, in the Measurement and Analysis Process Area description [4], identifies a series of steps that are usually taken to establish a measurement program. IEEE Standards 1045, Standard for Software Productivity Metrics [1], and 982.1, Standard Dictionary of Measures of the Software Aspects of Dependability [2], also provide a starting point for defining base measures that can be used to satisfy the organization's information needs. IEEE Standard 1045 provides details on measures that lead to a better understanding of the software process, and which may lend insight into improving it. It neither establishes industry software productivity norms, nor does it recommend what productivity measurements to take. IEEE Standard 982.1 provides a set of software reliability measures that can be applied to the software product as well as to the development and support processes. Like the IEEE Standard 1045, it is not prescriptive.

In overview, the establishment of a measurement program consists of the following steps. The business goals, objectives, and information needs are codified first. Based on that, and using, for example, the GQM methodology, the metrics and measures are defined so that they provide the necessary information to determine if the goals, objectives, and information needs are being met. The processes, services, and products that will yield the necessary measures and metrics are then determined. Along with that, methods for collecting and storing the data, and the methods of analysis, are specified. Some sample measures that can be used to form the basis of a systems and software measurement program are discussed below. These measures are:

- Size
- Defects
- Effort
- Duration
- Cost
- Customer satisfaction

4.1.1 Size

Anyone who has built a house, or had his house painted, is familiar with the concept of size and the use that is made of that. The cost of building a house is often quoted in so many dollars per square foot, as is the cost for painting one. The estimated time that it will take to complete the job is also based on the planned or existing size, multiplied by some historical productivity factor that is based on size (e.g., hours of painting per square foot and gallons of paint per square foot). Whether we are talking of building houses, airplanes, or developing software, size estimates and corresponding productivity numbers are an important factor for estimating cost, effort, and schedule.

To measure process performance, a measure of the size of the deliverable that the process produces is extremely useful. This provides the capability to investigate organizational productivity or quality. Common measures of software productivity are in terms of lines of code per day or function points per day. Manufacturing productivity can be measured in terms of the number of batches of parts produced per day or per week, where a batch is a given number of parts. Size can also be used as a parameter in measuring quality. The quality of software is often cited in terms of defects per thousand lines of source code (KSLOC) or defects per function point. It is also used in manufacturing, such as in soldering defects per thousand solder points.

In determining "size," we need a measure that is standardized across projects and independent of the methodology, technology, or development tools used to create the systems or software. Commonly cited measures of size include pages of documentation, number of objects, lines of code or function points, and number of manufactured units. For software, lines of code are typically measured without counting embedded comments in the source code and without special-purpose code used only for testing and debugging. The SEI guideline [5] for measuring lines of code is a standard that can be used to ensure that consistent definitions are used within the organization. As noted in Chapter 2, we emphasize again the importance of operational definitions, in this case to ensure that a line of code means the same thing to everyone involved. A function points is a dimensionless number that is representative of the extent of the functionality of the software—hence, a size measure. Guidelines for counting function points are provided in publications available from the International Function Points User Group (IFPUG). Function points also enable consistent size definitions within an organization, or within the software industry, for that matter, because they are independent of the programming language used for any project.

4.1.2 Defects

Organizations need to categorize and quantify defects resulting from their development, manufacturing, or service processes. This enables the organization to develop measures of product or service quality, essential for determining how well the processes are working for the organization. Each organization needs to clearly define defects for its own processes. IEEE Standard 1044, "Software Quality Measures" [7], can be an aid in defining defects for software in that it defines a framework for measuring software defects. Manufacturing defects are relatively easy to define in that tolerances for part acceptability are typically specified. Defects in services may not be as straightforward. The number of times a client has to call back customer support to get a problem resolved is one measure that is often captured in customer satisfaction surveys.

4.1.3 Effort

The amount of work effort involved in developing and supporting systems and software needs to be measured. Collecting these measures is helpful in estimating project cost and staffing. Effort is typically measured in staff-hours, staff-months, or staff-years. The tracking effort is always related to a process or a part of a process. For example, we can estimate and track effort by life-cycle phase or task.

4.1.4 Duration

A key measure is the calendar time required to complete a project, phase, or task. Collecting this data and maintaining it in a database is helpful in estimating future project schedules. This time can expressed as elapsed time in days, and may initially be calculated as the difference between the start date and the finish date. Later, duration may take into account such things as vacations, non-worked days, etc. to reflect the true elapsed time between a project's start and finish dates. Duration is tied to effort and may be correlated to other measures, such as software size, number of pages of documentation, or number of drawings.

4.1.5 Cost

Organizations need to measure the overall costs of developing and supporting system and software processes. Collecting these measures and maintaining them in a database are helpful for assisting in the estimation of project cost. Each organization must have a clear definition of what will be included in this measure and how it will be calculated (i.e., its operational definition). For example, the two major cost drivers in software development organizations are the human and computer resource utilization. Cost is tied to effort and may be correlated to other measures, such as software size.

4.1.6 Customer Satisfaction

Organizations need to focus on the needs of their customers. By measuring what is important to customers (e.g., response time to process a query or wait time on hold until a call to customer support is answered), organizations can target areas of their processes for improvement. It is also important to measure how satisfied the customer is with the products or services that the organization is providing.

A program to measure customer satisfaction must be carefully designed in order to preclude bias in the data. Designing a customer satisfaction survey questionnaire is not a trivial task. The wording of the questions, the scales for capturing responses, and the overall design of the questionnaire can introduce bias that cannot be corrected by statistical analysis. Professionals in planning, designing, and implementing customer satisfaction surveys need to be consulted in order to avoid producing ineffective or misleading reports. For more on the measurement of customer satisfaction and proper statistical analysis, see [11, 12].

4.2 Effectiveness of Process Measurement Programs

In this section we describe the attributes of an effective process measurement program and highlight positive and negative impact factors.

4.2.1 Attributes

The effectiveness of a measurement program is not contingent upon the existence of all the attributes included in this section. However, a process measurement program that lacks many of these attributes has a high risk of failure. The eleven attributes listed below represent the combined experience of the authors and the writings of general management consultants, such as Deming and Juran, as well as software process improvement specialists such as Watts Humphrey, the former Director of the Software Process Program at the SEI [10, 12, 13, 16]. For a comprehensive treatment of applications of statistics in quality and reliability in general, and measurement programs in particular, see articles such as "Six Sigma, Software Reliability and Data Collection"[20].

The attributes of an effective measurement program are:

1. *Measurement as a mission-critical function:* Measurement is a mission-critical function. The program should not be seen as overhead, and it does not require continuous justification.
2. *The measurement program is aligned with business objectives:* The program is aligned with business objectives and helps in the determination of how well the business objectives are being met. The value and benefits of measurement are understood by clients, sponsors, stakeholders, and other corporate departments.
3. *The measurement program is supported by management:* Support for the program by all levels of management is demonstrated in the following ways:
   - An empowered champion or sponsor leads the implementation and long-term support of the measurement program.

- Staff members responsible for the development and support of the measurement program are selected based on their qualifications, not by their availability.
- The importance of the program in the organization is demonstrated by the proactive participation of senior-level personnel.
- Management uses the measurement information to plan, control, and improve the processes and products of the organization.
- The focus is on quantifiable process improvement, not on measurement for the sake of measurement.

4. *Measurement is linked to decision making:* Appropriate measurement information is distributed to and used by all levels of management, as well as by team members. Measurement information is used to manage and estimate. It is also used as part of the decision-making process internally and with customers.

5. *Action plans are derived from reported measurements:* There are action plans for process improvement derived from measurement. Measurement provides evidence of the existence and magnitude of process improvements. This includes cost savings, increased quality, improved productivity, etc. In the case of processes under statistical process control, the action plans show the anticipated improvement in process performance baselines, and quality and process performance goals and objectives.

6. *The measurement program is integrated into the development effort:* The measurement program is integrated into the systems and software development and support processes. This means that the measurement tasks are included as part of the standard development process. In composing the process for a project, subprocesses are selected that will yield the necessary measures and metrics. Where possible, computer-based tools are acquired and incorporated into the development, test, and/or maintenance environments to facilitate the process of collecting the raw data. Raw data is collected, to the maximum extent possible, as a by-product of the developers' (or maintainers') daily activities, without requiring any added effort on their part.

7. *The measures are standardized and documented:* Accurate, repeatable, consistent data is maintained across all reporting entities (e.g., effort tracking and size tracking). Consistent definitions (operational definitions) of the measures are used across the organization to facilitate uniform understanding of the data. Procedures exist for data collection, storage, and analysis. A loss of or a change in personnel does not affect the quality and validity of data. Adherence to standards is audited. Continuous process improvement and the use of measurement data are included in an ongoing education program.

8. *The program uses a balanced set of metrics:* A balanced set of related metrics is used to gain a global and complete perspective. There is no reliance on a single metric for decision making. There is an appreciation of the value of sev-

eral compatible metrics. The inherent risk to decision making from reliance on a single metric is recognized, and no single metric is viewed as a panacea.

9. *The program is integrated with organizational initiatives:* The measurement program is integrated with initiatives such as continuous process improvement or Six Sigma. In proposing changes or improvements to the process, estimates are made of the impact on process performance baselines, as appropriate. Where possible, quantitative estimates are made of the proposed improvements in terms of quality and process performance goals and objectives. Proposed improvements are backed up by estimates of return on investment. Obviously, changes that cost more to implement than produce cost savings would not be changes to implement.

10. *The program is part of the culture:* Measurement is part of the corporate culture. Continuous improvement is demonstrated through measurement. For processes under statistical process control, actual improvements on process performance baselines or predicted quality and process performance objectives are monitored and compared to predicted improvement. The program remains sound despite turnover of personnel (at all levels), and it evolves to reflect changes in the business.

11. *Measurement focuses on processes—not people:* Measurement information is used to improve system and software development and maintenance processes, and not to measure individual performance. The program measures the process, not the people. Productivity may sometimes be mistakenly thought of as a people measure; however, productivity is really measuring the process in terms of its capability (in terms such as lines of code per day, number of cars assembled per day, or some such similar measure) *across the organization,* not with respect to one individual developer. We look at process improvement with respect to its ability to improve productivity across the board for the entire organization, not individuals. If measurement information is used inappropriately to measure individual performance, it can become the downfall of the measurement program. Nonetheless, it is possible, and even in some cases desirable, to assess individual performance data in the context of process improvement. We want to be able to understand if the workers in a peer group are performing within, above, or below process performance norms, and to see how the individual is performing relative to the peer group. To use this data in such a context in an individual's performance review requires identifying systemic, structural components of performance that are due to an individual's performance in a positive or negative way. Such analysis involves control charts as described in [11]. If such an analysis is properly conducted, one can identify the performance of individuals beyond expectations, within expectations, and below expectations. These individual measurements should focus on the individual's contribution to process performance (i.e., performance of that process relative to expected levels) rather than on the perspective of rewards or punishment for the individual.

4.2.2 Impact Factors

There are a number of factors that influence the effectiveness of measurement programs in organizations. Some have a positive impact, and others have a negative impact.

4.2.2.1 Positive Impact Factors

Factors with a positive impact can be grouped according to business value, awareness, and training:

1. *Business value:* Factors related to business value include:
   - Integrating the use of metrics into all levels of management decision making
   - Demonstrating to line management the business value of measurement by providing management information
   - Defining objectives for the measurement program
   - Ensuring the appropriate use and presentation of the data (measure processes, not people)
   - Communicating results promptly and appropriately
   - Establishing realistic management expectations for measurement
2. *Awareness:* The factors related to awareness include:
   - Establishing strong public and private sponsorship at all levels, including high-level, committed corporate sponsorship
   - Maintaining an awareness of the time required to conduct and implement measurement
   - Communicating measurement results to the relevant stakeholders; relevant stakeholders include the work staff performing the processes for which measurements have been obtained
3. *Training:* The factors related to training include:
   - Developing the measurement program with the same staffing consideration and disciplines as any other highly visible, high-risk project (phased approach, adequate resources, objectives, realistic time lines, etc.)
   - Communicating the purpose, benefits, and vision for measurement to all levels of the organization, including all levels of management
   - Including training and education on metrics, and data analysis for all measurement participants (for software projects, this would include methods for counting lines of source code or function points, as appropriate, or any other sizing approaches used by the organization)

4.2.2.2 Negative Impact Factors

Factors that organizations have found to negatively impact their measurement programs when improperly applied fall into the categories of management and communication:

1. *Management:* Factors related to management include:
   - Lack of ongoing commitment or support from management (e.g., under-funding, inadequately and inappropriately staffing, disinterest in the metrics provided, or lack of accountability)
   - A view of measurement as a finite project
   - Unclear goals and objectives for measurement
   - Misuse of measurement data for individual performance reviews
   - Failure to understand what is being measured and what the measurement results mean
   - Failure to act when data indicates that corrective action should be taken
2. *Communication:* Factors related to communication include:
   - Lack of communication of results and measurement information to the relevant stakeholders and the organization, as appropriate
   - Negligence in responding to questions and concerns from staff about measurement and the measurement program/process
   - Unmanaged expectations (e.g., unrealistic time frames for implementing a measurement program)
   - An expectation that measurement will automatically cause change

4.2.3 Measuring Impact

Kenett et al. [13] suggested evaluating the impact of improvement and problem resolution projects derived from measurement programs using a formula of Practical Statistical Efficiency (PSE). The PSE formula accounts for eight components and is computed as follows:

$$PSE = V\{D\} \times V\{M\} \times V\{P\} \times V\{PS\} \times P\{S\} \times P\{I\} \times T\{I\} \times E\{R\}$$

where:
   $V\{D\}$ = Value of the data actually collected
   $V\{M\}$ = Value of the statistical method employed
   $V\{P\}$ = Value of the problem to be solved
   $V\{PS\}$ = Value of the problem actually solved
   $P\{S\}$ = Probability level that the problem actually gets solved
   $P\{I\}$ = Probability level the solution is actually implemented
   $T\{I\}$ = Time the solution stays implemented
   $E\{R\}$ = Expected number of replications

These components can be assessed qualitatively, using expert opinions, or quantitatively if the relevant data exists. A straightforward approach to evaluate PSE is to use a scale of "1" for not very good to "5" for excellent. This method of scoring can be applied uniformly for all PSE components. Some of the PSE components can be also assessed quantitatively using measurements as opposed to subjective evaluations. P{S} and P{I} are probability levels, T{I} can be measured in months, and V{P} and V{PS} can be evaluated in any currency, such as euros, dollars, or English pounds. V(PS) is the value of the problem actually solved, as a fraction of the problem to be solved. If this is evaluated qualitatively, a large portion would be "4" or "5," a small portion "1" or "2." V{D} is the value of the data actually collected. Whether PSE terms are evaluated quantitatively or qualitatively, PSE is a conceptual measure rather than a numerically precise one. A more elaborated approach to PSE evaluation can include differential weighting of the PSE components and/or nonlinear assessments.

4.3 Measurement Program Implementation Steps

A measurement program does not magically appear. In establishing the scope and content of a measurement program, we must be cognizant of the fact that it must be implemented in stages. In Chapter 8 we deal with the mechanics of operating a measurement program. This section, however, focuses on establishing what to measure and report, responsibilities for measurement, and defining the stakeholders for the measurements. Table 4.1 lists phases and activities for implementing a measurement program. This is a suggested implementation. Each activity in the table is explained on the following pages.

Table 4.1   Measurement Program Implementation

Section	Phase	Activities
4.5.1	Plan/Evaluate	• Definition of goals, objectives, and benefits • Establishment of sponsorship • Communication and promotion of measurement • Identification of roles and responsibilities
4.5.2	Analyze	• Audience analysis and target metrics identification • Definition of software measures • Definition of the data collection, analysis, and data storage approach
4.5.3	Implement/Measure	• Education • Reporting and publishing results
4.5.4	Improve	• Managing expectations • Managing with metrics

This "project-based" approach to establishing a measurement program ensures that appropriate up-front planning is done before measurement begins. However, the exact order of the activities may vary from company to company. For example, some organizations need to establish sponsorship before defining goals. Once the "project" of implementing a measurement program is completed, the measurement program should be actively used to support management decision making. The most important steps in establishing a measurement program are to *clearly* define the goals and objectives.

It is critical that the defined goals of your measurement program apply directly to your organization's goals. This is the basis for determining what you will measure, what metrics you will use, and what data you need to collect based on the capabilities and maturity of your organization. The key is to look at the strategic and tactical goals of your organization and determine what you need to measure to provide management with information to track its progress toward its goals. For a discussion, with examples, of how to tailor these factors to organizational culture, see [10].

4.3.1 Plan/Evaluate Phase

4.3.1.1 Reasons for Implementation

One reason for measurement program failure is poor understanding by the organization of why it even began a measurement program. Measurement is then viewed as a panacea, and is performed for measurement's sake—not because it serves some specific purpose. A successful program must begin with a clear statement of purpose. The following list identifies some of the possible reasons for implementing a measurement program in an organization:

- Obtain the information to improve the organization's position in the marketplace (e.g., increase market share, move into new market segments, etc.).
- Establish a baseline from which to determine trends.
- Quantify *how much* was delivered in terms the client understands.
- Help in estimating and planning projects.
- Compare the effectiveness and efficiency of current processes, tools, and techniques.
- Identify and proliferate best practices.
- Identify and implement changes that will result in productivity, quality, and cost improvements.
- Establish an ongoing program for continuous improvement.
- Quantitatively prove the success of improvement initiatives.
- Establish better communication with customers.
- Manage budgets for development projects more effectively.
- Improve service to the customers of the systems that your organization produces.

4.3.1.2 Questions to Help Identify Goals

An effective measurement program helps determine if the organization's goals are being met. To determine if the strategic and tactical goals of an organization are being met, questions such as the following are asked:

- What can we do to improve our systems development life cycle and shorten the cycle time?
- Can we efficiently estimate product development costs and schedules? Are the estimates accurate?
- How fast can we deliver reliable systems or software to our customers? Does it satisfy their requirements?
- What is the quality of the products we deliver? Has it improved with the introduction of new tools or techniques?
- How much are we spending to support existing products? Why does one system cost more than another to support?
- Which systems should be re-engineered or replaced? When?
- Should we buy or build new software systems?
- Are we becoming more effective and efficient at product development? Why? Why not?
- How can we better leverage our information technology?
- Has our investment in a particular technology increased our productivity?
- Where are our funds being expended?

The key selling point for a measurement program is to provide a direct link between measurement data and the achievement of management goals. All levels of management need to understand what information they need, what is meaningful, and how the data will provide facts that will help manage and reach strategic and tactical goals. It is very important to obtain sponsorship and buy-in for a measurement program from managers who can and will commit to ensure its ongoing support and funding, at all affected levels in the organization. Buy-in is achieved by answering "What is in it for me?" for the entire organization. Facilitated sessions using an outside facilitator are often effective in helping the management structure identify, organize, and prioritize the goals and objectives.

4.3.1.3 Identification of Sponsors

An important step is to identify appropriate sponsors for the program. Because frequent initial reactions to measurement include apprehension and fear, upper management support is key to the success of the program. Consider including upper and middle managers, project leaders, and clients as sponsors and stakeholders.

Sponsorship has to be won. Usually it takes time to establish solid sponsorship. You can start with limited sponsorship and then build on it as the program progresses. To illustrate the point, a client at the division level did not wish to collect data on actual time spent on performing development tasks. It was not in the corporate culture to keep records of time spent. Reluctantly, the manager in charge of the division agreed to start some limited record-keeping, recognizing that there would be no way to determine if changes to the process were improving productivity without keeping such records. As time went on, the division manager began to see the benefits of keeping such records, and started requesting more detailed time record-keeping. In time, the project leaders began to also see the benefits.

To receive continual support and to expand that support throughout your organization, you must sell the measurement program to others in the organization, especially those who will be providing input. The following list includes some tactics that have been used to help establish buy-in for measurement:

- Provide education for the sponsors and stakeholders about measurement programs, metrics, and methods to effectively implement decision making and change.
- Use testimonials and experts to build credibility.
- Identify issues in product development and maintenance, and show how a measurement program can help. Use case studies from other organizations, if such data is available.
- Address concerns directly and realistically. The following concerns may be addressed:
  - How much resource time will it take to implement measurement?
  - How much will it cost?
  - Will individuals be measured?
  - Can we automate this process (automated counting, reporting, etc.)?
  - How can we maintain consistency?
  - Where can we get help?
  - How long will it be before we will see results?
  - What is in it for me?

Even with all the prerequisites in place (clear sustainable objectives for measuring, phased-in plan, audience analysis), your measurement program will face a number of obstacles before becoming a part of your organization. Several approaches can be used to sell a measurement program to the rest of the organization:

- Present the program as an integral component to quality and productivity improvement efforts. Demonstrate that the measurement program allows you to:
  - Track the changes in quality, productivity, or performance, resulting from improvements
  - Identify improvement opportunities

- Present the benefits of the metrics program to each staff level, describing in their own terms what is in it for them. These levels may include the professional staff, project managers, and senior management.
- Spread the word about measurement benefits as often and through as many channels as possible. There must be visible and demonstrable benefits, not just hype or industry claims. (For example, informal measurement discussions can occur around a brown-bag lunch.)
- Monitor the use of the data to ensure that it is used appropriately.
- Provide testimonials and concrete examples of measurement in practice, both within and outside your organization. Note that internal successes and testimonials from respected colleagues are often worth much more than external examples.
- Provide ongoing presentations to all levels.
- Address current user concerns. (For example, if users are complaining about the amount of money spent on maintenance, show comparisons of various software applications normalized by application size.)
- Address concerns and rumors about measurement candidly. Measurement fears (often from systems developers and analysts) are best dispelled by honestly addressing why measurement is needed and what benefits (e.g., better estimates) will accrue from its use.
- Depersonalize measurement (i.e., present data at a higher level of aggregation).

4.3.1.4 Identification of Roles and Responsibilities

When you establish a measurement program, the roles and responsibilities of participating individuals need to be defined and communicated. To establish the roles and responsibilities, the following questions need to be answered for your organization:

- Who will be responsible for defining what measurements to take?
- Who will decide what, how, and when to collect the measurement information?
- Who will be responsible for collecting the measurement information?
- How will the data be collected? What standards (internal or external) will be used?
- At which phases will the data be collected? Where will it be stored?
- Who will ensure consistency of data reporting and collection?
- Who will input and maintain the measurement information?
- Who will report measurement results? When?
- What will be reported to each level of management?
- Who will interpret and apply the measurement results?
- Who is responsible for training?
- Who will maintain an active interest in the measurement program to ensure full usage of the measurement information?
- Who will evaluate measurement results and improve the measurement program?

■ Who will ensure adequate funding support?
■ Will tools be provided to facilitate the measurement program?
■ Who will be responsible for evaluating the measurements being taken to ensure that the data being collected is still relevant?

Generally, organizations with a measurement program have established a central measurement coordinator or coordinators. For example, in software development organizations where they have established a Software Engineering Process Group (SEPG), some of the day-to-day management functions of the measurement program would be accomplished by it. Other functions (e.g., at the raw data collection and validation phase) may be performed by QA or a data coordinator. Still others may be performed by the engineering organization. No matter how it is organized, the following responsibilities need to be enumerated and assigned:

■ Collect the raw data.
■ Develop procedures for the collection, storage, and analysis of the data.
■ Establish a repository for the data.
■ Review measurements for accuracy, completeness, and consistency.
■ Provide measurement consulting assistance as necessary. For example, if a software development organization uses function points as the basis for size estimation, or if the organization counts lines of code, provide assistance, as needed, in performing counting. Similarly, a manufacturing department may need some assistance in interpreting control charts for signals indicating special causes of variation.
■ Distribute and support measurement reporting.
■ Consult with management on the analysis, interpretation, and application of the measurement information.
■ Maintain measurement data.
■ Collect and maintain attribute information for measurement analysis.
■ Maintain internal counting standards, requirements, and documentation for all collected measures.
■ Facilitate communication across the organization regarding the measurement program and its results.
■ Establish and maintain measurement standards, processes, and procedures.
■ Provide education, training, and mentoring on all aspects of the measurement program.

Depending on the size and scope of the measurement program, additional responsibilities listed below may need to be assigned to individuals in the organization:

■ Schedule and coordinate the measurements for development and/or maintenance (i.e., hardware upgrades, software application enhancement projects).

- Track all support changes for support reviews to keep baseline of measurements current.
- Submit measurement information to the coordinator.
- Schedule new applications or products to be counted after implementation.
- Ensure adherence to measurement policies, measurement standards, and requirements.
- Analyze, interpret, and apply resulting measurement information to improve development and support performance.

4.3.2 Analysis Phase

4.3.2.1 Analysis of Audience and Identification of Target Metrics

Once objectives have been set and sponsorship established, one of the next steps is to conduct an audience analysis and identify the target metrics. This activity should be conducted in conjunction with identifying roles and responsibilities discussed in the previous subsection. Work done here impacts the answers to the questions listed in the roles and responsibilities subsection (Section 4.3.1.4).

Conduct an audience analysis so that you will be sure to measure and track the appropriate data that will help your organization reach its goals. The first step is to determine which groups of people will require and use the measurement data (e.g., project managers, the COO, and directors). Anticipate needs to identify concerns and objectives, and work with management to recommend a relevant set of metrics. In applying these steps, one should:

- Select the most important requirements that will meet the goals of your measurement program.
- Set priorities and requirements for the metrics for each audience.
- Select only a few metrics to implement initially.
- Select realistic and measurable metrics by starting small and then building on success.
- Align audience needs with the overall objectives of measurement.

In identifying metrics, you should not:

- Choose metrics first and then create needs. A commonly made mistake is to select metrics based on what other companies are doing, and not on what the needs are for your organization.
- Try to satisfy all audiences and all requirements at once.

4.3.2.2 Definition of Metrics

After you have conducted an audience analysis and identified your initial set of metrics, you need to clearly define all the measures that will make up the metrics.

All measures and metrics must have operational definitions. An operational definition communicates what is being measured, its units of measure, what it includes, and what it explicitly excludes [4]. Identifying your metrics and defining your measures, using operational definitions, are important keys to ensuring that your data is collected consistently and that the metrics are uniformly understood by all concerned. You may wish to start from an existing set of standards. Enforcing the definitions and standards is very important for maintaining consistency. One needs to define all the component measures that will make up the metrics so that the reason they were captured, their meaning, and their usage can be clearly communicated to measurement participants and management. If measures are not clearly defined and understood by all individuals in your organization, your measurement data could be collected inconsistently and lose its reliability and usefulness.

For example, if the goal is to determine whether or not development productivity is improving, you need to collect data on development hours, project type, and project size, as a minimum. If you do not specifically define what is included in development hours, some developers may report overtime hours while others may not. Some may include analysis hours before beginning the project while others may not. It is obvious that if the hours are not captured consistently, the productivity rate (a metric combining work effort and size) will not be valid. Furthermore, if you do not also specifically identify the type of project for which the data was collected (e.g., command and control application versus information system), you may wind up comparing apples to oranges. Some products are more complex by their nature and require more hours of development time, given a comparable size.

There are many different approaches from which you can choose when first starting a measurement program. Possible options include:

- Establishing a baseline of all products (software applications, systems, etc.) in your organization as a function of application domain
- Running a measurement pilot project for one product or application
- Measuring activity against all installed products or applications
- Tracking only development projects
- Tracking defects, cost, and customer satisfaction
- Tracking only size and effort

The approach you select should be based on your own organizational goals and constraints for implementing a measurement program. For example, the following types of measures are often collected as part of an initial measurement program:

- Lines of code (for software projects)
- Work effort
- Defects
- Cost
- Customer satisfaction

How you track and collect measurement data depends on what you decide to collect. If you choose to start big and collect a large amount of data for the whole organization, you might want to consider investing in automated data collection tools and repositories. In addition to the types of data, the data collection approach you choose will depend on the scope of the measurement program you want to establish. One needs to establish standards and processes for data collection, storage, and analysis to ensure that the data will be usable and consistently collected and reported.

Whether you build or buy automated measurement tools that will store and analyze your data depends on your requirements. Even if you start small and decide to store the information in a spreadsheet, you will quickly amass a large amount of data and will need to obtain an electronic storage medium for collected data.

4.3.3 Implement/Measure Phase

4.3.3.1 Organizing for Just-in-Time Training and Education Processes

Education and training must be provided to those persons who will be involved in and affected by the measurement program. Such an effort should be geared toward providing these individuals an understanding of:

- Why measurement is necessary
- How it affects them
- How the information can help them manage
- Their responsibilities in the context of the measurement program

Different types of training or presentations are required to address the needs of different levels of personnel, depending on the depth of their involvement. For example, managers may require an executive summary presentation on how measurement applies to them, while system experts may require in-depth training on data collection and measurement. For organizations that intend to achieve a high maturity rating on the CMMI, training in process performance models, statistics, and statistical process control will also be needed. Training in the Six-Sigma methodology is also very helpful.

Particularly effective training incorporates realistic examples from the workplace of the participants. Exercises where hands-on exposure to the tools, techniques, and methodologies can be obtained are essential. Complementary to formal training sessions, a follow-up and support effort should be designed so that the participants "stay on course" and apply the skills they acquired during training. Monthly or quarterly follow-up meetings where problems, success stories, and difficulties in implementation are presented have also proved very effective.

When planning training and education programs, one should always keep in mind that learning is enhanced by positive reinforcements—and mostly fun! Using graphical user interfaces in conjunction with good case studies and simulations has helped achieve these objectives.

4.3.3.2 Reporting and Publishing Results

Whether you produce manual reports, develop your own reporting system, or purchase a measurement reporting package, you will need to report and publish the results. The following list includes suggested guidelines for publishing results:

- Make sure the people viewing the reports understand what the reports reveal so that the information is not misinterpreted and used to support poor decisions. This is another reason why operational definitions are important.
- Include an analysis and explanation with the reported data (i.e., mix graphs with text).
- Produce key statistics relevant to and usable by your audience.
- Provide relevant and actionable reports to managers.
- Use appropriate graphs to present your data (e.g., pie charts do not present trends well).
- Keep graphs simple (i.e., do not present multiple metrics on a single graph).
- Report source data in appendices.

4.3.4 Improve Phase

4.3.4.1 Managing Expectations

If the life span of your projects ranges from a few months to a year or more, consider how long it will take to collect enough data points to make your data statistically significant. Also, it will take a great deal of time to define the measures, set the standards for data collection, and train the individuals. And because it will take a great deal of time before the measurement program will reach its long-term goals, one should also concentrate on short-term goals and let all involved know when they can expect to see more immediate results. The short-term goals derive from the long-term goals. Keep your long-term goals in mind while developing your short-term goals. For example, over the first few months of a new metrics program, improvement in estimating accuracy could be a more immediate goal than shortening the development cycle. Once you are able to collect work effort and project size, you can begin to supplement this data with available industry-data delivery rates to help you in project estimating and shortening the development cycle. The amount of time and money that must be invested in measurement before payback is realized varies from organization to organization. Many organizations have found that long-term payback may not be realized until 2 years or more after the program is implemented.

4.3.4.2 Managing with Metrics

The staff responsible for a metrics program is not usually responsible for making management decisions. A measurement program provides input to the decision-making process, but decision making is outside of measurement programs. However, the most important process in a measurement program is to use the results to support management decision making. This is the central purpose of a measurement program, and it may be tied to continuous improvement, total quality management, and other organizational initiatives. You can manage with metrics by:

- Analyzing attributes and data
- Incorporating results into decision making
- Showing the cost-benefit

You need to analyze the attributes and determine which factors impact your processes. You also need to analyze the data to identify which processes are working well and which are not, which areas require further investigation (based on the data), and where opportunities for improvement exist. It is also necessary to analyze the data to determine how well the measurement program itself is working.

Incorporate the results and data into the management decision-making process in such areas as:

- Estimating the time and cost of new development and enhancements
- Allocating resources to support and development areas
- Improving quality and productivity
- Analyzing portfolios
- Making buy-versus-build decisions

Using measures to manage is how you can show the cost-benefit of a measurement program. This chapter concludes with case study examples of measurement program establishment and measures demonstrating the benefits of measurement programs, followed by a discussion of some advanced statistical and probabilistic techniques that are added into a measurement program by CMMI high maturity organizations.

4.4 Case Studies

In this section we give examples of effective and ineffective measurement programs. The actual examples will be amalgamations of actual cases in order to protect the identity of the actual organizations. In addition, we provide some examples of the benefits that can accrue from measurement programs. A number of measures exist that can be utilized for managing development or maintenance efforts. These give insight into the progress of the effort. This section includes five examples of such measures. Each of them, in and of themselves, does not necessarily tell a complete story, but if collected and utilized properly, provide effective warnings of potential problems.

4.4.1 Measurement Program Case Studies

4.4.1.1 An Effective Measurement Program

Division X of Company ABC had been collecting measures and generating metrics, but had been doing this in a very unorganized manner. There were corporatewide measures that were required, but these were financial in nature and oriented toward generating data to support quarterly reports to stockholders. Little was collected to support insight into project status or product quality. What was collected was determined in haphazard fashion, in that the measures that were collected were based on discussions with personnel from other divisions or from attendance at conferences.

Division X's business was defense oriented, and one of its Department of Defense (DoD) customers decided to require their contractors to implement a process that was equivalent to Maturity Level 3 on the CMMI. (Some DoD Program Offices are reluctant to risk having contract awards disputed over a contractor's maturity level rating, and will instead require a Maturity Level 3-like process, instead of requiring them to actually achieve that rating.) Division X began the process of implementing the CMMI and soon discovered that it would have to implement the Measurement and Analysis (MA) Process Area (PA) in order to achieve even Level 2. The MA PA requires that the organization derive its measures from the organization's business and information needs. Similar to what we have pointed out in this chapter, MA suggests using GQM as an approach for identifying the measures to collect and the metrics to calculate, and then to identify the processes from which these measures can be obtained. Division X identified two critical business needs (as well as other business informational needs): close tolerances on promised delivery dates and low residual defects in the delivered product. These were driven by customers' needs for better on-time delivery by Division X and better product reliability. Measurements were identified to yield this information, operational definitions were established, methods of data collection and storage were specified, and procedures were established for the analysis of the data.

At a later date, the organization decided to pursue achieving Maturity Level 5 on the CMMI. Subprocesses that were critical to achieving these business goals were identified, and were placed under statistical process control. A process performance model was constructed, and statistical data from the performance of these subprocesses was used to predict resultant quality and estimated delivery date. Process performance baselines were monitored for these subprocesses, and corrective action was implemented when special causes of variation were noted. Division X went on to achieve Level 5 on the CMMI. As a consequence of having implemented a Level 5 process, the Division was better able to predict delivery dates and reduce variability. Latent defects were also reduced in delivered products.

4.4.1.2 An Ineffective Measurement Program

Company Z performed an assessment based on the CMMI, and was rated at Maturity Level 1. It was clear to the assessment team that the organization did not have a measurement program in place. The organization decided to develop a strategic plan for process improvement, and to implement a measurement program as part of that plan. As it developed its measurement program, the organization began to identify goals for which the measurements would be used. This was done in a brainstorming session, but was not explicitly tied to any business goals or information needs of the organization. They were tied only to what struck the participants in the brainstorming session as important at that moment.

One of the objectives established at the brainstorming session was to improve the productivity of the organization by 7% the first year, and 20% by the following year. No baseline of productivity measures existed at that point in time, and no objective was established for the measurement program to determine such a baseline as the first step. Furthermore, no operational definitions for what constituted a productivity measurement were established.

Ultimately, the measurement program and the process improvement initiative failed because a great deal of meaningless data was being collected. No useful interpretations could be made of the data that was collected.

4.4.2 Beneficial Measurement Examples

In this subsection we present measurements that have proven useful for many organizations. The extent to which any measure used by one organization is useful to another organization depends on what that organization's information needs are. The specific measures used by organizations, and their operational definitions, will depend on the specific business goals and objectives of the organization that uses them.

4.4.2.1 Example 1: Staffing Profiles

Although this measure would appear on the surface to be one that is normally tracked, far too many organizations do not. Many organizations fail to utilize some of the most basic and obvious measures necessary for effective program management. Staffing profiles are one class of measure that would indicate if there is potential for overruns or underruns. It is simply a profile of the planned staffing versus the actual staffing (Figure 4.6). In this example, the actual staffing is lagging the planned staffing. This would indicate the potential for a schedule slip because the staffing levels are below that which was estimated as necessary to accomplish the project's objectives. Whether this was actually occurring is not known, on the basis of this chart alone, but it would indicate that an investigation into what was occurring was necessary. Other data would be needed, such as a milestone chart or earned value data, to see if progress was being maintained. Nonetheless, the chart served its purpose: to alert the project leader of a potential problem.

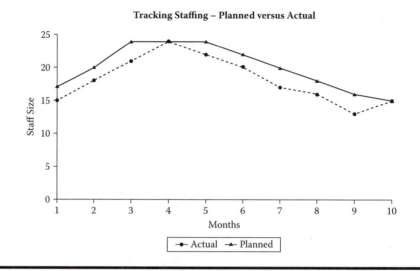

Figure 4.6   Planned versus actual staffing. Kenett, R.S. and Baker, E.R., *Software Process Quality: Management and Control,* **New York: Marcel Dekker, Inc., 1999.**

Clearly, if the staffing profile indicated that actual staffing exceeded the planned level, this would be an indication that there could be an overrun. As before, other data would be needed to confirm or reject the hypothesis. But as before, the chart would have served its purpose: to warn the project leader of a potential problem.

If supplemental information is added to these charts, expanded usage of these charts can be made for program management purposes. For instance, if we overlay onto these charts monthly data on staffing changes—that is, additions to and losses of personnel on the project—it is possible to focus more closely on the reasons for staffing shortfalls. If the changes to staffing are not negligible, the reasons for the shortfall may be due to a high turnover rate. This may be an indication of other problems, such as employee dissatisfaction. If the changes to staffing are negligible, it may be due to normal attrition experienced by all organizations. If we additionally overlay onto these charts monthly data on staffing levels by skill area or technical discipline (e.g., analysts, testers, quality assurance, etc.), it is possible to identify shortfalls in the technical disciplines necessary to implement the project's requirements.

4.4.2.2 Example 2: Software Size

As indicated earlier, software size is a measure essential for estimates of schedule, cost, and resources required for the accomplishment of a software project's objectives.

If a software development organization estimates the size of the software to be produced—either in terms of lines of code, function points, or Web pages; stores that data in a database; and then tracks the project against these estimates

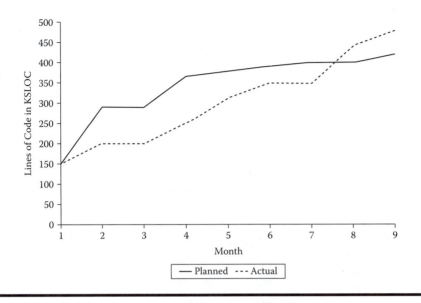

Figure 4.7 **Software size, over time. Kenett, R.S. and Baker, E.R.,** *Software Process Quality: Management and Control,* **New York: Marcel Dekker, Inc., 1999.**

(see Figure 4.7), it has the basis for performing rudimentary estimates of project schedule, cost, and required staffing. New start-up projects can be compared against other projects in the database for similarity; and the schedule, cost, and staffing estimates can be made on the basis of what that similar project experienced. Clearly, more accurate estimates can be made if the growth in size is also tracked, along with the related changes to requirements (if applicable), schedule, cost, and staffing. Refinements and improvements in the ability to estimate can be accomplished if supplemental information is tracked, such as the number of requirements to be implemented by the project.

In the beginning of a project, there would be a single number for the planned size of the software, for example (as is indicated in Figure 4.7), 150,000 lines of source code. However, during the course of a project, changes in scope can occur, causing either a decrease or increase in the planned lines of code. Sometimes, as the project matures, the project management team gets a better handle on what's required to implement the requirements, and the planned size of the software can change because of that. Due to these factors, the planned lines of code will not necessarily be a horizontal line, as is the case in Figure 4.7.

Tracking software size can also provide indications of development problems. If the software size grows during the term of the project, there must be a reason for it. It may be merely the poor ability to estimate (most developers often underestimate, sometimes by as much as a factor of 2 or 3). This can be resolved by recording and tracking estimates, as described in the previous paragraph. Often, however, there may be other causes. For example, perhaps the requirements are changing. If the requirements are changing in a controlled fashion, and the impact of these changes are evaluated in terms of cost and schedule prior to accepting the changes, then there is no problem. On the other hand, if the size is changing due to uncontrolled requirements changes, then there is cause for concern. This can lead to schedule slips, cost overruns, and customer unhappiness. Perhaps the change in size is due to poor understanding of the requirements by the developers. Corrective action may be required, such as additional staff, staffing changes to acquire the necessary skills, or training for the existing staff in the necessary skills. Clearly, additional data would have to be collected to drill down to the cause of the problem. But again, the measure served its purpose: to alert the project leader to a potential problem.

A word of caution: In the case of software size measured in terms of lines of source code, operational definitions for lines of code are absolutely essential, especially with regard to variables such as programming language, comments, and the like. Otherwise, no basis for comparison exists.

4.4.2.3 Example 3: Problem Report Tracking

Problem report tracking provides the capability to gain insight into both product quality and the ability of the organization to maintain the product. Problem report tracking can be done at the project level, product line level, or organizational level. At the project level, it provides insight into developmental issues. At the product line level, it provides insight into the quality of the delivered product and the effectiveness of the maintenance and support efforts. At the organizational level, it can provide insight into the quality of the products, in general, as well as the effectiveness of the maintenance and support efforts across all the organization's products.

Figure 4.2 shows, by week, the total number of problem reports that are still open. Figure 4.4 shows the number of open problem reports plotted by severity. In tracking problem reports over a product's development life cycle, there may be discontinuities as a function of life-cycle phase. For example, at the start of a new test phase, such as integration testing at each level, or acceptance testing, the number of problem reports opened would increase fairly rapidly, opening a gap between the number of problems reported and closed. If the problems are straightforward, we would expect the gap to decrease in short order. If not, that would be an indication of a possible problem. On the other hand, once a product has been fielded, one would expect the number of problem reports to diminish over time, until an update to the product was released.

Again, additional data helps to zero in on the problem, if one appears to exist. The difficulty may not be a difficulty at all. It may be that the problems reported are of little operational consequence and, as a result, their resolution can be deferred to a later date. Defining categories of problem severity, and overlaying that information on the chart as was done in Figure 4.4, helps determine if a problem exists.

To help in problem isolation, an additional set of charts that would be useful are bar charts of the number of weeks problem reports have been open at each severity level. For example, at the highest severity level, it would show the number open for 1 week, 2 weeks, 3 weeks, etc. If there are substantial numbers open for a long time, this could be an indication that the product is difficult to maintain, or that staff does not understand as well as it should the technical problem to be solved or corrected. Further investigation would be needed.

"Substantial numbers" and "open for a long time" are qualitative terms. There are no pat answers as to what is "substantial" or "long." This is based on an organization's own experience and the complexity of the product under discussion. This suggests that it is advisable to maintain records from past projects for use for comparison. Once again we see the importance of operational definitions for measures.

4.4.2.4 Example 4: Control Charts of Newly Opened Bugs, by Week

Processes in real-life environments often exhibit statistical variation. Much of that is what can be considered the "noise" of the process, in that variability exists in the performance of the processes. Although there might be tight controls on it, the process doesn't produce precisely the same results each time it is performed. Such variation poses significant challenges to decision makers who attempt to filter out "signal" from "noise." By signal we mean events with significant effects on the data such as short- or long-term trends and patterns, or one-time, sporadic spikes. The noise, in Statistical Process Control (SPC) terms, is called the "common cause of variation," while the signal is referred to as a "special cause of variation."

As an example, in software projects, peer reviews of the coding may be performed prior to unit-level testing or integrating the code into the application as a whole. One would not expect that the coding had occurred flawlessly for every unit of code; consequently, a variable number of problem reports would be written, peer review to peer review. It is in the context of such fluctuation that we want to determine if the number of opened problem reports, in a given peer review, exceeds the "normal" range of fluctuation. Another phenomenon we are interested in identifying is a significantly decreasing trend in the number of opened problem reports. Such a trend might be a prerequisite for a version release decision.

A primary tool for tracking data over time, in order to identify significant trends and sporadic spikes, is the control chart. Control charts have been applied to industrial processes since the 1920s, first at Western Electric, then expanding to U.S. industry, to Europe, and, with outstanding results since the 1950s, to Japan. Control charts are the basic tools in process improvement and SPC. Industrial statisticians developed several types of control charts designed to handle a variety of situations. These charts can be applied to individual data points such as the number of problem reports opened per week; to representative samples, such as complexity of randomly selected units; or to percentages, such as percent of lines of source code with comments. Sometimes data is collected simultaneously in several dimensions, such as response times to a routine transaction at several stations in a network. Such multivariate (as opposed to univariate data) can be analyzed with multivariate control charts. A comprehensive discussion of control charts is beyond the scope of this book; we focus here on interpreting real-life examples. The interested reader can find more information on control charts in the univariate and multivariate case in books such as Kenett and Zacks [11], Fuchs and Kenett [6], and the *Wiley Encyclopedia of Statistics in Quality and Reliability* [20].

All control charts consist of a graph tracking the relevant time-ordered data and three lines: an Upper Control Limit (UCL), a center line, and a Lower Control Limit (LCL).

The UCL and LCL in most cases represent the ±3-sigma limits of the variance for that control chart. The center line is the mean of all the values. Basic interpretation of control charts consists of monitoring the occurrence of four patterns, although there are additional patterns that can be applied as well (see [11, 15, 19]):

1. A single point below the Lower Control Limit or above the Upper Control Limit
2. Two out of three points in the region between either +2 and +3 sigma, or –2 and –3 sigma
3. Four out of five consecutive points between either +1 and +2 sigma, or –1 and –2 sigma
4. A run of eight points below or above the mean or center line

This is sometimes referred to as the 1-2-4-8 rule. Any one of these events indicates a nonrandom occurrence that justifies appropriate action. As noted above, other trends exist and are specified as signals for those who are implementing SPC using the Six Sigma methodology; however, the signals noted above are used most commonly by the general community of people performing SPC. As mentioned previously, these signals are causes for further investigation; for example, a point above the UCL might start an investigation as to the cause of this unusually high number of problem reports (for ease of interpretation, we are equating a defect to a problem report).

Figure 4.8 consists of a control chart of the number of new problem reports per week in a new software product from a company developing Rapid Application Development tools. The control charts in Figure 4.8 and Figure 4.9 were generated with Minitab version 15.1. We expect the number of reported valid problems to decrease over time, as latent defects are removed.

The I chart in Figure 4.8 is a control chart that tracks individual numbers and computes variability using a moving range of the individual values. In our case we use a chart of the number of new problem reports with standard deviation based on a moving range of size two, that is, the difference in absolute value of two consecutive values of the data (I control chart). Superimposed on the run chart tracking new problem reports are three lines:

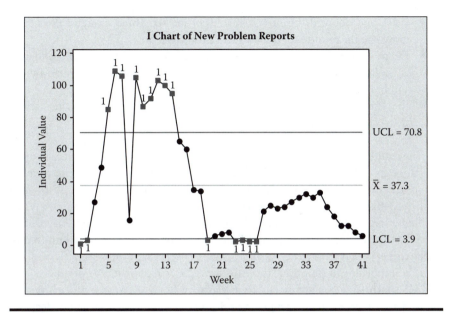

Figure 4.8   Control chart of new problem reports, by week.

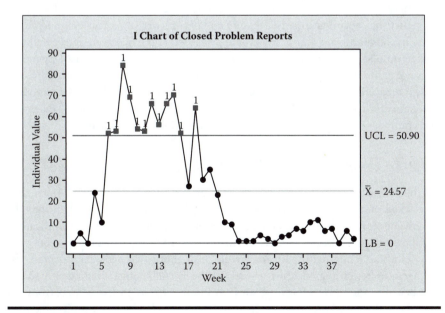

Figure 4.9   Control chart of closed problem reports, by week.

- The Upper Control Limit (UCL = 70.8) located three standard deviations above the mean number of new valid problem reports
- The mean number or centerline of valid problem reports (= 37.3) computed as the total number of reported new valid problem reports divided by the number of observation weeks
- The Lower Control Limit (LCL = 3.9) located three standard deviations below the mean number of new valid problem reports

The standard deviation of the data is computed by dividing the average moving range by 1.128. The I chart indicates a significant decrease in the number of reported bugs on week 14 and a gradual decrease toward the end. With such behavior, management could declare that the number of reported new valid problem reports has significantly dropped, thus allowing for version release. Overall, the average number of new bugs has dropped from about a hundred to about three.

It is important to point out that this example is representative of aspects of a CMMI Maturity Level 4 process. In a Maturity Level 4 process, process performance baselines are established based on historical data (which should be reevaluated periodically as more recent data is collected). The mean, UCL, and LCL used in the examples above are the process performance baselines that were established for beta testing, based on past history. Process performance baselines are used to control the critical subprocesses and implement corrective action to the process, as necessary. The critical subprocesses are derived from the organization's business and information needs. Process performance models are also established and are used to predict the capability to satisfy quality and process performance goals and objectives. In the discussion above, it was noted that when the observed defects fell to a predetermined number, the software can be released. Note that a reliability model does not *guarantee* that the *actual* number of latent defects is known with full accuracy. A reliability model merely *predicts* the expected number of latent defects based on data collected up to the point in time that the prediction is made. Percentiles and confidence intervals of these predictions can also be computed.

4.4.2.5 Example 5: Control Charts of Closed Bugs, by Week

The number of closed problem reports in Example 4 (see Section 4.4.2.4) can also be tracked on an I control chart (Figure 4.9). This provides some insight into the effectiveness of the diagnostic and corrective action process. The I chart indicates a significant decrease in the number of closed bugs on week 17, an increase on week 18, and a stable bug fixing rate after week 25 of about two per week. Overall, the average number of bugs closed per week has been 24.57, quite a bit lower than the number of reported new bugs per week. A lower capacity in fixing bugs from the rate of creation of new problem reports implies that a backlog of open problems is building up.

4.4.2.6 Example 6: A Cost-Driven Dashboard

A balanced scorecard [9] for a major system development project is presented in Figure 4.10. It presents six measures:

1. EVA: earned value, in tens of thousands of dollars, indicating what part of the planned progress has been achieved
2. Budget versus utilization: tracking planned burn rate versus actual burn rate
3. Government grants: reflects funding level for grants from the government (a major source of financing R&D projects)
4. Man-month versus budget: tracking of full-time equivalent employees (FTEs)
5. Milestone monitoring
6. Tracking of problem reports (PRs) and change requests (CRs)

4.5 Measurement and the CMMI High Maturity Practices

In this chapter we focused on the process of establishing measurement programs in organizations. Measurement programs are intended for a number of purposes, including, among other things, obtaining information about the state of an organization's business, the status of individual projects, and process implementation. From time to time, we have included various aspects of what are referred to in the CMMI as high maturity practices, but without a complete discussion of what these high maturity practices are and what they are intended to do. In this section we discuss that aspect.

The CMMI high maturity practices are a richer extension of the measurement programs being discussed in this chapter. They require a more complete quantitative knowledge of the organization's critical subprocesses in that the high maturity practices utilize statistical and probabilistic techniques; consequently, sufficient data must exist to apply statistical methods. A subprocess is defined in the CMMI as "[a] process that is part of a larger process. A subprocess can be decomposed into subprocesses and/or process elements" [4].

High maturity practices are used for the purposes of managing process performance, evaluating proposed and implemented process corrections, evaluating prospective process and innovative changes, and evaluating the actual implementation of the process changes and innovations after deployment.

Up to Maturity Level 3, knowledge of process performance tends to be more qualitative rather than quantitative. Measures will exist that provide information about the status of the various processes that have been implemented, but they don't provide the same kind of knowledge that exists at Maturity Level 4. At Maturity Level 4, the organization has collected various kinds of data on process status and

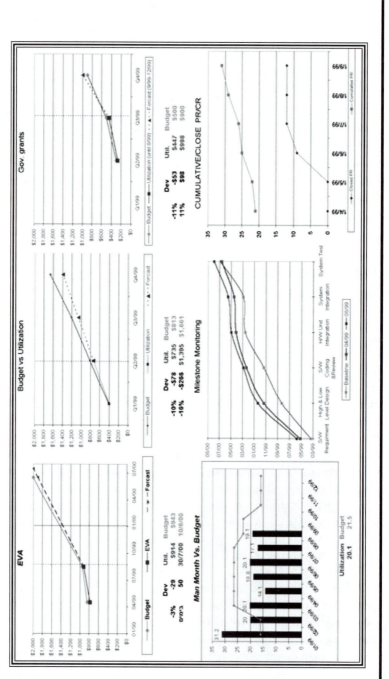

Figure 4.10 A sample project dashboard.

performance, and turns its attention to managing process performance with the intention of addressing the special causes of process variation. As previously noted, the special causes of process variation can be indicators of a problem in process performance—one that may require correction in order to maintain process performance within limits. At Maturity Level 5, attention focuses on reducing the common cause of variation (the "noise" of a normally running process) and improving the nominal level of process performance.

We discuss the two maturity levels in terms of each of the Process Areas (PA) that comprise them. This discussion demonstrates the logic involved in each PA and how that applies to the way those organizations manage and control their critical processes.

4.5.1 Maturity Level 4

There are two process areas at Maturity Level 4: Organizational Process Performance (OPP) and Quantitative Project Management (QPM).

The focus of OPP is to establish which processes should be controlled, define what measures to take for these, and establish process performance baselines (PPBs) and process performance models (PPMs). The determination of which processes to control is based on their link to the business needs of the organization. Using a methodology like Goal-Question-Metric (GQM), measures that link to the business needs, goals, and objectives can be established, and the processes that will provide that data can be identified.

Once the critical processes and subprocesses are identified, process performance baselines can be established. These are often in the form of a control chart generated from relevant data across all the projects that have implemented that process. A process performance baseline must be based on stable processes; that is, all special causes of variation must have been removed. A PPB generated for an unstable process establishes a baseline that is unreliable. The PPB for any particular process is used by the projects for managing that process, and in the absence of any specific process performance requirements specified by a project's internal or external customer, becomes the "voice of the customer," that is, the specification for performance of that process.

Process performance models, on the other hand, are used to estimate or predict achievement of quantitative project quality or process performance objectives and goals that cannot be measured until much later in the life cycle. An example would be defect density at the time of delivery, if this were specified as a quality requirement. Examples of PPMs include Monte Carlo simulations* and reliability models, among others [11, 18, 20]. These are also intended to be used by the projects as a means of making early determinations of whether or not their quality or process

---

* A pioneer in the development of Monte Carlo simulations was Stanislaw Ulam, who used Monte Carlo simulations to evaluate complex integrals that arise in the theory of nuclear chain reactions while working on the Manhattan Project in the early 1940s.

performance goals and objectives will be achieved. They can also be used for other purposes, as will be discussed later. All PPMs must relate the behavior of a process or a subprocess to an outcome. As such, they must include factors (for example, the mean, UCL, and LCL) from measurements of that subprocess to conduct the prediction. That enables projects to take corrective action on the performance of the included subprocesses if the prediction looks unfavorable. By including these subprocesses, it enables the project to identify to what subprocess to apply corrective action.

Recall from Chapter 1 that we set up an example of how GQM can be used to define the measures that need to be taken to achieve the goal of improving the quality of the software at delivery by 5%. In defining these measures, we established that we need to know data about nominal defects by phase and by the activities that occur in these phases, and the variance of these defects. We also wanted to know about the effectiveness of the QA audit process by looking at the nominal level of noncompliances detected per audit and its variance. Based on that, PPBs can be established for each of the development processes implemented during these phases, and a PPB can be established for the QA audit process. Having done that, we can do a regression analysis to see which factors have a significant influence on delivered quality and set up a Monte Carlo simulation as a PPM. Projects can then use this simulation by inserting actual values recorded during the early phases of development to see if the required quality objectives will be met.

Quantitative Project Management uses this information generated from implementing OPP (i.e., PPBs and PPMs) to manage their projects. Projects identify the critical processes and subprocesses that they must control, and the measures that provide insight into these objectives. These are based on what the organization has identified as critical (see the discussion above on Organizational Process Performance), as well as any needs that are unique to the project, for example, a quality requirement specified by the customer that is different from those specified by the organization. Based on this information, the project manager "composes" the project's process. That is, the project manager selects the processes and methodologies that will produce the required results. If there are alternative methods of achieving the required outcomes, PPMs can be used to evaluate the best approach to take. For example, there can be several different life-cycle models that can be applied to meet a delivery schedule constraint. A Monte Carlo PPM that evaluates schedule factors by life-cycle phase can be used to determine which model will meet the constraint with the highest probability. In composing the process, the project manager defines which processes and subprocesses are to be statistically managed.

The project then manages the critical processes and subprocesses in accordance with the measures thus identified. Control charts are used to identify if these critical processes and subprocesses are stable and capable. By capable, we mean that the process performance (variance) is inside the specification limits. By stable, we mean that there are no special causes of variation (elsewhere we discussed the signals that identify potential special causes of variation). When a special cause of variation is noted, corrective action is implemented to remove it. Some of the techniques involved with the implementation of the Causal Analysis and Resolution (CAR) process area can be applied to identify and remove the special cause of variation. The PPMs are used with actual data recorded during execution of these critical processes and subprocesses to estimate the quality of the delivered product based on the data recorded to date. They can also be used to evaluate the efficacy of proposed corrective actions.

The results of implementing QPM are recorded in the organizational process database. Periodically, at the organizational level, the results from all the projects are aggregated and analyzed to determine if the organizational PPBs need to be updated. In some cases, constants or correlation factors utilized in the PPMs may also need to be updated.

4.5.2 Maturity Level 5

As noted earlier, at Maturity Level 5, we are focusing on reducing the common cause of variation, that is, the "noise" of any performed process. We do this in three different ways: (1) by reducing or eliminating the occurrence of the most frequently uncovered errors and defects that the noise of the process produces, (2) by making changes to the process itself to make it more efficient and effective, and (3) by introducing innovations that have the effect of making the process more efficient and effective. The two process areas at Maturity Level 5—Causal Analysis and Resolution (CAR) and Organizational Innovation and Deployment (OID)— accomplish this. We are also reliant at Maturity Level 5 on the PPBs and PPMs we established at Maturity Level 4 to help us in this process.

CAR is nominally a project-level PA, in that it is the project that keeps day-to-day track of the performance of the processes and has the data to characterize the types of problems and defects encountered and the frequency of their occurrence. Nonetheless, it is also incumbent upon the project to assess if any change implemented should be made to the organizational standard process, and also to initiate a request to make that change if it seems that the organizational standard process should change.

By implementing CAR, the project classifies the various types of problems and defects and counts the frequency of its occurrence. Although the process may be performing normally, and there are no special causes of variation, certain types of defects or errors may still be occurring more frequently than others within the "noise" of the process. Using tools like a Pareto chart, the project can determine if a given type or types of defect is or are occurring much more frequently than others. If the project determines that any type or types of defect are occurring at a much greater frequency, the project, with the participation of appropriate stakeholders, can then implement a causal analysis, using tools like the Ishikawa (fishbone cause and effect) diagram. Having identified the most likely cause or causes, these stakeholders can propose likely fixes. A quantitative estimate should be made as to the effectiveness of the proposed solution. If possible, an estimate of the expected mean, and upper and lower control limits for the fix should be made, and that information input into the appropriate PPM to estimate the effect on the ultimate result in terms of process performance or product quality. Clearly, if the change is going to have negligible impact, the change should not be made—that would not be cost effective.

After the change has been implemented, measurements should be made of the implemented solution to see if the change performed as well as estimated. In some cases, a statistical test of the significance of the difference in performance of the process before and after the change should be made. If the change proves effective and would change the way that the standard process is implemented, then the project should initiate a change request for that through the mechanisms provided by implementing OID.

Many of the techniques described above for the implementation of CAR can also be applied for addressing special causes of variation. We alluded to this in our discussion of QPM. Performing a causal analysis, proposing solutions, and estimating their expected benefits, for example, are useful concepts for dealing with special causes of variation.

Although CAR is a project-level PA, in some organizations it may be more effective to implement it as an organizational-level activity or a hybrid of organization- and project-level activity. If an organization only has small short-term projects, the projects will not last long enough or produce enough defects individually (hopefully!) to make effective use of CAR at the project level. Similarly, if the organization had a mix of large, long-term projects and short-term small projects, the long-term projects should implement CAR at the project level, but the data for the much smaller projects may perhaps be aggregated and analyzed at the organizational level when it comes time to review the PPBs to determine if they should be updated.

OID is implemented at the organizational level (as is apparent from its name). By implementing OID, the organization collects, evaluates, and implements potential changes to the process from two different sources: (1) internally requested changes and (2) innovative technological or methodological changes in the marketplace that have been investigated by (perhaps) the SEPG (or equivalent organizational entity). The SEPG (or equivalent entity) reviews the proposed changes and, where the data is available, estimates the effect of the change on the process. As with CAR, estimates of the change in the PPBs should be made, and the effect on the resultant process performance or quality should be made, using the appropriate PPM. (This may not be doable for every change, due to the nature of the change.) When possible, the change should be piloted to evaluate if the proposed change will produce the anticipated results. A low risk environment should be selected as the host for the pilot, such as a start-up project that is not mission critical. The pilot should be handled like a complete rollout, in that a process description, standards, procedures, and training materials should be prepared and training in the changed process performed. The extent to which this is done, however, should be consistent with piloting a change, as opposed to a complete rollout. In addition, the implementation of the process improvement pilot should be managed like a project, and should have a project plan. Measurements of the pilot process should be taken and compared against estimates. Feedback should be collected so that if the pilot is successful, the organization will know what corrections to make to the materials, if needed. If the pilot is successful, the new process will be rolled out.*

When the decision is made to roll out a changed process, similar to what has been described above for a pilot; the rollout should also be handled as a project, with the same kinds of considerations: plan, process description, standards, procedures, training materials, measures, and training. The actual deployment should be measured against the estimates to ensure that the new or changed process meets expectations. As with CAR, it may be necessary, in some cases, to do statistical tests of significance in order to determine if the new or changed process significantly improves the process or ultimate product quality.

As with CAR, the results of OID are also entered into the organizational process database. As appropriate, new PPBs may be generated or PPMs updated.

4.5.3 *High Maturity Summary*

As we have seen, the high maturity practices greatly expand the capability of an organization to assess the extent to which their processes are effective in terms of capability and stability. It permits the projects within the organization to not only manage and control their critical processes and subprocesses quantitatively, but also permits the organization to quantitatively evaluate methods of improving its processes to determine if the changes make any statistically significant improvements in capability, and return on investment.

4.6 Chapter Summary

Previous chapters provided an introduction to software measurement programs. In them, we elaborated on the question *Why measure?*, discussed key definitions, and the Goal/Question/Metric method. In this chapter we presented a discussion on the effectiveness of process measurement programs. Attributes of an effective program were listed together with positive and negative impact factors. A four-step implementation plan was discussed in detail so that practitioners can tailor this program to their own organization-specific needs. We included case studies that provide real-life examples of effective and ineffective measurement programs and how measurements can be beneficially used in development and maintenance

---

* The CMMI practices do not indicate that these steps be performed for a pilot; but as a practical matter, handling pilot projects to test out changed processes in this manner makes good sense, based on experience.

organizations. Finally, we discussed the CMMI high maturity practices in relation to how they enrich a measurement program and enable us to improve our processes.

References

1. ANSI/IEEE Standard 1045, Standard for Software Productivity Metrics, IEEE Standards Office, Piscataway, NJ, 1992.
2. ANSI/IEEE Standard 982.1-2005, Standard Dictionary of Measures of the Software Aspects of Dependability, IEEE Standards Office, Piscataway, NJ, 2005.
3. Basili, V.R. and Weiss, D., A Methodology for Collecting Valid Software Engineering Data, *IEEE Transactions on Software Engineering*, 1984, p. 728–738.
4. CMMI Product Team. CMMI for Development, Version 1.2 (CMU/SEI-2006-TR-008), Pittsburgh, PA: Software Engineering Institute, Carnegie Mellon University, 2006.
5. CMU/SEI-92-TR-20 Software Size Measurement: A Framework for Counting Source Statements, 1992.
6. Fuchs, C. and Kenett, R.S., *Multivariate Quality Control: Theory and Applications*, New York: Marcel Dekker Inc., 1998.
7. IEEE Standard 1044, Standard Classification for Software Anomalies—Description, IEEE Standards Office, Piscataway, NJ, 1993.
8. ISO/IEC 15939, Software Engineering—Software Measurement Process, International Organization for Standardization, Geneva, Switzerland, 2002.
9. Kaplan, R.S. and Norton, D.P., The Balanced Scorecard as a Strategic Management System, *Harvard Business Review*, January–February, 61–66, 1992.
10. Kenett, R.S. and Albert, D., The International Quality Manager: Translating Quality Concepts into Different Cultures Requires Knowing What to Do, Why It Should Be Done and How to Do It, *Quality Progress,* 34(7), 45–48, 2001.
11. Kenett, R.S. and Zacks, S., Modern *Industrial Statistics: Design and Control of Quality and Reliability*, San Francisco, CA: Duxbury Press, 1998. (Spanish edition, 2000; 2nd paperback edition, 2002; Chinese edition, 2004.)
12. Kenett, R.S., The Integrated Model, Customer Satisfaction Surveys and Six Sigma, *The First International Six Sigma Conference,* CAMT, Wrocław, Poland, 2004.
13. Kenett, R.S., Coleman, S.Y., and Stewardson, D., Statistical Efficiency: The Practical Perspective, *Quality and Reliability Engineering International,* 19, 265–272, 2003.
14. McGarry, J., Card, D., Jones, C., Layman, B., Clark, E., Dean, J., and Hall, F., Practical Software and Systems Measurement: A Foundation for Objective Project Management, Version 4.0c, Department of Defense and the U.S. Army, March 2003.
15. Motorola University, *Six Sigma Green Belt Program, Participant Guide*, Volume 2, Version 3, undated.
16. Ograjenšek, I. and Kenett, R.S., Management Statistics, in *Statistical Practice in Business and Industry*, Coleman, S., Greenfield, T., Stewardson, D., and Montgomery, D., Editors, New York: Wiley, 2008.
17. Rockart, J.F., Chief Executives Define Their Own Data Needs, *Harvard Business Review*, 1979, p. 81–93, 1979.
18. Rubinstein, R.Y. and Kroese, D.P., *Simulation and the Monte Carlo Method, 2nd ed.*, New York: John Wiley & Sons, 2007.

19. Wheeler, D.J., *Understanding Variation: The Key to Managing Chaos, 2nd ed.,* Knoxville, TN: SPC Press, 2000.
20. Ruggeri, F., Kenett, R.S. and Faltin, D., Editors in Chief, with Ruggeri, F., *Wiley Encyclopedia of Statistics in Quality and Reliability,* New York: John Wiley & Sons, 2007, online edition Wiley InterScience, 2008.

IMPROVEMENTS AND TESTING OF SYSTEMS AND SOFTWARE

Chapter 5

System and Software Development Process Improvement

Synopsis

Chapter 2 introduced generic process improvement methodologies, including Six Sigma and Enterprise Knowledge Development. In this chapter we focus on the application of these methodologies to continuous process improvement for systems and software development. Specifically, we present a full methodology for assessing and prioritizing improvement projects; describe how such projects are managed, including a description of roles and responsibilities; and conclude with a section on After Action Reviews for mapping lessons learned from current improvement projects for the benefit of future organizational improvements.

5.1 Implementing System and Software Process Improvement

Implementing system and software process improvement effectively and comprehensively requires the development of an overall process improvement strategy. Process improvement does not occur overnight, and cannot be implemented using short-term planning. Process improvement requires long-term commitment on the part of all parties involved: management, the developers, and the stakeholders. Many small steps are involved in process improvement. The first, and most crucial step is generating awareness of process change. There must be an acknowledgment by all parties that process improvement is required in order to remain competitive. Accompanying that acknowledgment must be a willingness on the part of management to commit the necessary resources to accomplish rational process change, and a commitment to implementing it.

5.1.1 Developing a Long-Range Action Plan for Process Improvement

In Chapter 2 we discussed the fundamentals of the improvement process. As mentioned above, the first step of this journey is problem recognition. This may be accomplished by performing an appraisal, as described in Chapter 3. In this chapter we use the CMMI-based appraisal methodology as the basis for obtaining a baseline from which to begin the process improvement effort. An appraisal of the current state of the practice leads to a set of findings and a corresponding set of recommendations for process improvement. These become the basis for selecting process improvement projects (the second step) and developing the action plan for process improvement (see Figure 5.1).

The strategy for implementing the recommendations needs to be addressed in a plan that accomplishes the actions as a series of small projects. The plan should identify the resources (including personnel, capital outlays, and software and hardware tools) needed to execute the projects, the schedule, associated tasks, project responsibilities, and measures that will be utilized to indicate success or failure against which

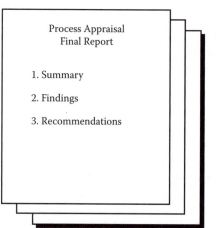

Figure 5.1 Structure of a process appraisal final report. Kenett, R.S. and Baker, E.R., *Software Process Quality: Management and Control*, New York: Marcel Dekker, Inc., 1999.

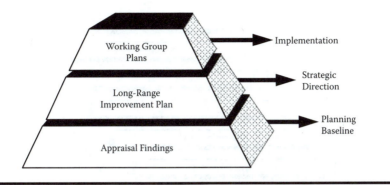

Figure 5.2 Action planning strategy. Kenett, R.S. and Baker, E.R., *Software Process Quality: Management and Control*, New York: Marcel Dekker, Inc., 1999.

the process improvement project results will be evaluated. At the top level we have an overall plan that identifies and prioritizes the individual projects. Rough, order-of-magnitude costs for executing these projects are estimated, together with a schedule that shows the phasing of each of these projects. Once the long-range plan has been approved, detailed implementation plans for each of the approved projects can then be developed. These plans will contain a more refined cost estimate and schedule for performing each of the projects. Figure 5.2 illustrates the relationships.

The long-range action plan should be one that is commensurate with the overall business objectives of the organization. An approach for developing such a plan has been developed by Frank Koch, a principal and co-founder of Process Strategies, Inc.* A description of the approach used by Process Strategies follows. Figure 5.3 illustrates the process.

In developing this plan, a team is organized, consisting of the personnel who participated in the appraisal, as well as any other personnel who have a vested interest in the outcome of the planning effort. This team is the focal point for the planning process. In Chapter 2 we discussed the concept of the Quality Council. We pointed out that sometimes the role of the Quality Council is split between senior management and the Software or System Engineering Process Group (SEPG). Members of that Quality Council (the SEPG) will typically participate in this team. As we show later, senior management also has an important role to perform as a strategic steering committee.

To expedite carrying out the various analyses that need to be performed, and to achieve agreement on priorities, the planning process is typically conducted in a workshop mode, led by a facilitator.

General Structure. Typically, the development of the strategic plan for process improvement occurs in a facilitated workshop setting. Participating in the workshop

---

* Printed by permission of Frank Koch, Process Strategies, Inc., Box 104, Walpole, ME, 04573-0104.

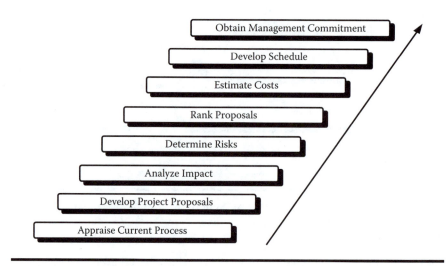

Figure 5.3 The long-range action planning process. Kenett, R.S. and Baker, E.R., *Software Process Quality: Management and Control,* **New York: Marcel Dekker, Inc., 1999.**

are a facilitator and the process improvement planning team. The process improvement planning team is generally comprised of five to seven people, typically those who are members of the SEPG, as well as those who have a vital interest in process improvement in the organization. They should be senior people, respected within the organization, who have good insight into what will or won't work with respect to process improvement. Once the team has completed the workshops, the process of implementing specific process improvement projects begins. These projects will generally be staffed by small teams of experts within the technical discipline associated with the process in question. Usually these experts work on these process improvement projects on a part-time basis. A project leader or manager is assigned to each team, and the process improvement project is managed no differently than any other project in the organization. To ensure that the quality improvement effort will succeed, and that the process improvement eventually implemented will endure, it is necessary to obtain the commitment of management to provide the needed resources, including the commitment of time for the personnel having the required skills and capabilities to implement the project, including skills as change agents. Accordingly, the project descriptions identify members and assignments for the implementation team.

The planning workshops are each of 2 days duration, and occur generally over a 4- to 6-week period. There are specific objectives established for each workshop session. They are as follows:

- Workshop 1:
    - Define business goals and critical success factors (CSFs), and weighting factors for the CSFs.

- Address findings:
  - Develop process improvement project solutions for the findings.
  - Select 12 to 14 for further development.
- Make assignments for writing process improvement project proposals.
- Begin developing risk factors and associated weights.
■ Workshop 2:
  - Review drafts of process improvement project proposals.
  - Finish risk factor identification and weighting.
■ Workshop 3:
  - Weight the criticality of specific CMMI Process Areas.
  - Score proposals against all the CSFs, risk factors, and CMMI PAs.
  - Develop initial rank order, based on scores.
  - Factor in contingencies and estimated costs and benefits.
  - Develop final proposal rank order.
  - Develop presentation to senior management on recommended approach.
  - Present briefing to senior management to obtain their concurrence.

The details of what occurs in these sessions are discussed in the following paragraphs.

Select Improvement Projects. The selection of an improvement project begins with an informal analysis or a structured appraisal, which determines the current status of the state of practice within the organization. The results of the analysis, or appraisal, are a set of findings that characterize the strengths and weaknesses of the processes currently operating in the organization. The weaknesses suggest the need for improvement. As a consequence, there is usually a need to nominate improvement projects. The action planning that results from the appraisal establishes priorities for process improvement and for the selection of improvement projects.

Nomination of the process improvement projects begins with a brainstorming session. In brainstorming sessions, the approach is to avoid squelching suggestions, and allowing the participants to make suggestions freely. Typically, the tendency will be to nominate one project for each weakness found. The result is a long list of nominations, which must be winnowed down to a manageable few. This is accomplished by noting overlaps between suggested projects and combining them into a single project, as well as looking to see where a single project could take care of more than one weakness found during the appraisal. In most cases, the result of this exercise is to produce a list of 12 to 14 process improvement projects that will be prioritized by some of the later workshop activities.

Develop Project Proposals. The strategic direction for process improvement is set based on the recognition of an existing problem within the organization. The problem could be loss of market share or degraded quality in delivered products. Some general underlying causes for the problem can be determined, by comparison against a standard benchmark, such as the CMMI. If the organization desires to improve productivity and product quality, the CMMI and the appraisal findings help to set the strategic direction. It identifies the CMMI PAs and associated practices that must be implemented to reach that goal. Near-term priorities for accomplishing the appraisal's recommendations should be based on the most critical process issues currently facing the organization; however, the priorities must be consistent with the findings of the appraisal. A Level 1 organization, for instance, should not normally be addressing Level 3 issues in their action plan (unless one of them was a particularly critical issue for the organization) because there are Level 2 issues that need to be addressed first. Even within the Level 2 issues, some may be a more burning issue for the organization than others.

While the CMMI is one factor that helps set the strategic direction, other factors enter into the picture as well: alignment with the organization's business goals and risk. We will look at that further in the next section.

Based on the findings of the appraisal, a set of recommended solutions to the process issues identified in the appraisal report is proposed. This begins the process of project selection. Typically, numerous projects are proposed, as we previously discussed; and through discussions and nominal group techniques, the list is winnowed down to a manageable size: something on the order of a dozen or so. All these projects cannot be implemented at once. Some sort of prioritization must occur. This occurs over the course of the workshops, as we describe in the following sections.

As noted above in the description of the workshop sessions, writing assignments for the process improvement proposals are made during the first workshop. Each member of the process improvement planning team is assigned two or three of the aforementioned proposals to write. The project descriptions are each contained in a one- or two-page summary (see Figure 5.9) that identify the project, provide a brief description of it, describe the expected benefits, identify the personnel who will be responsible for performing the project, identify the costs and associated resources, and define the duration of the project. In the description of the project will be an identification of how the process will change. It will identify the methods that will be investigated, address implementation in a trial application, identify the effort required to establish a set of trial standards and procedures, training of the personnel in the trial application in the new methods, incorporation of feedback into the methods and trial standards and procedures, and re-trial, if necessary. The estimated costs and associated resources address these activities. These plans, at this stage, are high level and do not provide sufficient content to initiate, execute, and manage the process improvement projects. When the proposals are accepted, a more detailed plan will be written for the accepted projects.

Analyze Business Impact. In arriving at the highest-priority improvement projects, three factors must be considered: business objectives impact, risk, and benefits. When we look at impact, we are looking at the impact of the projects on the overall strategic business objectives of the organization. We also look at the number of PAs affected by the project because an action to be implemented can impact more than one PA. For example, suppose there is a recommendation to implement configuration management to accomplish the baselining of requirements documentation and instituting change control over them. Suppose there is another recommendation concerning subcontractor management that addresses having a solidified set of requirements for the prospective subcontractors before the subcontract is let. Here, too, the concept of baselining is involved. A single project to implement a well-thought-out change management process can have an impact on the business objectives, as well as implementing process improvement based on both of these PAs.

In evaluating the business impact, the beginning point is the statement of the organization's business objectives. These are then characterized by a set of critical success factors that help determine if these objectives are being met. These should be expressed in terms that relate to the development organization. It is best to limit this list to a maximum of about seven critical success factors (CSFs). Generally, this list is developed in a joint session consisting of cognizant senior management and the members of the process improvement planning team. If the business objectives have already been clearly and publicly stated by management in some documented form, the focus is then on the development of the CSFs. Sometimes we find that the business objectives have not been articulated for the organization, and the first order of business then becomes to state those objectives. The intent of the activity is to reach consensus on the CSFs affecting the development organization, and to establish weights for each of the factors. It is also the intent to enlist senior management participation in this activity, thus establishing their commitment to the process improvement activity. Recall from Chapter 2 that senior management's active involvement is essential for institutionalizing process improvement.

A number of techniques exist to attain a consensus within a group on the CSFs and their weights. Techniques such as Australian balloting may be used quite effectively. (A discussion of these techniques is outside of the scope of this book.) Each of the projects is then scored against these CSFs, in terms of how well they will support achieving them. For each project, a normalized score is calculated, which is later used for rank-ordering the projects. Figure 5.4 illustrates the process.

Analyze Process Improvement Model Impact. A second impact analysis is then performed, in which each of the projects is evaluated against its impact on achieving the next level of capability on the CMMI. A simpler weighting is used here, based on a High–Low scale. Recall that some projects can impact more than one PA. Assuming that there is no pressing requirement to implement a PA at a maturity level higher than the one above the level at which the organization was rated, those PAs associated with the next level of maturity receive the highest weighting, while those that are associated with higher levels of maturity receive lower weighting. The objective is generally to get to the next higher level of maturity. Accordingly, as a first cut, all PAs at the next level of maturity receive the identical (higher) rating, while the PAs at the next level of maturity receive an identical lower rating. The project is then scored against each of the affected PAs, and a project normalized score is calculated, as previously described.

As a refinement, a second cut at the CMMI impact analysis may be performed. At the next level of maturity, some PAs may be more important than others, based on the appraisal findings. For example, an organization that was rated at Level 1 may

Process Improvement Proposal Impact Analysis for Project "C"			
Critical Success Description	Factors Weight	Impact Score	Wtd. Score
CSF 1	6	5	36
CSF 2	4	1	4
CSF 3	10	9	90
CSF 4	2	6	12
CSF 5	8	4	32
TOTALS:	30		174
NORMALIZED			5.8

Figure 5.4   Impact analysis based on business objectives. Kenett, R.S. and Baker, E.R., *Software Process Quality: Management and Control,* **New York: Marcel Dekker, Inc., 1999.**

have a pressing need to implement the Requirements Development PA, which is at Maturity Level 3, not at Level 2. Consequently, they may decide to give that PA a higher weighting than any of the Level 2 PAs. The issues that are presented as appraisal findings have been identified by consensus and reflect the most pressing issues for the organization. Accordingly, the PAs associated with those issues will receive a higher weighting. As before, weightings are established in a consensus-gathering session, again using techniques such as Australian balloting. Figure 5.5 illustrates the process. In performing this analysis, keep in mind that the cumulative sum of the weighted scores for a given PA across all projects cannot exceed 100%. The significance of a score of 100 is that the PA is totally achieved. A score in excess of 100 means that the PA has been more than achieved, which is not logically possible.

Analyze Risk. Risk refers to the difficulty associated with implementing the proposed plan. Is implementing the process improvement project a gamble, or are the results reasonably predictable? In determining this, a number of factors are considered, grouped into three categories: project size, structural issues, and technology (see Figure 5.6).

Project size, as a risk category, refers to the magnitude of the project in terms of staff-hours to implement it. In general, the smaller the number of staff-hours to perform the project, the less the risk.

The category of structural issues can include a number of factors, such as:

■ The number of functional groups within the organization involved in the project.
■ The complexity of the process improvement project.

Process Improvement Proposal Impact Analysis for Project "C"			
Process Area		Impact	Wtd.
PA	Weight	Score	Score
Project Planning	6	9	54
Project Monitoring & Control	6	7	42
Requirements Management.	6	2	12
PPQA	6	1	6
Configuration Management.	6	1	6
Integ. Project Management.	2	5	10
TOTALS:	32		130
NORMALIZED SCORE:			4.1

Figure 5.5   Impact analysis based on key process areas. Kenett, R.S. and Baker, E.R., *Software Process Quality: Management and Control,* **New York: Marcel Dekker, Inc., 1999.**

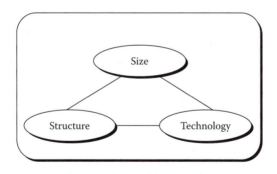

Figure 5.6   Project risk categories. Kenett, R.S. and Baker, E.R., *Software Process Quality: Management and Control,* **New York: Marcel Dekker, Inc., 1999.**

■ The experience in the affected process domain of the people assigned to develop the solution: Are the people assigned to develop the solution novices in this area?

■ The experience in the affected process domain of the people who will ultimately use it. In other words, will a great deal of training be required to familiarize the users with the practices involved?

■ The anticipated resistance of the organization to change in this area.

The category of technology issues can includes factors such as:

- The maturity of the proposed process within the system, software, or hardware engineering discipline. Is the project attempting to implement a cutting-edge solution?
- The availability of training in using the new methodology.
- The complexity of tools or other aids that will be acquired to implement the solution.
- The maturity of the tools and other aids. Is the organization going to acquire Version 1.0 of the tool (which is always a risky proposition)?
- Experience in the use of the tools or aids. How experienced are the users with the tools or aids used to support this process?
- Impacts on existing interfaces.

The risk evaluation is performed in a manner similar to that of the impact analysis. In the case of risk, however, a tailored set of risk factors is defined. The list of risk factors shown here is a generic list. They may not all be applicable for some organizations. Other, additional factors may be. The list needs to be refined for each organization so as to be applicable to that organization's specific environment.

For each project, a score is determined for each of the risk factors, based on guidelines (see Table 5.1), and a normalized score is calculated, based on the sum of the weighted scores for each of the factors. Note that there is one major difference. In the case of the impact analysis, the greater the impact, the higher the score. In the case of risk, the higher the risk, the lower the rating. Projects with the lowest risk receive the highest score.

Rank-Proposed Projects. Once the impact and risk analyses have been performed, the projects are ranked according to total score. The general equation for calculating the total score is:

$$\text{Total Score} = (\text{Weight}_1)(\text{Business Objective Impact}) + (\text{Weight}_2)(\text{PA Impact}) + (\text{Weight}_3)(\text{Risk})$$

where the impacts and the risk for each project are the normalized scores developed in the manner described in the paragraphs above, and weights 1, 2, and 3 are determined by consensus. To illustrate how the rankings are developed, some organizations may consider all three items of equal importance. Under those circumstances, the equation would reduce to:

$$\text{Total Score} = (\text{Business Objective Impact}) + (\text{PA Impact}) + (\text{Risk})$$

Another organization might consider the business objective impact three times more important than PA impact and risk twice as important as PA impact. Under those circumstances, the equation would reduce to:

Table 5.1   Examples of Risk Factor Scoring Guidelines

Risk Factor	Parameters	Score
Project size	Less than 160 staff-hours	10
	160–400 staff-hours	5
	More than 400 staff-hours	1
Number of functional groups	1–2 Groups	10
	34 Groups	5
	More than 4 groups	1
Process maturity	Vintage	10
	Middle-aged	5
	Bleeding edge	1
Process developer experience with this type of process	Expert	10
	Intermediate	5
	Beginner	1
Complexity of tools for supporting process	Simple	10
	Moderate	5
	Very	1
Resistance to change	Minimal	10
	Somewhat	5
	High	1

Source: Kenett, R.S. and Baker, E.R., *Software Process Quality: Management and Control,* New York: Marcel Dekker, Inc., 1999.

Total Score = 3(Business Objective Impact) + (PA Impact) + 2(Risk)

Each proposed project is thus scored, in turn.

A tentative ranking is now established on the basis of the scores recorded for each project, with the project achieving the highest score ranking the highest. A further refinement of the ranking is then made after the conclusion of the next step.

A cost-benefit analysis (e.g., return on investment) is not likely to be performed by organizations to support the ranking process. As a rule, Level 1 and 2 organizations will be unable to accurately forecast tangible benefits to be achieved from the improvement projects. These organizations typically will not have collected the

Process Improvement Cost Summary				
Rank	Project	Cost	Cumulative Cost	
1	Project A	$23,000	$23,000	
2	Project F	55,000	78,000	
3	Project D	100,00	178,000	Available Funding
4	Project C	62,000	240,000	Level = $185,000
5	Project E	15,000	255,000	
6	Project B	37,000	292,000	

Figure 5.7 Identifying candidate projects. Kenett, R.S. and Baker, E.R., *Software Process Quality: Management and Control,* New York: Marcel Dekker, Inc., 1999.

data and metrics to support such projections. Such analyses are feasible for Level 4 and 5 organizations, and may likely be achievable for Level 3 organizations, or organizations close to Level 3.

Estimate Cost and Schedule. After the proposals are ranked, the next step is to estimate the schedule and cost for each project by itself. In this step, the intent is not to develop fine-grained costs or schedules, but to get an overall rough order-of-magnitude estimate, in order to get a general idea of what the commitments would be for each project. Knowing the staffing and financial resources available for the near-term (e.g., the remainder of the fiscal year), the team can then identify the candidate projects for the near-term plan, based on priority and available resources. Figure 5.7 illustrates the methodology. Considering the fact that the projects have been prioritized, and in all likelihood there will be interdependencies between them, the next step is to develop an overall strategic schedule for all the projects that reflects their priorities and interdependencies. Figure 5.8 is an example of such a schedule. It is an example of a schedule for calendar year 2005, showing that the projects overlap 2 fiscal years, and the output from Project E becomes an input for Project B, which, as the plan indicates, would be performed in calendar year 2005. Also, the schedule indicates that Projects B and F become inputs to Project D. Consequently, the priority of the projects can change as a result of the interdependencies between projects. Projects A, F, and B may now become the projects recommended for the first year.

Another consideration is impact and visibility. For low CMMI Level 1 organizations, management may not have much credibility where it relates to process improvement. Past experience may prove that management "talks the talk" but doesn't "walk the walk." Considering that, a short-term project having low risk, high visibility, and some nontrivial benefits may be a better candidate for a high-priority, initial project, even though it may have had a much lower ranking.

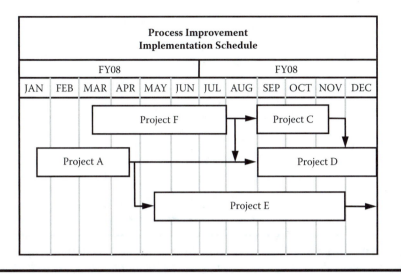

Figure 5.8   Scheduling projects. Kenett, R.S. and Baker, E.R., *Software Process Quality: Management and Control,* **New York: Marcel Dekker, Inc., 1999.**

Obtain Management Commitment. The next step is to formalize the plan and present it to management to obtain their concurrence and commitment. Earlier, we spoke of the necessity for senior management involvement in order to institutionalize process improvement. We discussed this in relation to their participation in the generation of the critical success factors. Once again, their participation is necessary, this time in approving the strategic plan and authorizing the initiation of the first process improvement projects.

The overall plan for process improvement is prepared and presented to the steering committee, which includes senior management. We discussed this in Chapter 2 when describing the function of the Quality Council. Figure 5.9 illustrates the typical content of the plan, which is comprised of the process improvement proposals. Recall that these are high-level plans and do not contain the detail necessary to implement the proposed process change or innovation, but are rough order-of-magnitude estimates of cost, benefit, and schedule. For the proposals selected for implementation, the next step would be to produce a detailed plan sufficient to manage its implementation. For organizations implementing Six Sigma, a Team is an appropriate format for such a plan.

An introductory section in the plan presented to management or the steering committee should explain the purpose of the plan, describe the methodology utilized in developing the content, and briefly summarize the findings and recommendations contained in the appraisal report. (It would refer the reader back to the appraisal report for more detail concerning the findings and recommendations.) The remaining content of the plan is a one- to two-page summary of each of the projects that we discussed earlier. This plan, after review and approval

Figure 5.9 Structure of a process improvement proposal. Kenett, R.S. and Baker, E.R., *Software Process Quality: Management and Control*, New York: Marcel Dekker, Inc., 1999.

by the steering committee, is signed by them, signifying their concurrence and commitment to providing the personnel and other resources. Once the strategic plan has been approved, the detailed plans for the individual projects are then developed. We discuss the implementation of the detailed plans in more detail in Section 5.2.

In Chapter 2 we discussed the relationship between the SEPG and the Quality Council. The SEPG is the focal point for implementing process improvement and managing change within the organization. Typically, it will be the responsibility of the SEPG to coordinate and manage the implementation of the action plan. They should track each project against its planned schedule, funding, and task performance. Most project plans require a pilot application of the change in the process. Once the trial application has shown that the new methods will work, the SEPG becomes responsible for communicating this to the organization as a whole, providing education and training to the organization at large in the new methods, and promulgating the standards and procedures necessary to implement them. Senior management is also involved in this process. Senior management, acting as a steering committee, meets regularly with the SEPG to discuss progress and status, and monitor the project to closure. Senior management also authorizes the official rollout of the changes in the process. After a pilot application shows the efficacy of the change, and feedback from the pilot application has been incorporated into the standards, procedures, and training materials, as appropriate, the SEPG makes a report to the steering committee. The steering committee then decides if the change should be rolled out, and, if it should, provides the authorization. The steering committee also periodically reviews the status of the action plan and decides on which improvement projects to initiate next.

5.1.2 Work Environment Factors Affecting Process Improvement Plan Implementation

In developing the detailed plans, it is important to remember that there likely are barriers to successful implementation of process improvement. If these barriers did not exist, it is quite likely that process improvement efforts would have already been further advanced. The plan must take these barriers into account, and address the mitigation of their effect.

The next several subsections discuss some of these barriers.

5.1.2.1 Facilities

Richard Templar in *The Rules of Management* states, "Get your team the best, the very, very best, and then let it get on with its job" [11]. The physical environment in which employees work can have a major impact on productivity. Studies show that the amount of floor space allocated to an employee can have a significant effect on employee productivity, thus moderating the effect of other actions taken to improve productivity through process improvement. Capers Jones [3] cites a study by DeMarco showing that the productivity of people who worked in offices having 78 square feet or more were in the high quartile. On the other hand, those who had 44 square feet or less were in the low quartile. An analysis conducted by TRW on its company-funded Software Productivity Project [8] showed that personnel productivity on this project was 42% higher than that predicted for such a project. These personnel had been supplied with private offices having 90 to 100 square feet, floor-to-ceiling walls, carpeting, soundproofing, and ergonomically designed chairs. They were provided with adequate desk area, lighting, and storage. Each office had a terminal or a workstation, which was networked to computers, file servers, and printers. Roughly 8% of the productivity gain was attributed to the comfort of the work environment.

In addition to the floor space, personnel require adequate shelf space for reference material and storing other materials (even personal items) considered important by the developers. A minimum of 20 feet of shelf space is recommended [1].

Noise levels also adversely affect productivity. Noise can be especially bad in facilities where partitions do not extend to the ceiling, floors are not carpeted, and/or several people share offices. If an office is shared, people dropping in to visit with an office-mate can be very distracting, especially if the conversation is not work related. Offices with low partitions, if the office area has not been properly soundproofed, are susceptible to all sorts of ambient noise (people walking down the corridor, hallway conversations, etc.). Even with good soundproofing, low partitions can still allow in a fair amount of ambient noise. The positive effects on productivity and quality of providing personnel with the latest and greatest in tools, techniques, methodologies, and training can still be blunted by failing to provide an agreeable work environment.

There are other facility/environmental factors to consider. These include, among other things, the availability of [1]:

■ Adequate lighting. Numerous studies indicate that fluorescent lighting *lowers* productivity.
■ Free parking close to the facility.
■ Availability of food service and vending machines. One site with which we are familiar provides free pretzels, cookies, and potato chips of various flavors to all work areas. Because of this and other environmental considerations, this company achieves significant output from its employees, as well as a great deal of creativity.
■ Wall space for notes—a minimum of 10 square feet. This can be in the form of corkboards, chalkboards, or whiteboards. The availability of Post-its™ (or equivalent "stickies") is a great help in note-keeping and replacing "to-do" files.

5.1.2.2 Software Tools to Facilitate Process Implementation

Development organizations sometimes rely on software tools as a replacement for process. They believe that if they acquire a tool, software or otherwise, they have acquired a process. Software tools should be acquired only to serve and facilitate an established process or one that is being piloted. Reliance on tools without having an underlying process in place is very risky. Until an organization has defined, established, and codified the methodology it uses for developing a product, software tools will be of little or no value. The tools will provide more capability than the organization can assimilate or use effectively. Accordingly, organizations functioning at very low levels of process maturity are not able to realize significant benefits from them. Consequently, organizations that see tools as a panacea for their productivity and quality problems without addressing the basic process problems first will find that tools are an obstacle, rather than an aid, and will have wasted considerable sums of money.

Software tools for development and testing can be either a godsend or a hindrance, depending on if they are used, the context in which they are used, and how they are used. We sometimes find companies that do not readily acquire tools. They are reluctant to expend the capital funds necessary to acquire them. Consequently, the process they have put in place becomes slower and more cumbersome to implement. As an example, imagine implementing an entire measurement program using spreadsheets, with attendant manual collection of raw data, manual inputting of data into the spreadsheets, and manual execution of formulae. Imagine how much easier it would be, and how much more readily a measurement program would be accepted, if a software tool was available to facilitate most of the effort, and to make the raw data collection as unobtrusive as possible to the developers.

To compound matters, many companies, when they acquire tools, are often reticent to contract for the associated training. The non-acquisition of training is treated as a cost savings. In actuality, it is costing these organizations more by not acquiring the necessary training. Tools do not teach the user how to use them: trainers do. Without the training, the tools either go unused or are used inefficiently and ineffectively. Any productivity and quality gains that could have been realized by the use of these tools are quickly negated.

5.1.2.3 Personnel Resources

Management attitude toward personnel strongly affects how the workforce performs and the success of process improvement. In Japan, when people enter the workforce, they tend to stay at one company for their entire working career. Layoffs tend to be uncommon. (Although both of these conditions are starting to change, this is still the general rule [10].) There is a strong feeling for the group, and people work together for the good of the organization. Maintaining motivation and having an adequate, experienced staff is much less of a problem than it is in the Western World (although temporary employees are becoming more common).

In the Western World, people want to be rewarded for their individual contributions. If they feel that they are not, they feel free to leave their employer and go to work for another. Management, on the other hand, wants to have a free hand in reducing the workforce as economic conditions change. They want to be able, when they see fit, to shift personnel from project to project to balance costs, and to sometimes replace experienced personnel with less-experienced personnel. (Note that management can sometimes accomplish the same economic results by process improvement, rather than by arbitrary staff reductions or sudden changes in staffing.) If personnel do not feel valued, performance and acceptance of change will suffer.

Most personnel like to have training and want more of it. We hear this quite frequently during the course of process appraisals. Some like training in disciplines related to their current jobs. Others look at training as a way of getting a better job somewhere else. Management's views of training, on the other hand, are sometimes unclear. Training is often one of the first things cut out of the budget when companies fall on hard times. The need to reduce the cost of doing business when economic downturns occur is certainly understandable; however, many companies practice cut-and-slash without taking a careful look at their business practices. This was alluded to in the previous paragraph. Improving business processes can often effect significant cost savings without the necessity for cutting personnel or drastically reducing the training programs, which can result in improved staff performance. If cutting and slashing is perceived as indiscriminate, this can severely overtax the surviving staff and cause serious morale problems.

How companies use their training budgets varies considerably. In some companies, training is sparse. Other companies provide a great deal of training. European,

Japanese, and especially Indian companies provide a great deal of training—far more than that provided by companies in the United States. (A recent article in *Businessworld* [9] points out that in India, human resources budgets now focus on plan-for-growth, not plan-for-cost; consequently, training budgets are increasing at an annual rate of 20% to 25% in some companies.) Some companies succeed in providing for both individual growth as well as growth of the organization. Others provide only for growth of the organization. Still others succeed in providing the right kinds of training, but at the wrong time—either too far before training in the discipline is needed, or too late, well after the people who needed it have limped along without it.

Intelligent utilization of personnel resources is essential for process improvement. Process improvement requires motivated personnel if the process improvement effort is to be seen as something other than the fad of the week. Experienced, skilled personnel must be employed in the planning and implementation of the process improvement efforts. Training in the new skills and disciplines to be implemented must be provided. Management must be seen as being behind the effort (more on that in the next section).

Process improvement planning must take these factors into account.

5.1.2.4 Management Policies

Process improvement starts at the top. Management commitment to process improvement must exist or no improvement will occur. Management must be in for the long haul. There must be a recognition that process improvement takes time, and that changes will not occur overnight. If management is unable to recognize that fact, or is not willing to stay the course during the time it takes for process improvement to take hold, it is better not to begin.

Management must provide the resources, establish the priorities, and provide encouragement for implementing process improvement. All are essential ingredients. This fact has been recognized by the SEI in the structure of the CMMI. Each PA is comprised of Specific and Generic Goals and corresponding Specific and Generic Practices. The Generic Practices (GPs) address providing an infrastructure to ensure that the performance of the activities of the process area (Specific Practices) will occur in a consistent manner from project to project. The GPs include such things as promulgating the necessary policies within the organization requiring it to perform the good practices encompassed by the PA, specifying management responsibility for seeing to it that the policy is consistently implemented, providing the necessary resources to implement the practices, such as funding and training, and management oversight of the regular performance of the PA.

In general, if these activities are not properly managed for process improvement, they can discourage the improvement efforts. Management must be willing to make capital outlays to pay for the necessary staff to oversee the process improvement efforts and to acquire the software tools and hardware associated with the process improvement projects. This does not mean that management should sign a blank check. Reasonable budgets should be established, and the projects should be required by management to live within their budgets and schedules. Priorities should be established to take care of contingencies. By setting reasonable priorities based on sound process improvement objectives, looking for "silver bullets" can be avoided. Too often, management squanders precious financial resources on software or hardware, which is looked upon as a quick solution.

The foregoing assumes, of course, that management has a genuine commitment to quality. This means that schedule is not the primary motivating factor in the organization. In too many organizations, quality assurance, configuration management, and test (among other good practices) fall by the wayside when projects fall behind. Process improvement efforts will fail, and personnel will be demotivated if management is perceived as paying lip service to quality. If schedule is overriding, quality cannot be perceived as important.

5.2 Managing Improvement Projects

The improvement projects should be managed no differently than any other project. As part of the initial action plan development, rough order-of-magnitude estimates are defined for each project. Prior to the start of each project, a detailed plan for that project should be drawn up. A detailed definition of the tasks to be performed should be established. Based on the task definitions, detailed estimates of the labor hours to perform the tasks should be made, together with the schedules for their performance. If possible, identify the specific individuals who should perform these tasks, because it will be essential to obtain their management's commitment for them to support the process improvement projects. They should be individuals who have specific expertise in the affected disciplines, as well as being respected within the organization. This is necessary to facilitate acceptance of the proposed changes within the organization. Establish detailed cost estimates for each task, including nonrecurring costs. This could include the cost for tools, outside consultants, training, books, etc., as well as the costs for printing draft procedures, standards, training materials, and the like for the pilot projects.

Earlier in this chapter we discussed the structure of the process improvement projects. We addressed the need to characterize the existing process, collect measures of the current process, and analyze the data. In addition, we discussed the need for measures to determine if the process changes provide significant benefit. By doing this, we provide data essential for the proper management of improvement projects. Consequently, in developing the plans for the projects, the detailed tasks should address (among other things) the following:

- What does the current process look like? This may involve developing a model of the current process. For instance, if the process improvement project is geared toward improving the system requirements definition process, it may be very useful to develop a model of the process currently used to develop these requirements.

■ What are the inputs to and outputs from the current process? How do they relate to the quality issue?
■ What measures do we have or can we generate to determine the quality of the inputs and outputs?
■ How are we going to collect and analyze the data?
■ What changes do we need to make to the process?
■ What measures do we have to collect in order to determine if the change is beneficial?

The plan should also include provisions for pilot applications of the process change, as described previously. This involves the following:

■ Identifying and selecting an ongoing or new development or maintenance project in which to conduct a pilot application of the change
■ Producing draft standards and/or procedures to define how the process change is to be implemented in practice
■ Developing and implementing draft training materials for use in training the practitioners in the required practices
■ Utilizing the measures we cited in the previous paragraph to evaluate the efficacy of the change
■ Applying feedback and lessons learned from the pilot application for overhauling or fine-tuning the draft materials (standards, procedures, training materials) before rollout to the organization as a whole
■ Releasing the changed process to the organization-at-large, using the revised standards, procedures, and training materials

The plan, when completed, should be submitted to the SEPG for review and comment. Feedback from the SEPG review should be incorporated into the plan. When satisfied with the plan's content, the SEPG will then submit it to the senior management steering committee for review and approval (as discussed previously). The steering committee, when it finds the plan acceptable, must provide the specific resources required. This includes adequate budget, as well as personnel assignments. The steering committee should be the focal point for obtaining the commitment of lower-level managers to provide the specific individuals identified in the plan to support the project. If the specified individual can't be allocated to the project, then a suitable substitute should be negotiated between the steering committee and the designated project leader.

The project leader should manage the performance against the approved plan. This includes managing the expenditure of funds, tracking the schedule, and tracking the performance of the tasks. Where variances exist, they should be investigated; and if corrective action is necessary, it should be implemented. The commitment of personnel should also be monitored. Where commitments of individuals to the project are not being kept, that problem must be worked up through the chain to the SEPG to the steering committee.

Successful process improvement projects must follow the same good project management practices that any competent project manager would follow in managing a development project. For more on characteristics of improvement projects, see [2, 4–7].

5.3 Roles and Responsibilities and Specifying Mission Statements

Improvement projects, in system and software development environments, as in other organizations, are carried out by improvement teams. The process improvement project, if not initiated as a result, for example, of a CMMI-based appraisal or an ISO audit, has generally been initiated as the result of the efforts of a champion. The champion is a person (or persons) who saw the necessity for a process improvement and advocated for it. The champion may in some cases be included in the improvement team, if he/she has the necessary expertise (and can be made available to support the project), but has no special role as a result of advocating the change.

To distinguish such teams from teams working in other organizational contexts, we will call them Process Improvement Project Teams. The infrastructure for operating such projects involves a Steering Committee or Quality Council or SEPG, as described in Chapter 2, a Sponsor, a Team Leader, and a Team Facilitator. The Sponsor is a member of the Steering Committee or Quality Council or SEPG assigned to a specific improvement project to represent the interests of the Steering Committee or Quality Council. The Team Leader is the project manager of the process improvement project and is primarily responsible for getting the project to achieve its mission. The Team Facilitator plays a professional role supporting the team leader and the team and provides expertise in methodologies and tools. The facilitator is responsible for the team's work standards and proper use of tools. In the Six Sigma context, facilitators are called Black Belts.

In the remainder of this section we take an operational view of process improvement, identifying roles and responsibilities and describing how to design, launch, and maintain process improvement initiatives in the context of system and software development. We begin by describing the roles and responsibilities of the key actors in the improvement process.

- ■ The Sponsor:
    - – Oversees the team vis-à-vis the Steering Committee, committed to the success of the project
    - – Assists the team in case of difficulty (disagreement on progress, external non-cooperation, data availability)

- Attends team meetings from time to time as needed and in coordination with a GQM facilitator
- Participates in periodic follow-up meetings with team leader and a GQM facilitator
- Supports team leader in preparation of periodic report
◼ The Facilitator:
  - Is responsible to the Steering Committee for the professional aspects of the improvement project
  - Continuously reports to the Steering Committee
  - Focuses the team members on the project's objectives and methodology
  - Oversees progress according to the schedule
  - Is a knowledge resource of the methodology and the improvement process
◼ The Team Leader:
  - Implements good project management practices
  - Is responsible to the Steering Committee for the improvement project's progress
  - Assures attendance by the team members
  - Makes assignments
  - Facilitates group problem solving
  - Follows up on commitments
  - Acquires outside support, when needed

Where an organization has implemented Six Sigma, the Team Leader may be a certified Black Belt.

Improvement teams are driven by a mission statement. Some examples of mission statements from improvement initiatives in software development are listed below:

◼ Project Management Improvement Team mission statement:
  The purpose of the Project Management Improvement Team is to define and implement measurable improvements to the project management processes used throughout the system product life cycle. These changes will improve the ability to plan and trace projects so that execution of programs meet committed plans.
◼ Documentation Improvement Team mission statement:
  The purpose of the Documentation Improvement Team is to define and implement measurable improvements to the documentation used to support the processes used throughout the system product life cycle. These changes will achieve improved quality, time to market, and productivity by reducing the number of design-related bugs.
◼ Engineering Improvement Team mission statement:
  The purpose of the Engineering Improvement Team is to define and implement measurable improvements to the development processes. These changes will

> *achieve improved product quality, time to market, and improve productivity by reducing the number of bugs that are found within the product at initiation of Alpha test and by reducing the number of regression-related bugs.*

■ Release Management Improvement Team mission statement:
> *The purpose of the Release Management Improvement Team is to define and implement measurable improvements to the Change Control processes and development environment. These changes will achieve improved software product quality, time to market, and productivity by reducing developer errors in delivering code to final product.*

In addition to specific improvement teams such as the ones described above, it is not uncommon to also form a support or infrastructure improvement team. The mission of such a team is listed below:

■ Support Improvement Team mission statement:
> *The purpose of the Support Improvement Team is to define and implement measurable improvements to the configuration management, quality assurance, and causal analysis processes and environment. These changes will achieve improved software product quality, time to market, and productivity by reducing and avoiding defects in the delivered product.*

■ Testing Methodology Improvement Team mission statement:
> *The purpose of the Testing Methodology Improvement Team is to define and implement measurable improvements to the test strategy, methodologies, and procedures. These changes will achieve improved software product quality, time to market, and productivity by improving early identification of system flaws.*

The mission statements are composed by the Steering Committee. To launch an improvement project, management must also consider the proper team composition and act accordingly. In the next section we present several criteria for forming such teams.

5.4 Criteria for Team Composition

Team members are selected by the sponsors and team leader using professional criteria such as:

■ Employees with potential for advancement
■ Employees considered professionals in their job
■ Employees with organizational capabilities and opinion leaders
■ Employees with high commitment
■ Employees with the ability to critique current practices

- Employees with leadership capability
- Employees who can be objective and can come to the table without hidden agendas

5.5 Launching Improvement Teams

Properly launching improvement teams requires preparation and commitment. Usually the launch is executed during a meeting of the Quality Council or SEPG in the presence of the team leader, its facilitator, and the team members. On that occasion, the problem and mission statement prepared by the Quality Council or SEPG are handed out to the team leader, and the background for choosing the specific improvement areas is explained. In addition to specifying the problem statement, management must also set a mission statement for the team. These two documents serve to set expectations between management and the team. Because improvement teams sometimes meet over several months, these documents often serve as a reminder of what the project is about and helps keep the improvement effort focused. We saw in Section 5.3 several examples of mission statements that are characteristic of process improvement projects in systems and software development environments.

5.6 Managing Improvement Teams

Like any other initiative, the progress of process improvement projects needs monitoring and management attention. The most common monitoring activity is to hold a monthly meeting of the Quality Council or SEPG where a progress review of all process improvement projects is carried out. In such meetings, one can spend more time on specific projects and conduct a quick review of other projects. Projects under comprehensive review might be represented by all team members. The status of other projects is typically provided only by the team leader and team facilitator. Status, schedule, budget, risks, and issues (technical, management, coordination, etc.) are generally included in the topics covered.

5.7 After Action Reviews

An After Action Review (AAR) formalizes the process of learning from past experience in general and from improvement teams in particular. Its purpose is to carefully analyze a project once it has ended, and identify what went well and what went poorly so the team can do better on the next project. A major purpose of the AAR is to collect lessons learned. It can also serve to give closure to a project. This is particularly important if people have spent quite a bit of time and energy on the current project and are being rolled into another project right away or are being disbanded. By definition, an AAR should take place after a project is finished. However, the process of analyzing work completed and learning from that experience is valuable at any stage of a project, so major milestones are also appropriate times to apply the principles of an AAR.

While there are no rules dictating when to schedule an end-of-project AAR, it's best to conduct it between 2 and 5 weeks after a project is complete. If you schedule it too soon, people may focus their discussion too heavily on the latter part of the project, or they might still be too close to the project emotionally and mentally to be objective in their analysis. If you wait too long, people will forget the details of the project or be so removed that what was actually a very frustrating part of the project may not now seem like a big deal.

There are many ways to organize the AAR. You can have a group meeting to discuss the issues, or you can organize an AAR team to research the issues and present their results. While some AAR meetings are day-long events, it's usually better to keep them to 4 hours or less, even if it means breaking an AAR into two different meetings. Keep in mind that much of the work can be done outside the meeting. Invite anyone who was involved in the project—even if only for a small percent of the time—to the AAR. If inviting everyone makes the group too large, however, consider conducting mini-AARs, dividing people into groups that worked on specific parts of the project (e.g., user education, development, testing, or program management). Representatives from each of these mini-AARs could then attend the final AAR. Another alternative is to have people send in their comments on the agenda items to a few key players and have the key players conduct the AAR. Ultimately, however, it's better—if at all possible—to gather everyone together for the AAR so they can speak for themselves.

A sample AAR agenda is:

■ Timeline for project
■ Players and the percentage of their time spent on the project
■ What went badly
■ What should be done differently
■ What went well
■ Recommendations for the future

As long as all these elements are included in some form in the AAR process, it doesn't matter whether they all take place during the actual meeting. The timeline can be developed by one or two people ahead of time and distributed to others for review. Each person involved can report on his/her involvement, role, percent of time spent on the project, dates involved, and so forth. Typically, the timeline acts as a vehicle for focusing people on the entire project, rather than just the end or just when they were involved. Send out the agenda at least a week prior to the AAR with instructions for people to begin thinking about the project with the agenda topics in mind (what went well, what went poorly, and what should be done differently).

You may want people to send a list of "good" and "bad" parts of the project to the person organizing the AAR a day in advance. That list can then become the new agenda. Otherwise, it is perfectly appropriate for people to simply think about the agenda and bring up their good and bad points during the AAR. You should also have some of the team review old AARs from your group, if they exist, prior to the meeting. Also, review some of the AARs that exist for other projects. It can be very helpful to review AARs from projects that were similar to your own.

To make an AAR most effective, you need to have a neutral person facilitate the meeting. This means you need someone who was not involved in the project and who has no investment in the outcome of the discussion. The facilitator will be in charge of the process in the AAR, making sure people stay on topic, no one gets personally attacked, the agenda items are covered, everyone is equally involved in the discussion, and everything gets done in the time allotted for the meeting. The facilitator also needs to understand the overall purpose of the AAR and have various techniques available for use during the actual meeting if straight discussion isn't working.

It is essential to record what is being discussed as it is being discussed. The best way to record the AAR discussion is to do it in such a way that everyone can see what's being written. That can be done by taping large sheets of flipchart paper on all the walls in the room so the recorder can simply move from piece to piece, writing what is being said. Using whiteboards is also effective, although you may run out of space, and the recorded information must be later copied to a more permanent medium. Recording the discussion by writing it down is an effective approach because everyone involved in the discussion focuses on the recorded information instead of on each other, which keeps the discussion on track. It is also an excellent facilitation tool because it offers an alternative way to handle ideas that arise that are not related to the current topic of discussion. The recorder simply writes off-subject ideas (sometimes referred to as the "parking lot") on another sheet of paper. That way, good ideas don't get lost, but they don't defocus the team either. Another benefit of recording comments on sheets of paper is that it prompts people to be more concise in what they say and to avoid repeating each other as often, because they can see the comments. The recorder should be organizing the information, as well as recording it, by topic. One good way to do that is to use different locations and different colors, depending on the topic. For example, red could be for negative comments, while green could be for positive ones, and yellow for things needing to be changed. If it's possible to arrange, the recorder should also be someone who was not involved in the project being critiqued. Often, the facilitator and the recorder are the same person.

Open the AAR meeting by setting everyone's expectations for what the meeting is supposed to accomplish and how it will be conducted. Ensure that everyone present understands that it is not a chance to get back at people, place blame, or simply vent. The purpose is to come up with recommendations for future projects based on experience with the previous project. It is also important that people understand that the comments made during the AAR will not be reflected in their performance reviews. It needs to be a productive working discussion. The facilitator should then explain his/her role in keeping the discussion on topic, and protecting people, and lay some ground rules—for example, don't interrupt and don't attack another person. This will make it easier for the facilitator to jump in during the discussion to refocus people when they need it. The recording function should also be explained, and everyone should be encouraged to correct the recorder when appropriate.

The first topic of discussion should be the timeline, looking at who was involved in the project when, and the amount of time each person was involved. If this information was generated ahead of time, it should be prominently displayed in the room, allowing the recorder to simply add to it appropriately. If this information was not generated ahead of time, use several sheets of paper to create a timeline for the project and to begin filling in dates as people remember things. Either way, working on the timeline is an effective way to get all members immediately focused on the project and to remind them of the entire time frame to be discussed.

Next, it is usually best to discuss what went poorly because people are usually anxious to get to that anyway, and you can then end the meeting with what went well. Psychologically, it has the effect of ending the meeting on a positive note. If you haven't collected a list of topics from people ahead of time, your first task should be to get a list of everything that went wrong. This can be done by brainstorming, which involves having people say whatever comes to mind as they think of it.

Brainstorming is a technique that can be used to develop the list. Another nominal technique to use involves having everyone list individually what they think went wrong, and then discuss those items, one at a time. The technique requires that each person read one item from his/her list, after which the next person does the same, and this continues until everyone has read and has had recorded everything on his/her list. Storyboarding is another technique that can be used. It involves having everyone write their lists on index cards (one item per card), which are then piled together and sorted by topic until they form a comprehensive list. Pick the technique likely to work best for the people involved. If you have people likely to dominate a brainstorming session, you might want to use the nominal technique. If you have people worried that others know what they say, use storyboarding. If you collected the list of what went wrong ahead of time, go over it now and ask for any additions. Once you have the list generated, you need to prioritize the list so that you discuss the most important issues first, in case you run out of time. One simple technique is to have everyone vote on their top three issues. This can be done two ways: by having people mark their top three items using little round stickers and placing them on the list, or by simply voting by raising their hands. Then, simply start with the most important issue and open up the floor for discussion. This is when the role of the recorder really kicks in. Almost everything that is said should be recorded on the flipchart paper. As discussion about why this issue was a problem winds down, ask the group what signals or flags there were that might have warned something could be a problem. This is an important question because these are the things the team needs to be on the lookout for during its next project.

Also, wrap up each issue with what the group would have done differently to make it more successful. For each issue, then, the discussion should flow as follows:

- Why this was a problem/what went wrong?
- What were the signs that should have warned us?
- What should we have done differently?

The part of the AAR that deals with things that went well is generally much easier to facilitate. You simply need to list what people thought went well and record why it was successful. Note that some of the things listed will contradict what was discussed during the "what went poorly" part, which is okay. It's important to record that there were differences of opinion.

Discussing things to change for the future is a nice way to summarize the AAR. Looking back at the information recorded for both the "good" and "bad' discussions, summarize what the group would recommend for future projects. Sometimes this is done after the meeting by the person writing the AAR document.

The team should prioritize each item and issue that was identified as something that worked well or didn't work well, based upon effectiveness at reaching the goals of the team. The recommendations should focus on only the high-priority items and issues. In such an analysis, modern text mining techniques such as *Association Rules* can be used to scan large amounts of textual records (see [12] and other papers in the quoted volume). For an application of Association Rules to the analysis of aircraft accident data and IT operational risk data in a financial institution, see [13].

Problems, items, and issues that didn't work well should be studied carefully. Alternatives should be developed to address those problems. If best practices exist to address those problems, they should be listed as recommended solutions. For problems that have no known best practice solution, the team should determine the best alternative to solve the problem. A summary of all findings and recommendations should be presented to the management team and to the project team. All additions, revisions, and corrections should be recorded for inclusion in the final report. The team should select which recommendation will be accepted for implementation in the development process for the next milestone or for the next project.

Some organizations maintain a "lessons learned" database, in which information like this is organized and stored for future use by similar such projects. One client company organizes its lessons learned into general categories with subcategories of "the good, the bad, and the ugly." That helps in data retrieval and to be able to look for potential problems for various types of projects of interest. Data quality in such databases is clearly a critical issue. If the records are screened with some types of records consistently omitted or the record includes incomplete information, the resulting bias can affect future use and have a misleading effect. The track from data to information and knowledge is nontrivial and requires careful planning and maintenance (see [14]).

Documenting the AAR is very important, both for future reference and for the people involved, so they have a summary of the meeting. Decide ahead of time who will be responsible for writing up the document. The document should include the timeline, names of people involved, list of recommendations, and the information recorded from the discussion. There should be very little deleted from the information that was recorded during the meeting or from e-mail collected by the AAR team, because that most accurately reflects what the group said. Make sure everyone gets a chance to review the document before it is published so they can propose needed changes.

5.8 Summary

In this chapter we elaborated on various principles discussed in Chapter 2 regarding structuring process improvement programs and process improvement teams, and presented specific applications of these principles to system and software process improvement. We described the application of CMMI appraisals to the implementation of process improvement planning. In this way, findings from CMMI appraisals become the input for the activity of strategic planning for process improvement. A formal methodology for planning process improvement was discussed in detail, with a definition of roles and responsibilities.

In the next two chapters we discuss the design of a test strategy and beta programs that precede official release of a software or system version. In Chapter 8 we expand the discussion to the analysis of field data for establishing baselines of current performance, and measure actual improvements achieved by specific improvement initiatives.

References

1. Bliss, R., The Other SEE, *CrossTalk*, January 1994.
2. Godfrey, B. and Kenett, R.S., Joseph M. Juran, A Perspective on Past Contributions and Future Impact, *Quality and Reliability Engineering International*, 23, 653–663, 2007.
3. Interview with T. Capers Jones, *CASE Strategies*, Vol. II, No. 9, September 1990.
4. Juran, J.M., *Managerial Breakthrough, 2nd ed.*, London: McGraw-Hill, 1995.
5. Juran, J.M., *Juran on Leadership for Quality—An Executive Handbook*, New York: the Free Press, 1989.
6. Juran, J.M., The Quality Trilogy: A Universal Approach to Managing for Quality, *Proceedings of the ASQC 40th Annual Quality Congress*, Anaheim, CA, May 20, 1986.
7. Kenett, R.S. and Albert, D., The International Quality Manager, *Quality Progress*, 2001, p. 45–51.

8. Putnam, L.H. and Myers, W., *Measures for Excellence: Reliable Software On Time, Within Budget*, Englewood Cliffs, NJ: Yourdon Press, 1992.
9. Rajawat, Y., Where Are the People?, *Businessworld*, January 29–February 4, 2008.
10. Sayonara, Salaryman, *Economist.com*, January 3, 2008.
11. Templar, R., *Rules of Management: The Definitive Guide to Managerial Success,* Upper Saddle River, NJ: Pearson P T R, 2005.
12. Kenett, R.S. and Salini, S., Relative Linkage Disequilibrium: A New Measure for Association Rules, In P. Perner (Ed.), *Advances in Data Mining: Medial Applications, E-Commerce, Marketing, and Theoretical Aspects*, ICDM 2008, Leipzig, Germany, Lecture Notes in Computer Science, Springer Verlag, Vol. 5077, 2008.
13. Kenett, R.S. and Salini, S., Relative Linkage Disequilibrium Applications to Aircraft Accidents and Operational Risks, *Transactions on Machine Learning and Data Mining*, 1(2), 83–96, 2008.
14. Kenett, R.S., From Data to Information to Knowledge, *Six Sigma Forum Magazine*, November 2008, p. 32–33.

Chapter 6

System and Software Testing Strategies

Synopsis

This chapter deals with the design and implementation of a testing strategy in system and software development. Such testing is necessary in order to increase the probability that the software or system developed by the development organization meets customer needs and performs according to specifications. The goal of such efforts is to ensure that the delivered product will gain acceptability of users and to provide feedback and insights to developers so that they can improve and monitor development processes. Organizations at Level 5 of the CMMI are able to make efficient use of such feedback, while organizations at low maturity levels are using testing as a final attempt to prevent problems from reaching the customer.

We begin this chapter with a section on the design of a test strategy, followed by sections on test strategy implementation, the analysis of test data, and the assessment of testing effectiveness and efficiency using STAM—the Software Trouble Assessment Matrix. We include in this chapter a discussion on the newly evolving field of *software cybernetics* as a method for facilitating the monitoring and control of a testing program. We conclude the chapter with an introduction to software reliability models. The mathematical details of the software reliability models are included in an appendix to this chapter; they require a proper mathematical background and can be skipped without loss of continuity.

6.1 Designing a Test Strategy

6.1.1 Background

A test strategy describes how testing will be carried out in the context of an over-all development strategy. The most common test strategies implement a V-model approach (Figure 6.1). The model maps the testing activities to the various development activities, beginning with the requirements definition and on through to design implementation. The concept of the V-model applies to software and system development. In some cases, the V-model is tailored to fit the culture of the development organization. In this chapter we provide one such example.

The V-model applies, in one form or another, to almost all development strategies, including waterfall [7], JAD/RAD [5, 36], rapid prototyping [4, 36], spiral model incremental development [7], and Scrum or Agile development [2, 6, 9, 30, 37]. In all cases, testing is conducted to ensure that the design and implementation process produces a product that meets the requirements. The case study in Chapter 10, where more details on the approach are provided, is based on an application of Agile Scrum development. Here in this chapter we focus on the design and implementation of a generic test strategy based on the V-model.

The test strategy of a specific project should take into consideration testing constraints derived from time-to-market and quality improvement goals. The design of a test strategy should be documented in a system Test Plan that covers the entire

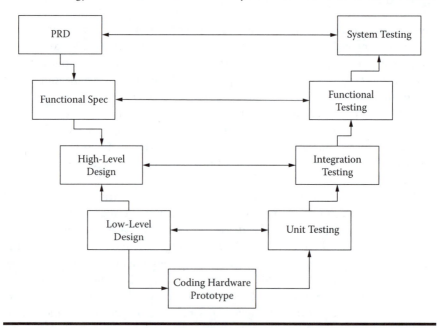

Figure 6.1 V-model of software and system development: from product requirements definition (PRD) to system testing.

test program. A Test Plan design starts by mapping the organizational structure and characteristics of the test teams to the development organization and corresponding activities. In terms of organizational structure, the main alternatives are test teams reporting to the project managers, or test teams reporting to an independent testing or quality manager. In any case, the test strategy should clearly define how the execution of testing at all levels will establish that the deliverable product meets customer and user requirements. The Test Plan should identify the organizational responsibility for all tests, such as unit tests, feature, integration, and regression testing. It is usually the responsibility of the product team to provide estimates of actual product quality. A test strategy should specify the quality objectives for the test team as derived from the quality strategy of the whole organization and quality objectives for the product itself. Achieving such objectives requires, in many cases, focused improvement efforts. A testing improvement team mission statement states its roles and responsibilities. Typical testing team mission statements are:

■ Improve customer satisfaction by implementing test scenarios that cover as much as possible the way that the product will be used (as derived from the requirements and operational concepts and scenarios) and detecting severe bugs as early in the test phases as possible.
■ Improve time-to-market by involving test engineers in the design phase, testing as early as possible, implementing automation test solutions, and reporting real-time status, including release readiness assessments.

With respect to the organizational structure, the test team typically provides testing services to the product development teams as well as services to other groups such as project management, configuration control, and documentation. The advantages of a functional, independent test team are that:

■ Test team resources are efficiently distributed and can be easily reassigned to different products.
■ Test engineers are working according to one testing approach, which gives them the ability to test faster and better.
■ Knowledge and experience are shared.
■ The complexities of testing issues are handled professionally and not left to the responsibility of busy development managers.
■ Biased evaluations of testing results that could be introduced by developers who were involved in the development of the product performing the testing are eliminated.

The main disadvantage of such a structure is that the transfer of product technical knowledge from development to testing is hard to achieve. When testing is integrated within development, such transfer is easier. Figure 6.2 presents a typical organizational structure of a functional testing organization.

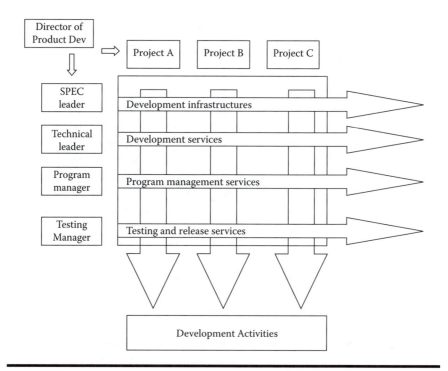

Figure 6.2 A typical organizational structure, including the role of the test team.

The V-model defines testing activities with respect to the design activities and emphasizes the parallelism of testing, design, and coding (see Figure 6.1). Design and development begins with the production of the Product Requirements Document (PRD), which captures the requirements for the system or software from the customer's or user's perspective. The development team needs to analyze the PRD and create the functional specification documents, which are lower-level, more detailed requirements from which design and implementation can proceed. Based on the functional specifications, a system architecture is developed, detailed design is performed, and lower-level components, such as hardware or units of software code, are produced. At this point, when something tangible exists, testing can begin.

The levels of testing have analogs in the various levels of the design and development effort. Testing at even the lowest levels can also impact other levels. For example, when a defect is detected in system testing, fixing it requires development engineers to go all the way back to the customer requirements, and to reviewing the PRD and related design documents. Organizations typically work on the premise that it is better to detect problems in early phases rather than later, implying that it is cost efficient to invest efforts in document reviews and involve the test team in early phases. Figure 6.3 presents a chart supporting this claim that is based on proprietary data from projects conducted by KPA Ltd. (www.kpa.co.il).

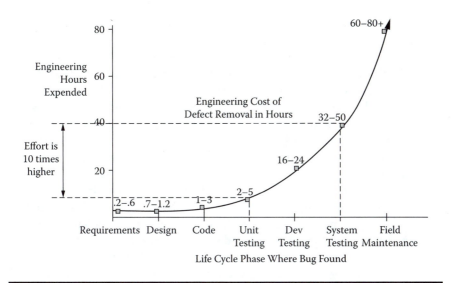

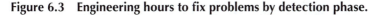

Figure 6.3   Engineering hours to fix problems by detection phase.

A standard V-model includes six phases toward product release:

1. Product definition
2. Design
3. Implementation and unit testing
4. Internal verification (alpha testing)
5. External validation (beta testing)
6. Installation and maintenance

Tailoring the V-model and testing strategy requires consideration of time-to-market requirements, and quality and performance results. Some guidelines for setting up the test strategy include:

■ *Product definition:* Marketing activities or customer and user needs should be determined prior to approval of the product plan, and combined within the PRD document. The testing phase of the PRD in the V-model is performed by system testing.

■ *Design phase*: This phase of the V-model includes high-level design and low-level design. During these activities, test engineers need to participate in design reviews in order to gain knowledge and better prepare for the testing phase. This knowledge will contribute to quality and time-to-market.

■ *Implementation and unit testing*: Prototype development for hardware, and coding and unit testing activities, is typically carried out by the development team, sometimes with tools or test environments provided by test engineers.

At that phase, the test team should be part of the detailed design or code reviews for enhancing the test engineer's knowledge of the product.

■ *Internal verification (alpha test):* This phase is sometimes called "feature testing" or verification testing, and includes verifying that the features are properly working according to function specifications. Integration testing is also usually part of this phase and is sometimes described in a special integration Test Plan. A major concern of integration testing is evaluating the integrity of the interfaces between the component parts. Sometimes, feature testing and integration testing are combined into one test activity such that both objectives are accomplished.

■ *External validation (beta testing):* At this stage, an external team or specific customers initiate external validation. There is a distinct difference between validation and verification. Validation testing is carried out to determine if the requirements, as stated in the requirements specifications, have been implemented in the completed product and to test the implementation of features in the system. Sometimes this effort is called "system testing." Another external activity is beta testing, which is designed to test the product in a real environment, beyond the development and test teams' ability to emulate. Beta testing is very important for the improvement of product quality. (More on beta testing is presented in Chapter 7.)

Problem reports, failures, or bugs* derived from testing are typically classified by level of severity, which is typically based on cost impact, system performance impact, or user impact. For example, severity classes for a telecommunication network system might be based on the ability of the network to continue to handle calls when a failure is encountered. A typical generic definition of severity classes includes:

■ *Severity 1—Catastrophic:* Reasonably common circumstances cause the entire system to fail; or a major subsystem to stop working; or if on a network, other devices are disrupted, and there's no work-around.

■ *Severity 2—Severe:* Important functions are unusable, and there's no work-around, but the system's other functions, and if on a network, the rest of the devices work normally.

■ *Severity 3—Moderate:* Things fail under unusual circumstances, or minor features don't work at all, or things fail, but there's a low-impact work-around. This is the highest level for documentation bugs.

■ *Severity 4—Minor:* Things fail under very unusual circumstances, and recover pretty much by themselves. Users don't need to install any work-arounds, and performance impact is tolerable.

---

* We use the term "bug" to mean any problem uncovered during testing, whether it is hardware, software, or system testing.

- *Severity 5—Cosmetic:* No real detrimental effect on system functionality.
- *Severity 6—Comment:* A comment or observation determined useful to document but not really a bug; a minor editing comment affecting documentation.

Problem reports are usually managed in a dedicated problem reporting system. Once entered into the system, problem reports are classified according to their status. Sample categories representing the status of a problem reports include:

New	(N)	A new bug report that has not been classified yet.
Un-reproducible	(U)	Problem cannot be reproduced by the test engineer.
Junked	(J)	Bug report is discarded because it does not describe a problem that requires changes to hardware, software, or documentation.
Information required	(I)	Bug reports that needs additional information to determine the cause of the problem. Typically, this is easily retrieved information, such as a trace or dump, or a better description of the problem and its symptoms. This state is also used when a diagnostic or special image has been provided to the customer in order to gather more information to continue the analysis. In either case, development engineering cannot make progress until the required information is provided.
Duplicate	(D)	This bug report describes the same problem documented in another bug report.
Closed	(C)	Bug report is valid, but a conscious decision has been made not to fix it at all or in all releases. Normally, only a development engineering manager moves a bug report to this state. This state is not available in all projects.
Re-open	(R)	A bug report that was handled in an earlier release is re-opened due problem re-occurrence

Beyond their status in the problem reporting system, bugs or problem reports are classified by type, following a diagnostic assessment. One example of a classification of bugs by type for software—in this case, the development of TEX—is provided by Donald Knuth [29]. The categories are:

A—Algorithm: a wrong procedure has been implemented.

B—Blunder: a mistaken but syntactically correct instruction.

C—Cleanup: an improvement in consistency or clarity.

D—Data: a data structure debacle.

E—Efficiency: a change for improved code performance.

F—Forgotten: a forgotten or incomplete function.

G—Generalization: a change to permit future growth and extensions.

I—Interaction: a change in user interface to better respond to user needs.

L—Language: a misunderstanding of the programming language.

M—Mismatch: a module's interface error.

P—Portability: a change to promote portability and maintenance.

Q—Quality: an improvement to better meet user requirements.

R—Robustness: a change to include sanity checks of inputs.

S—Surprise: changes resulting from unforeseen interactions.

T—Typo: Change to correct differences between pseudo-code and code.

For more information and a statistical analysis of this data, see [24].

In 1994, Chillarege established a relationship between the statistical methods of software reliability prediction and the underlying semantic content in the defects used for the prediction (see [12]). He demonstrated that the characteristics of the defects, identified by the type of change necessary to fix defects, had a strong influence on the maturity of a software product being developed. This opened the door to the possibility of measuring product maturity via extracting the semantic content from defects and measuring their change as the process evolves, and he developed at IBM a methodology labeled Orthogonal Defect Classification (ODC). ODC is a technique that bridges the gap between statistical defect models and causal analysis. Specifically, a set of orthogonal classes, mapped over the space of development or verification, is established to help developers provide feedback on the progress of their software development efforts. This data and its properties provide a framework for analysis methods that exploit traditional engineering methods of process control and feedback. ODC essentially means that we categorize a defect into classes that collectively point to the part of the process that needs attention, much like characterizing a point in a Cartesian system of orthogonal axes by its (x, y, z)-coordinates. Although activities are broadly divided into design, code, and test, in the software development process, each organization can have its variations. It is also the case that the process stages in several instances may overlap while different releases may be developed in parallel. Process stages can be carried out by different people and sometimes by different organizations. Therefore, for classification to be widely applicable, the classification scheme must have consistency between the stages. Without consistency it is almost impossible to look at trends across stages. Ideally, the classification should also be quite independent of the specifics of a product or organization. If the classification is both consistent across phases and independent of the product, it tends to be fairly process invariant and can eventually yield relationships and models that are very useful, thus forming a good measurement system that allows learning from experience and provides a means of communicating experiences between projects.

One of the pitfalls in classifying defects is that it is a human process, and thus is subject to the usual problems of human error and a general distaste if the use of the data is not well understood. However, each of these concerns can be handled if the classification process is simple, with little room for confusion or possibility of mistakes, and if the data can be easily interpreted. If the number of classes is small, there is a greater chance that the human mind can accurately resolve among them. Having a small set to choose from makes classification easier and less error prone. When orthogonal, the choices should also be uniquely identified and easily classified (see [38]). We discuss STAM in Section 6.4; it is an efficient method for tracking defects that precedes but has many similarities to ODC that was developed by Kenett (see [20]).

The plan for performing internal testing should account for the time it takes to test all features as described in the functional specification, and possible adjustments and retesting that may occur after the first tests. Retesting can occur at various levels of test. At the lowest level, it can occur at the unit of code or hardware component level after a test failure and correction of the problem. At the very minimum, the steps that failed should be repeated after the fix has been made; and preferably, a complete retest should be performed, depending on the size and complexity of the unit of code or component under test.

When testing occurs at levels higher than the unit or component level (e.g., integration testing), a regression test strategy must be in place. By regression testing we mean that a subset of the original test cases that have been completed must be rerun. Using software as an example, when a test failure is observed, it will be traced back to the unit or units of code that were the cause of the fault. That unit of code will be sent back for correction to the developer responsible for that unit (or the person designated to fix it). As indicated above, when the correction has been made, the unit affected will be retested to determine if the problem has been fixed. It is not safe to assume that the fix that has been made absolutely cures the problem. The unit of code may function well as a stand-alone unit of code, but when integrated with other interfacing units of code, it may cause some unforeseen problem. Consequently, a regression test strategy that looks at the interfaces between the lines of code that have been fixed and other units that use the results of that execution or input into that execution, will define the test cases that need to be rerun.

Regression testing should be performed at each level of test below the level at which the problem was discovered. Referring back to Figure 6.1, if a fault is discovered at system test, after the fix is made to the unit of code or hardware component, there will be a retest of the unit or component. Regression testing of the integration test cases will then be performed, followed by regression testing of the functional testing, followed by regression testing of the system test cases completed prior to

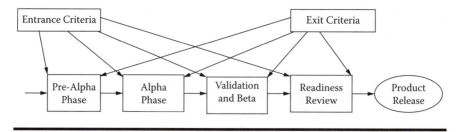

Figure 6.4 High-level test process flowchart.

the failure. After all that regression testing has been performed, the remaining system test cases to be completed can be run.

Sometimes organizations short-circuit regression testing completely, or shorten it severely, with a significant increase in risk exposure. Organizations at CMMI Maturity Level 5 can assess the level of the risk so as to make informed decisions. Organizations at lower maturity levels simply take a gamble, sometimes with severe consequences.

Testing is an integral part of the product development effort and therefore should be synchronized with the development activities. Moreover, test engineers should be fully familiar with the requirements for the product under test. Professional test engineers are systems or software engineers with several years of experience in development and testing so that they can properly evaluate the quality of the product. In many cases, however, the software industry employs in such positions students with no experience as test engineers. The result is ineffective and inefficient testing with negative impact on time-to-market, product quality, and performance. Testing, like any other process, should include "Entrance" and "Exit" criteria that are documented in the test strategy. We describe below a testing process with three main phases: "Pre-Alpha," "Alpha," and "Validation and Beta" (see Figure 6.4). We use the terms "defects," "test failures," and "bugs" as equivalents (see [3]).

6.1.2 "Pre-Alpha": Early Testing before Feature Freeze

The following "Entrance" criteria are typically used for this phase:

- The product has been approved by appropriate management functions.
- The Product Requirement Documents, Functional Specifications, and Test Plan documents are approved.
- The development team agrees with the testing team to start an early phase of testing. Starting from this point, bugs are reported by the testing team as part of that agreement.
- A Build Plan or Development Product Version Plan is ready and approved by development team.
- A Test Plan based on the Build Plan or Development Product Version Plan is ready and approved by team.

- A Quality Scorecard based on Quality Graphs is ready.
- The development and test teams agree to follow the process of bug reviews.

The following "Exit" criteria are typical for this phase:

- The test team completed the tests according to the Test Plan as defined for this phase.
- The development and test teams agreed that the "Pre Alpha" test phase is completed. In case some features are not ready according to the original Build Plan or Development Product Version Plan, the development and test teams can decide on waivers.

6.1.3 "Alpha": Feature Freeze

The following "Entrance" criteria are usually used in this phase:

- There are alpha Go/No-Go acceptance criteria.
- The development and test teams are following the Build Plan or Development Product Version Plan.
- All features are covered by a Feature Test Plan document.
- Test automation is ready (as part of the plan) and follows the Build Plan or Development Product Version Plan readiness.

The following "Exit" criteria for this phase are:

- All features have been tested according to the Feature Test Plan document.
- The Quality Graphs display a clear positive trend:
  - Number of new reported bugs is going down.
  - No new code or components were added for the last test version.
  - Development team is closing the gap on "open bugs."
  - Testing team in good rate of "verification" trend.
- All Severity 1 and Severity 2 defects have been resolved. Some Severity 3 defects are allowed as long as there is no significant impact on functionality or performance with respect to customer requirements.

6.1.4 "Validation and Beta": Development Is Complete

The following "Entrance" criteria are being used in this phase:

- No or only a small portion of unresolved bugs are left.
- No new components or code are added.
- According to the Quality Graphs:
  - The number of new reported bugs is low.

- No new Severity 1, Severity 2, and Severity 3 defects have been detected in the last alpha build.
- All tests have been covered according to plans.

The following "Exit" criteria are used for this phase:

- ■ "Exit" criteria are meeting the Go/No-Go requirements.
- ■ In addition:
  - All Severity 1, Severity 2, and Severity 3 bug fixes have been verified by the test team.
  - No Severity 1, Severity 2, and Severity 3 defects have been detected during that phase. If such bugs are detected, then extension of this testing phase should be considered. As a result of such bugs, for software, code can be added or modified, thus overriding the "Development is Complete" requirements; however, fixing defects in the software should not be used as unrestricted permission to continue trying to perfect the code. Such changes often wind up creating new defects while making negligible improvements in either functionality or performance.

For more references on the V-model and the design of a test strategy, see [11, 17, 23–25].

6.2 Implementing a Test Strategy

As already discussed, test engineers must be involved in the design phase; participate in design reviews, documentation, and project plans; be knowledgeable in the implementation of the design; and use this information in creating test documentation and designing the specific project test strategy.

In follow-on releases of a product, test engineers need to bring information from "customer cases" representing how customers actually used the product. To organize such data, test engineers can, for example, conduct surveys, interview customers, and benchmark competitors' products. A knowledge of customer cases is essential for achieving time-to-market improvements.

6.2.1 Ten Steps for Time-to-Market and Quality Improvement

The following ten steps represent what test managers and developers should do to help their organizations achieve time-to-market and quality requirements:

1. Test managers should be involved in the management decision on the future of the product, in order to plan current test resources.

2. Test engineers should be involved in design reviews.
3. Test engineers should be involved in developing usage scenarios.
4. Development teams should prepare a software Build Plan or a Product Development Version Plan synchronized with test team planning, documentation, marketing, etc.
5. Test teams should prepare Test Plans that are based on the software build or product development version.
6. Test managers should be involved with the internal decisions of the development teams.
7. Test teams should plan for implementation of automatic testing tools, which will reduce testing time and improve quality by releasing test engineer resources for planning additional test scenarios. This also includes stress tests for completed features designed to determine performance and capacity limitations, and to detect memory leaks and product malfunctions in early phases.
8. During test execution, test managers should adjust Test Plans by monitoring test results, using quality data.
9. During test execution, test managers should report testing coverage and show-stopper bugs, using quality graphs.
10. During test execution, test managers should look ahead and anticipate the behavior of the product and plan accordingly.

6.2.2 *Build and Development Product Version Plans*

A Build Plan or Development Product Version Plan is typically prepared by the development team in conjunction with the Test Plan and the development activities that are required to achieve a successful release of the product.

The development team needs to plan development activities by weeks that correspond to the plan, so that, for example, each week the development team submits a software build or a version of the evolving system to testing. Such an approach helps the test and development teams focus on common objectives.

While development and testing teams are planning the execution phase, the development team is planning its development phases as a function of builds or evolving system version. As previously noted, the development team needs to analyze the PRD and create the functional specification documents. When the development team fully understands the product requirements, a Build Plan or Development Product Version Plan is designed. Development and test teams review the relevant plan together. After completing the review process, the test team creates a Test Plan based on the Build or Development Product Version Plan. The development team should approve the Test Plan. Figure 6.5 presents the flowchart of the Build or Development Product Version Plan and Test Plan development process.

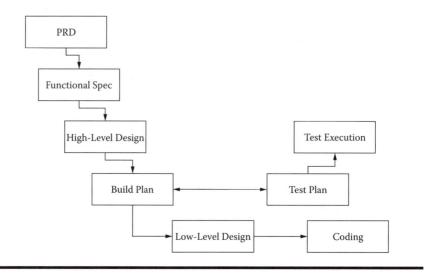

Figure 6.5   Flowchart of the Build Plan and Test Plan development process.

6.2.3 Main Phases in the Test Plan

The test team's main objective is to initiate and perform testing on the product in a minimal time frame with maximum feature coverage and functionality interactions, recognizing that today's systems are so complex that it is impossible to achieve 100% feature coverage. The test team needs to cover all features designated for testing during the test period and to perform additional testing tasks like regression testing, performance testing, and stress testing. The testing team should learn as much as possible about the product while the development team is in the early implementation phase, so that it can start preparing automation tools and test scripts. While the development team is working on design and early code or component implementations, the testing team can start testing early features. Moreover, the ability to initiate early stress tests will provide early feedback to the development team about the quality of features. Such early warnings make fixing the high-severity bugs easier in the early phases of development. Such tests are not a substitute for unit-level or component testing, which are white-box, bottom-up tests, as opposed to the Test Plan, which is black-box and top-down, being driven by functional specifications. For ease of distinction, white-box testing is testing performed to verify that the design has been implemented correctly, while black-box testing is testing performed to determine if the requirements have been properly implemented. Figure 6.6 presents the main phases of the Test Plan.

The right test strategy will stimulate quality improvements and initiate early development and testing process control. We measure the test process using quality metrics and quality graphs derived from problem reports and build or development version information. Using this information we can estimate the level of product quality and how long it will take for the team to meet quality objectives.

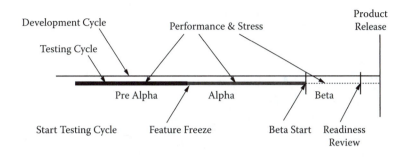

Figure 6.6   Main phases of a Test Plan.

6.2.4 *Monitoring and Controlling the Test Process*

One of the main strengths in implementing a testing process reflecting a clear test strategy is the achievement of proper integration of tools supporting test engineers with their testing tasks. Tracking the life cycle of defects or problem reports during the test process enables the test team to adjust testing tasks and to reorganize testing resources according to test results. Moreover, test engineers have the responsibility to alert management to negative trends and to predict the post-release quality of the product.

Typical charts that are used during the test process for tracking development and test team performance and product quality include:

- New defects per evolving product version
- New Severity 1 to Severity 3 defects per build
- Total defects by status
- Total defects by severity
- Accumulated submitted defects versus accumulated defects closed by date
- Accumulated lines of code for software and accumulated defects per evolving product version
- Customer found defects by priority by product version
- Customer found defects by severity by product version

The following subsections provide examples of data collected and analyzed during the product life cycle to monitor and control the test process.

6.2.5 *Defects per Build*

Defects are counted over fixed time intervals such as day, week, or month. The tracking of defects can be done effectively using a "control chart" (see Chapter 4, Section 4.5.1). The chart for tracking defects is a c-type control chart for attributes. The c-chart control limits are calculated using the Poisson distribution assumption. We present such a chart in Figure 6.7 for data from release 6.4 of

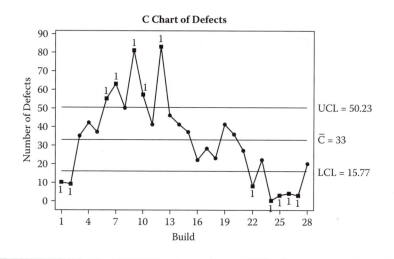

Figure 6.7   Control chart of all new defects per build, release 6.4.

a telecommunication system. Over 28 consecutive software builds, the problem reporting system collected 924 confirmed problem reports of defects. The average number of new defects per build was 33, and the upper (UCL) and lower (LCL) control limits of the c-chart are

$$UCL = \bar{c} + 3\sqrt{\bar{c}} = 33 + 3 \times 5.744 = 50.23$$

$$LCL = \bar{c} - 3\sqrt{\bar{c}} = 33 - 3 \times 5.744 = 15.77$$

where:

$$\bar{c} = \frac{T}{K} = \frac{924}{28} = 33$$

T = Total number of defects = 924
K = Number of subgroups (builds) = 28

For the theory and application examples of control charts, see Kenett and Zacks [26].

Builds 1, 2, 22, and 24 through 27 showed an unusually low number of defects. This is as expected because initially bugs are "hidden" by other bugs and it takes time to uncover them. Later, toward the end, we expect fewer bugs because we are getting ready for code freeze. We do notice, however, a general downward trend in the number of reported defects following the peak in build 12.

If we want better performance in tracking defects, we can analyze the same data with a *cumulative sum* (CUSUM) sequential procedure designed to signal weaker changes in distribution. CUSUMs are more responsive to weak changes

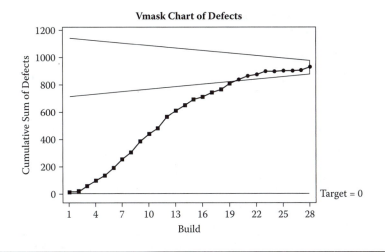

Vmask Chart of Defects

Figure 6.8   Cumulative sum control chart of defects per build, release 6.4.

than the control charts presented above. Control charts are, however, more responsive to abrupt changes than CUSUM charts (see [26]). A CUSUM tracks cumulative differences relative to a target value from the data being monitored. In our case, the CUSUM tracks differences in the number of defects from a target value of "0." The procedure can be seen as analogous to tracking the status of a bank account, which combines deposits with expenses over time. The cumulative sum is analyzed by positioning a "V-mask" on the last observation. The mask is determined by two parameters, h and k. The parameter h is the width of the mask at its right-hand side; the parameter k is the slope of the V-mask arms. The values of h and k determine the rate of false alarms and the time to detection once the software has reached acceptable quality levels. The build number falling outside the mask is identified as a point of change, where the system has reached an adequate quality level. Figure 6.8 presents such a chart with a V-mask positioned at build 28. The fact that the V-mask does not include observation 19 indicates that from build 19 onward, we experience a significant decrease in the number of defects. For more on CUSUM procedures, see [26]. The procedure was set up using version 15.2 of the Minitab™ software with a target value of "0" and specifications for the mask dimensions (h = 4.0 and k = 0.5) that ensure a false alarm rate of 1/370 and a reaction to change of about four observation times.

Tracking defect data over time can also result in fitting a model that describes the behavior of the data (see, e.g., [14, 18, 19, 21, 25, 31, 39, 40]). If one fits a model to the number of defects per build, one can derive predictions and use the model for future planning. Take, for example, the data from release 6.4 at the 23rd build. We fit to the data from the 28 builds a two-parameter Weibull distribution with parameter estimates presented in Table 6.1 [26]. The software used for this analysis was version 15.2 of Minitab. The data with the Weibull fit is presented in Figure 6.9.

Table 6.1   Parameter Estimates from Weibull Fit to Release 6.4 Data

Distribution: Weibull				
Parameter Estimates				
			95.0% Normal CI	
Parameter	Estimate	Standard Error	Lower	Upper
Shape	2.23624	0.0604642	2.12082	2.35794
Scale	12.4038	0.197628	12.0224	12.7972

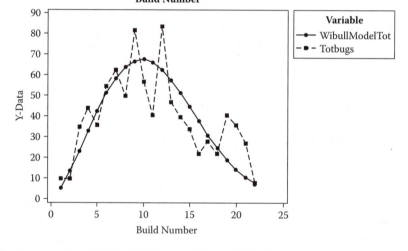

Scatterplot of WibullModelTot, Totbugs versus Build Number

Figure 6.9   Defects per build with Weibull fit, release 6.4.

The model can be described as estimating the percentage of total defects accumulated at a specific build number. These estimates are presented in Table 6.1 where "percentile" corresponds to the build number. From Table 6.1 we can determine, for example, that 4% of the defects are still expected to be found after build 21, and that at build 25, we will have covered 99% of defects potentially uncovered by the Test Plan.

6.2.6 Defects by Severity/Status

It is often very helpful to evaluate defects by severity. A piece of code or a hardware component that has a large number of Severity 6 bugs presents very little concern with respect to functionality or performance, whereas a piece of code or a hardware

Table 6.2   Predictions at Build 23 Based on the Weibull Fit to Data from Release 6.4

Table of Percentiles				
			95.0% Normal CI	
Percent	Percentile	Standard Error	Lower	Upper
0.1	0.565082	0.0507193	0.473927	0.673770
1	1.58551	0.0989960	1.40288	1.79191
2	2.16654	0.117614	1.94786	2.40977
3	2.60317	0.128958	2.36230	2.86860
4	2.96738	0.137034	2.71059	3.24849
5	3.28639	0.143226	3.01732	3.57944
6	3.57392	0.148183	3.29497	3.87648
7	3.83803	0.152271	3.55090	4.14839
8	4.08392	0.155710	3.78986	4.40080
9	4.31517	0.158650	4.01516	4.63759
10	4.53435	0.161192	4.22917	4.86154
20	6.34240	0.175127	6.00828	6.69510
30	7.82237	0.180546	7.47640	8.18436
40	9.18541	0.183773	8.83220	9.55276
50	10.5287	0.187575	10.1674	10.9028
60	11.9282	0.194341	11.5533	12.3153
70	13.4773	0.207332	13.0770	13.8899
80	15.3452	0.232704	14.8959	15.8082
90	18.0106	0.287973	17.4550	18.5840
91	18.3746	0.297132	17.8014	18.9663
92	18.7712	0.307515	18.1781	19.3837
93	19.2087	0.319438	18.5927	19.8451
94	19.6989	0.333360	19.0562	20.3632

Continued

Table 6.2 Predictions at Build 23 Based on the Weibull Fit to Data from Release 6.4 (*Continued*)

Table of Percentiles				
			95.0% Normal CI	
Percent	Percentile	Standard Error	Lower	Upper
95	20.2598	0.349989	19.5853	20.9575
96	20.9213	0.370504	20.2076	21.6602
97	21.7377	0.397093	20.9731	22.5301
98	22.8278	0.434618	21.9916	23.6957
99	24.5552	0.498352	23.5976	25.5516
99.9885	33.2543	0.880277	31.5730	35.0252

component that has a significant number of Severity 1 and 2 bugs will cause a lot of rework, and may possibly delay delivery of the system. Figure 6.10 is a basic bar plot of defects by severity in release 6.4. A similar analysis can be performed for defects by status, as a means of monitoring and controlling defect correction efficiency. To avoid being repetitive, we focus in the subsequent subsections on defects by severity.

In Figure 6.10 we see that in release 6.4, bugs of Severity 3 and 4 make up most of the reported defects (87%). Severity 1 defects make up 0.1%, and Severity 2 defects make up 3.1% of total reported defects.

As we were testing release 7.0, at a certain point in time, we accumulated 131 defect reports. If the defects present themselves with the same severity distribution as in release 6.4, we would expect to see, in these 131 reports, 0 reports of Severity

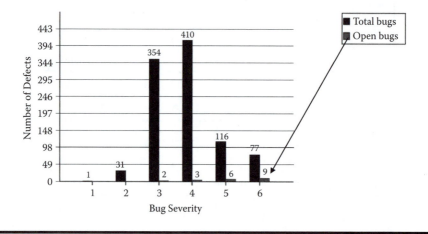

Figure 6.10 Total defects by severity of release 6.4.

1, 4 reports of Severity 2, 47 reports of Severity 3, and 54 reports of Severity 4. These expected numbers are computed very simply as the product of 131 and the relative proportions in release 6.4. By comparing the number of defects actually reported with the expected number of reports, we derive an efficient statistical test to help us determine if the release 6.4 distribution of bug severity still prevails in release 7.0. The M-test presented in [22] was designed to do just that. It is based on a standardized residual computed by dividing the difference between observed and expected values, by the standard deviation of the expected count. The result of this computation is a statistic labeled "Z." Using the Bonferroni inequality to account for multiple comparison effects, critical values are determined so that individual cell differences can be assessed as significant or not.

Let
N = Total number of defects in the new release (release 7.0)
K = Number of defect categories
n_i = Number of defects of severity i , i = 1, ..., K
p_i = Proportion of severity i in baseline release (here it is release 6.4)

Specifically, the M-test consists of five steps:

1. Compute the expected number of defects of severity i, $E_i = N_{pi}$, i = 1, ..., K.
2. Compute $S_i = \sqrt{(Np_i(1 - p_i))}$, i = 1, ..., K.
3. Compute the adjusted residuals

$$Z_i = \frac{(n_i - E_i)}{S_i}, \quad i = 1, \ldots, K$$

4. For a general value of K and a 10%, 5%, or 1% significance level is determined by comparing Z_i to the critical values $C_{0.1}$, $C_{0.05}$, and $C_{0.01}$ corresponding respectively to significance levels of 10%, 5%, and 1% provided by Kenett [22]. For K = 6, the critical values are: $C_{0.1}$ = 2.12, $C_{0.05}$ = 2.39, and $C_{0.01}$ = 2.93.
5. If all adjusted residuals Z_i are smaller, in absolute value, than C, no significant changes in Pareto charts are declared. Cells with values of Z_i, above C or below $-C$ are declared significantly different from the corresponding cells in the baseline.

Table 6.3 presents the defect distribution, by severity on release 6.4 and release 7.0, and the values of Z. We observe that release 7.0 shows a significant decrease in defects of Severity 5 at the 1% level. For more on the M-test, with applications to software data, see [22].

In Figure 6.11 we see the accumulated number of problem reports submitted versus accumulated closed and open by date in release 6.4. We observed the number

Table 6.3 M-Test Applied to Defects in Release 7.0, Relative to Release 6.4

Severity	6.4	%	Expected	Actual 7.0	Z
1	1	0.001	0	1	2.38
2	31	0.031	4	4	−0.05
3	354	0.358	47	52	0.93
4	410	0.415	54	63	1.54
5	116	0.117	15	4	−3.09
6	77	0.078	10	7	−1.04
Total	989	100	131	131	

of open bugs slowly decreasing toward the end, and stability in submitted reports is characteristic of the final stretch.

In Figure 6.12 we see the accumulated lines of code and accumulated defects per build of release 6.4. This data display allows us to assess the number of defects reported with respect to the work performed.

6.3 Software Cybernetics

6.3.1 Background

Cybernetics was defined by mathematician Nobert Wiener in 1948 as control and communication in the animal and the machine [41]. The essential goal of cybernetics is to understand and define the functions and processes of systems that have goals, and that participate in circular, causal chains that move from action to sensing to comparison with desired goal, and again to action. Studies in cybernetics provide a means for examining the design and function of any system, including social systems such as business management and organizational learning, including for the purpose of making them more efficient and effective. The emerging interdisciplinary area of software cybernetics (and, by extension, systems cybernetics) is motivated by the fundamental question of whether or not, and how system and software behavior can be controlled. System and software behavior includes the behavior of development processes, maintenance processes, evolution processes (from development to operational status), as well as that of software operation or software execution itself [8]. In general, software cybernetics addresses issues and questions on (1) the formalization and quantification of feedback mechanisms in software processes and systems, (2) the adaptation of control theory principles to software processes and systems, (3) the application of the principles of software theories and engineering

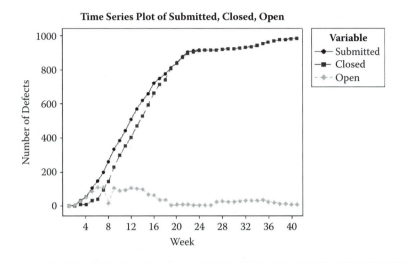

Figure 6.11 Accumulated submitted versus accumulated closed-by date, for release 6.4.

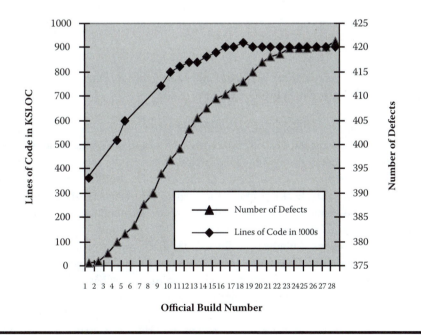

Figure 6.12 Accumulated lines of code and accumulated defects per build of release 6.4.

to control systems and processes, and (4) the integration of the theories of software engineering. The basic elements of software and systems cybernetics are:

■ *Processes:* The software and systems development process, the software and systems maintenance process, the software and systems evolution process, and all the related artifacts delivered during such processes, including the delivered software or systems.
■ *Control:* A finite or infinite sequence of actions that are taken and applied to the controlled object to achieve a given a priori goal for the controlled object in a systematic and repeatable manner.
■ *Control system:* A system that comprises at least two components: controlled object and controller. The controller delivers control signals or actions to the controlled object, which forces the controlled object to achieve a desired goal.
■ *Control of software and systems:* Here, the software or system serves as the controlled object and controls are applied to the software or system. The actions taken may include:
  – The identification or estimation of internal information and parameters of the software or system or software or system processes for determining whether or not they function properly
  – Whether or not a software or system process will achieve its intended objective so that a controller can transform the software or system or modify appropriate parameters in a process so that desired objectives are achieved on schedule

The control process must be quantifiable and hence repeatable under similar circumstances for similar controlled objects.

■ *Open-loop control:* Control (the sequence of actions) is determined offline without feedback; the responses of the controlled object to the applied actions are not utilized for determining subsequent actions taken on the controlled object.
■ *Closed-loop control:* The behavior of the controlled object is monitored and used in an online feedback control strategy to enforce a desired objective; that is, real-time information is used to determine subsequent control actions.
■ *Feedback control:* Closed-loop control.
■ *Adaptive control:* An advanced form of feedback control; during the process of adaptive control, not only the actions to be taken are derived and updated online, but also the controller is updated online on the basis of the responses of the controlled object to actions taken already.

Various related works have been reported in various conference proceedings, research journals, and edited books under different umbrellas, including feedback control of the software test process [9, 33], incremental software control [32], agile development [10], among others. Software and systems cybernetics explore the interplay between software or systems and control, and are motivated by the fundamental question of whether or not and how software or system behavior can be controlled. This new discipline is receiving increased attention and promises to play a substantial role in modern complex systems such as service-oriented architectures (SOAs) and Web services.

6.3.2 Modeling of Incremental Software and System Control

The study of software and system development and maintenance processes has evolved from prescriptive and descriptive modes into predictive modeling and decision support. Techniques range from statistical process control to system-dynamics models and "what-if" scenario evaluation, including techniques from the engineering discipline of control theory. The CMMI high maturity practices involve the use of statistical and probabilistic methods for purposes of quantitatively managing and controlling software and system development, services, and acquisition processes [13]. For example, a high maturity organization might use as a process performance model a Monte Carlo simulation to predict the probability of achieving a quality goal in the deliverable product, using as inputs the variance in the number of defects found during peer reviews at various stages of development. A high density of defects found during a code review, for example, could result in the model predicting a low probability of achieving the quality goal, thus necessitating corrective action for the design implementation (coding) process.

A framework for constructing a complete software development process model, including software coding and testing phases, was proposed in [32]. The proposal includes a *state model* formulation of the individual software development activities that can be composed into a model of software development in general, honoring any scheduling dependencies. In this context, the following types of work items flow between activities: (1) completed features for testing, (2) completed test cases for execution, (3) failure reports from new test cases, (4) failure reports from regression test cases, (5) feature change requests, (6) test case change requests, (7) new test case specifications (i.e., when test strategy deficiencies are identified), (8) corrected feature code for testing, and (9) corrected test cases for re-execution.

Scheduling constraints allow for the execution of a development activity to be made dependent on the cumulative progress of another development activity. For example, the test execution activity may be made dependent on the progress of the feature coding activity—limiting testing to those completed features. Scheduling constraints are implemented by synthesizing a controller to regulate the requested outflow rate of the queue for the schedule-constrained activity; the control law is constructed to limit the cumulative outflow of the controlled queue to the set of work items for which the scheduling dependencies have been satisfied. The scheduling constraints for a development activity are specified as an inequality in the total cumulative outflow outputs of the related development activities' queues. The study considers a project with two releases, say 10, then 20 features; there are 70 test cases for the first 10 features, then 130 more test cases once the remaining 20 features are completed and integrated into the product. The study demonstrates that the complex behavior of schedule-constrained software development activities can be captured through a state model. Furthermore, the simulation process for such a model is relatively simple, opening the way for iterative as well as deterministic decision-support applications. For example, an idle developer is recognizable as a discrepancy between the productive capability of the workforce and the outflow rate of the associated queue. The ultimate goal of such studies is to develop the capability to optimize software development processes through the techniques of modern control theory; by developing a framework for building general software process models within the formalism of state modeling. For more details on this application of cybernetics to software development, see [32].

6.3.3 Modeling of Test Processes

A formal theory of system testing is proposed in [11]. The model assumes that a testing process exists, and that the test and debug cycles alternate or occur concurrently and are not based on a predetermined schedule but, rather, as and when the need arises. Specifically, a product or a component of the product P could be in one of three states: Test, Debug, or End. When in Test, P is executed until a failure is detected. At this point, if a Debug mode condition is true, then P enters the Debug state; otherwise, it remains in Test. The Debug mode condition could be, for example, "The number of failures detected exceeds a threshold." As another example, it could be "A failure that will prevent succeeding tests from running as planned." P remains in the Debug state until errors are found and fixed, when it returns to the Test state. However, while P is in Debug, another version of it could be in the Test state. This is when we say that testing and debugging are taking place concurrently. This test-debug cycle can be handled effectively by our model; other scenarios are possible with some care in collecting the needed data. In [11] the author solves a process control problem that arises within the context of the system test phase.

Consider a product P under test and assume that the quality and schedule objectives were set at the start of the system test and revised before the test process had started. For a scheduled target date or a set of checkpoints leading to a target date, the quality objective might be expressed in a variety of ways: (1) as the reliability of P to be achieved by the target date, (2) as the number of errors remaining at the end of the test phase, or (3) as the increase in code coverage for software or functionality for hardware. Further, the quality objective could be more refined and specified for each checkpoint.

We assume that a test manager plans the execution of the test phase to meet the quality and schedule objectives. Such a plan involves several activities, including the constitution of the test team, selection of the test tools, identification of training needs, and scheduling of training sessions. Each of these activities involves estimation. For example, constituting the test team requires a determination of how many testers to use. The experience of each tester is another important factor to consider. It is the test team that carries out the testing activity and, hence, spends the effort intended to help meet the objectives. The ever-limited budget is usually a constraint with which to contend. During the initial planning phase, a test manager needs to answer the following question: How much effort is needed to meet the schedule and quality objectives? Experience of the test manager does help in answering this question. The question stated above is relevant at each checkpoint. It is assumed that the test manager has planned to conduct reviews at intermediate points between the start and target dates. The question might also arise at various other points—for example, when there is attrition in the test team. An accurate answer to the above question is important not only for continuous planning, but also for process control and resource allocation. A question relevant for control is: How much additional test effort is required at a given checkpoint if a schedule slippage of, say, 20% can be tolerated? This question could be reformulated in many ways in terms of various process parameters. A few other related questions of interest to a test manager are:

1. Can testing be completed by the deadline and the quality objective realized?
2. How long will it take to correct the deviations in the test process that might arise due to an unexpectedly large number of reported errors, turnover of test engineers, change in code, etc.?
3. By how much should the size of the test team be increased if the deadline is brought closer without any change in the error reduction objectives?

To answer these questions, the model proposed in [11] is based on the application of the theory of feedback control using a state variable representation. The model allows comparison of the output variables of the test process with one or more setpoint(s). Such a comparison leads to the determination of how the process inputs and internal parameters should be regulated to achieve the desired objectives of the Test Plan. When applied to the system test phase, the objective of feedback control is to assist a test manager in making decisions regarding the expansion of or reduction in workforce and the change of the quality of the test process. The control process is applied throughout the testing phase. Although it does not guarantee that testing will be able to meet its schedule and quality objectives, it does provide information that helps the test manager determine whether or not the objectives will be met and, if not, what actions to take. For more on such models, see [10, 11].

6.4 Test Strategy Results

In this section we present and compare results from release 6.4 and release 7.0 of a telecommunication product. The examples emphasize the benefits of using quality graphs as monitoring tools. The examples are based on the following charts:

- Accumulated lines of code and accumulated defects per build
- New defects per build
- New Severity 1, Severity 2, Severity 3 defects per build
- New defects by date

The objective here is to compare two releases of a specific product and identify what are possible causes for the significant differences.

6.4.1 Accumulated Lines of Code and Accumulated Defects per Build

In this analysis we track accumulated lines of code and accumulated defects per build in release 7.0. From Figure 6.13 it is apparent that a substantial amount of

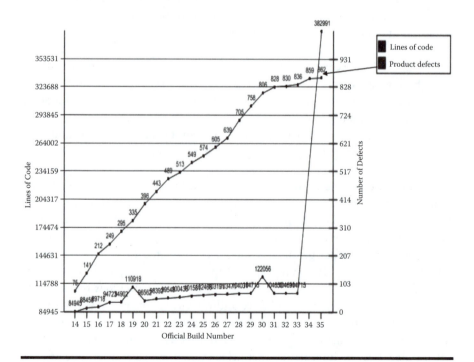

Figure 6.13   Accumulated lines of code and accumulated defects per build of release 7.0.

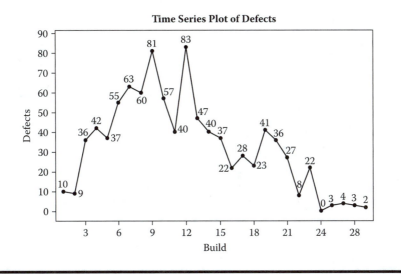

Figure 6.14 New defects per build, release 6.4.

lines of code was added in build 34. Moreover, the profile of accumulated defects does not stabilize up to build 35. In comparison with Figure 6.10 presenting the same data for release 6.4, we see some potential problems in release 7.0 due to a lack of control and instability.

6.4.2 New Defects per Build

Figure 6.14 is a chart of new defects per build for release 6.4. The equivalent chart for release 7.0 is shown in Figure 6.15. The patterns of defects build-up appear

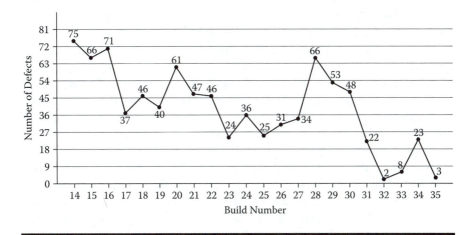

Figure 6.15 New defects per build, after build 14, release 7.0.

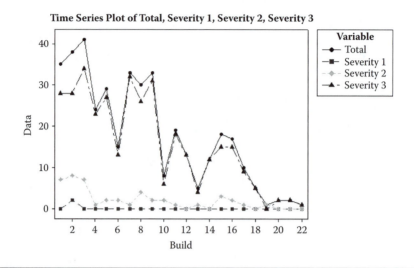

Figure 6.16   New defects of Severity 1 through 3, per build, from build 14, release 7.0.

similar. In release 6.4, the pattern from build 14 to build 24 is encircled. In both cases, toward the end, there is a peak in the number of defects. This occurred in build 28 for both releases, and in build 34 in release 7.0.

6.4.3 New Severity 1 through Severity 3 Defects per Build

The chart in Figure 6.16 represents new defects of Severity 1 through 3 per build, starting with build 14. It reflects the same trend that was observed at the total defect per build in release 6.4 (see Figure 6.6). The chart for release 6.4 (Figure 6.6) reflects a good level of synchronization between testing and development. When looking at Figure 6.10, representing the accumulated lines of code and accumulated defects per build in release 6.4, we can observe that for the last build, no code was added. Figure 6.16, in conjunction with Figure 6.13, describe a different situation in release 7.0. Here we had new code added at the last build and the reports on new defects did not stabilize. Both these facts are reasons for concern, and additional precautions were taken during product release, such as adding staff to the post-installation support group.

A key objective in tracking detects is to be able to determine, with a given level of confidence, if the system is ready for release. To reach such a decision, we need to model the software reliability trends. Section 6.6 discusses a variety of software reliability models that can be used for this purpose. However, before fitting such models, we discuss a methodology designed to assess our testing effectiveness and efficiency.

6.5 Assessing Testing Effectiveness and Efficiency

6.5.1 The Software Trouble Assessment Matrix (STAM)*

Some process improvements are driven by the need to stay one step ahead of the competition. Other process improvements are the result of attempts to learn from current and past problem reports. In the latter case, prerequisites to such improvement efforts are that the processes have been identified, and that process ownership has been established (see Chapters 2 and 3). Typical development processes consist of requirements analysis, top-level design, detailed design, coding (for software) or component prototype development (for hardware), and testing. A causal analysis of faults classifies faults as being attributable to errors in any one of these activities. For such an analysis to be successful, it is essential that there be agreement on the boundaries of the processes represented by these activities. In particular, the entrance and exit criteria for each process must be clarified and documented (see Section 6.2). Such definitions permit effective data collection and analysis. The Software Trouble Assessment Matrix (STAM) is a method introduced in [20] to analyze data derived by answering three questions:

1. Where were errors detected in the development cycle?
2. Where were those errors actually created?
3. Where could the errors have been detected?

STAM is a methodology that enables an organization to answer these questions, and is equally applicable to systems or software development. The STAMs are compiled by determining the phases in which each error is created, detected, or could have been detected. The term "error" is used here in the sense of problem report, defect, or bug. The total number of errors is then accumulated for each successive phase. The area under the curve is determined by adding again the cumulative totals. This gives a figure that represents the area under the cumulative frequency curve.

Three measures are easily computed from the data collected in a STAM analysis. Let P = Number of phases, E = Total (Cumulative) number of errors, and TS1 = Area under the curve of accumulated errors by detection phase, TS2 by earliest detection, and TS3 by creation phase. Three ratios are calculated: the Negligence, Evaluation, and Prevention ratios. The definitions and formulae for the negligence, evaluation, and prevention ratios are as follows:

■ *Negligence ratio* = 100 × (TS2 − TS1)/TS1. This ratio indicates the amount of errors that escaped through the inspection process filters. In other words, it measures inspection efficiency.

---

* The contribution to this section by Alberto Coutino-Hill, Director, Quality and Delivery, Xwave, is gratefully acknowledged.

- *Evaluation ratio* = 100 × (TS3 − TS2)/TS2. This ratio measures the delay of the inspection process in identifying errors relative to the phase in which they occurred. In other words, it measures inspection effectiveness.
- *Prevention ratio* =100 × TS1/(P × E). This ratio is an index of how early errors are detected in the development life cycle relative to the total number of reported errors. It is a combined measure of performance of the development and inspection processes. It assesses the organization's ability to generate and identify errors as early as possible in the development life cycle.

A typical error analysis begins with assessing where errors could have been detected and concludes with classifying errors into the life-cycle phases in which they were created. This procedure requires a repeat analysis of each recorded error. As previously noted, the success of an error causal analysis is highly dependent on clear entrance and exit criteria for the various development phases.

Using the negligence, evaluation, and prevention ratios, developers can better understand and improve their inspection and development processes. They also can use the STAM to benchmark different projects within their companies and against those of different companies.

The results of a STAM analysis for data from release 6.4 of the telecommunication system introduced in Section 6.2.5 are presented in Table 6.4 and Table 6.5. Table 6.4 shows the number of errors per phase, plotting the phase where the error could have been detected versus the phase where the error was detected. For example, there were three (3) errors that could have been detected during analysis, using the current procedures that are in place, that were not detected until the design phase. ·

In Table 6.4, two sets of cumulative frequency are calculated while one is calculated in Table 6.5. The final values calculated (TS1, TS2, and TS3) represent the total area under the cumulative frequency curve for each metric collected. TS1 represents a failure profile of where the errors were detected; TS2 represents a failure profile of where the errors could have been detected; and TS3 represents a failure profile of where the errors were created. As explained previously, these values are used to calculate the negligence, evaluation, and prevention ratios.

6.5.2 A STAM Analysis of Release 7.0

This section presents the application of STAM analysis for data from release 7.0 of the telecommunication system introduced in Section 6.2.5. The results are presented in Table 6.6 and Table 6.7.

6.5.3 Comparing Releases with a STAM Analysis

The negligence, evaluation, and prevention ratios are calculated for each release using the formulae discussed in Section 6.5.1. The values determined for the area of

Table 6.4 Software Trouble Assessment Matrix (Part 1)

Phase Where Errors Could Have Been Detected	Phase Where Errors Were Detected								Total per Phase	Cumulative Total	Cumulative Area
	PI	A	D	C	UT	IT	ST	AT			
Project Initiation (PI)	22								22	22	22
Analysis (A)		54	3				1		58	80	102
Design (D)			65	5					70	150	252
Coding (C)				87		3	2		92	242	494
Unit Test (UT)					48	4	2	1	**55**	**297**	**791**
Integration Test (IT)						26			26	323	1114
System Test (ST)							35	2	**37**	**360**	**1474**
Acceptance Test (AT)								22	**22**	**382**	*1856*
Total per Phase	22	54	68	92	48	33	40	25			**TS2 = 1856**
Cumulative Total	22	76	144	236	284	317	357	382			
Cumulative Area	**22**	**98**	**242**	**478**	**762**	**1079**	**1436**	*1818*	**TS1 = 1818**		

Table 6.5   Software Trouble Assessment Matrix (Part 2)

Phase Where the Error Could Have Been Detected	Phase Where the Error Was Created							
	PI	A	D	C	UT	IT	ST	AT
Project Initiation (PI)	22							
Analysis (A)	1	55						
Design (D)	1	2	69					
Coding (C)			3	91				
Unit Test (UT)				5	50			
Integration Test (IT)						26		
System Test (ST)			4				33	
Acceptance Test (AT)								20
Total per Phase	24	57	76	96	50	26	33	20
Cumulative Total	24	81	157	253	303	329	362	382
Cumulative Area	24	105	262	**515**	**818**	**1147**	**1509**	*1891*

TS3 = 1891

Table 6.6 Software Trouble Assessment Matrix (Part 1) for Release 7.0

Phase Where Errors Could Have Been Detected	Phase Where Errors Were Detected									Total per Phase	Cumulative Total	Cumulative Area (S2)
	PI	A	D	C	UT	IT	ST	AT	I			
Project Initiation (PI)	8									8	8	8
Analysis (A)		122	1							123	131	139
Design (D)			97	7						104	235	374
Coding (C)				142		1				143	378	752
Unit Test (UT)					66					66	444	1196
Integration Test (IT)						10				10	454	1650
System Test (ST)							27			27	481	2131
Acceptance Test (AT)								19		19	500	2631
Implementation (I)									7	7	507	*3138*
Total per Phase	8	122	98	149	66	11	27	19	7			**TS2 = 3138**
Cumulative Total	8	130	228	377	443	454	481	500	507			
Cumulative Area	8	138	366	743	1186	1640	2121	2621	3128	**TS1 = 3128**		

Table 6.7 Software Trouble Assessment Matrix (Part 2) for Release 7.0

Phase Where the Error Could Have Been Detected	Phase Where the Error Was Created								
	PI	A	D	C	UT	IT	ST	AT	I
Project Initiation (PI)	8								
Analysis (A)		123							
Design (D)			104						
Coding (C)			2	141					
Unit Test (UT)					66				
Integration Test (IT)						10			
System Test (ST)		1	1	1			24		
Acceptance Test (AT)								19	
Implementation (I)									7
Total per Phase	8	124	107	142	66	10	24	19	7
Cumulative Total	8	132	239	381	447	457	481	500	507
Cumulative Area (S3)	8	140	379	**760**	**1207**	**1664**	**2145**	**2645**	*3152*

TS3 = 3152

Table 6.8   Negligence, Evaluation, and Prevention Ratios for Release 6.4

STAM Quality Assurance Metrics for release 6.4			
Negligence Ratio:	2.09%	Inspection Efficiency:	97.91%
Evaluation Ratio:	1.89%	Inspection Effectiveness:	98.11%
Prevention Ratio:	59.5%		

Table 6.9   Negligence, Evaluation, and Prevention Ratios for Release 7.0

STAM Quality Assurance Metrics for release 7.0			
Negligence Ratio:	0.32%	Inspection Efficiency:	99.68%
Evaluation Ratio:	0.45%	Inspection Effectiveness:	99.55%
Prevention Ratio:	68.6%		

cumulative frequency of errors (TS1, TS2, and TS3) are used in these formulae to calculate the ratios. The ratios for release 6.4 are shown in Table 6.8, and the ratios for release 7.0 are shown in Table 6.9.

The *negligence ratio* is calculated by comparing when errors were detected versus when errors could have been detected. For release 6.4, about 2% of the errors escaped through the inspection processes that are currently in place. This indicates that the inspections (walk-throughs and test results reviews) were about 98% efficient. That is, 98% of the errors were detected as soon as they could have been detected.

The *evaluation ratio* is calculated by comparing when errors are created versus when errors could have been detected. Fewer than 2% of the errors were created but could not be detected until a later phase. This indicates that the inspections were over 98% effective during release 6.4.

For release 7.0, less than 1% of the errors escaped through the inspection processes that are currently in place. This indicates that the inspections (walk-throughs and test results reviews) were over 99% efficient. That is, 99% of the errors were detected as soon as they could have been detected.

Less than 1% of the errors were created but could not be detected until a later phase. This indicates that the inspections were over 99% effective during release 7.0.

As indicated from the above ratios, the last two releases of the software have been very successful. The ratios also indicate a degree of improvement from one release to the next. This can also be seen from the prevention ratio. This ratio assesses the ability to identify errors as soon as possible in the life cycle. This is an index of how many errors are detected early in the life cycle relative to the total number of errors. While it is impractical to expect this ratio to be 100%, improvement can be observed as this value increases. This ratio increased by over 9% from release 6.4 to release 7.0.

6.6 Software Reliability Models

Software reliability is the probability of execution without failure for some specified interval, called the mission time (see [26]). This definition is compatible with that used for hardware reliability, although the failure mechanisms in software components may differ significantly. Software reliability is applicable both as a tool complementing development testing, in which faults are found and removed, and for certification testing, when a software product is either accepted or rejected. When software is in operation, failure rates can be calculated by computing the number of failures, say, per hour of operation. The predicted failure rate corresponding to the steady-state behavior of the software is usually a key indicator of great interest. The predicted failure rate may be regarded as high when compared to other systems. We should keep in mind, however, that the specific weight of the failures indicating severity of impact is not accounted for within the failure category being tracked. In the bug tracking database, all failures within one category are equal. Actual interpretation of the predicted failure rates, accounting for operational profile and specific failure impact, is therefore quite complex. Predicted failure rates should therefore be considered in management decisions at an aggregated level. For instance, the decision to promote software from system test status to acceptance test status can be based on a comparison of predicted failure rates to actual, as illustrated below. As early as 1978, the now-obsolete MIL-STD-1679(Navy) [34] document specified that one error per 70,000 machine instruction words, which degrades performance with a reasonable alternative work-around solution, is allowed; and one error causing inconvenience and annoyance is allowed for every 35,000 machine instruction words. However, no errors that cause a program to stop or degrade performance without alternative work-around solutions are allowed.

These reliability specifications should be verified with a specified level of confidence. Interpreting failure data is also used to determine if a system can be moved from development to beta testing with selected customers, and then from beta testing to official shipping. Both applications to development testing and certification testing rely on mathematical models for tracking and predicting software reliability. Many system and software reliability models have been suggested in the literature (see [1, 16, 21, 25, 26, 28, 31, 35, 39], among others). In the appendix to this chapter we review the classical reliability models that apply to software and overall system reliability, and provide a detailed discussion of the classical software and system reliability models, the assumptions underlying them, and the equations used in performing the reliability calculations. The models discussed in the appendix include the Markov, nonhomogeneous Poisson process (NHPP), and

Bayesian models. For some of the models, the mathematics are quite complex and require an advanced level of knowledge of mathematics to fully understand them. In each software reliability model, there always exist some parameters that must be estimated based on historical data. Advanced methods for designing test cases of Web services in service-oriented architectures using group testing and application ontologies have been proposed [4]. They are based on statistical experimental design techniques described in [26] and introduced in Chapter 7 for designing beta tests. In Chapter 7 we also dwell on the topic of assessing system usability using customer usage statistics [15, 27]. Such analysis is critical for making improvements in modern systems and software applications.

6.7 Summary

This chapter dealt with methods to plan and improve system and software testing processes. A comprehensive discussion of how to design a test strategy was presented, together with examples of how the strategy was applied. Basic tools for the analysis of defect data, such as simple run charts and control charts, were presented. More advanced tools such as CUSUM charts and the M-test were also introduced. CUSUMs provide efficient, ongoing monitoring of processes and defect data so that decisions on release adequacy can be reached safely and in a timely manner. The M-test is a special statistical method used to compare frequency distributions across projects or time windows. The chapter also discussed the Software Trouble Assessment Matrix (STAM) as a technique to assess the effectiveness of the test strategy. In particular, the STAM provides three metrics—the evaluation, the negligence, and the prevention ratios—that are used to compare data from various releases. The chapter concluded with a review of the basic models of software reliability and a brief introduction of a promising young discipline—software cybernetics. Mathematical details of the software reliability models are presented in the appendix to this chapter and can be skipped without loss of continuity.

References

1. Almering, V., van Genuchten, M., Cloudt, G., and Sonnemans, P., Using Software Reliability Growth Models in Practice. *IEEE Software*, 24(6), 82–88, 2007.
2. Ambler, S.W., *Agile Modeling: Effective Practices for Extreme Programming and the Unified Process*, New York: John Wiley & Sons, 2002.
3. ANSI/IEEE Std. 729-1983, IEEE Standard Glossary of Software Engineering Terminology, IEEE Standards Office, P.O. Box 1331, Piscataway, NJ, 1983.
4. Bai, X. and Kenett R.S., Risk-Based Adaptive Group Testing of Web Services, *Proceedings of the 33rd Annual IEEE International Computer Software and Applications Conference, COMPSAC 2009*, Seattle, WA, 2009.

5. Basili, V.R., Briand, L.C., and Melo, W., A Validation of Object-Oriented Design Metrics as Quality Indicators, *IEEE Transactions on Software Engineering*, 22, 751–761, 1996.

6. Binder, R.W., Modal Testing Strategies for OO Software, *IEEE Computer*, 1996, p. 97–99.

7. Boehm, B.W., A Spiral Model of Software Development and Enhancement, *Proceedings of the International Workshop on Software Process and Software Environments*, California, 1985.

8. Cai, K., Cangussu, J.W., DeCarlo, R.A., and Mathur, A.P., An Overview of Software Cybernetics, *Eleventh International Workshop on Software Technology and Engineering Practice*, Amsterdam, IEEE Computer Society, 2004, p. 77–86.

9. Cangussu, J.W. and Karcich, R.M., A Control Approach for Agile Processes, *2nd International Workshop on Software Cybernetics—29th Annual IEEE International Computer Software and Applications Conference (COMPSAC 2005)*, Edinburgh, U.K., 2005.

10. Cangussu, J.W., A Software Test Process Stochastic Control Model Based on CMM Characterization, *Software Process: Improvement and Practice*, 9(2), 55–66, 2004.

11. Cangussu, J.W., DeCarlo, R.A., and Mathur, A.P., A Formal Model for the Software Test Process, *IEEE Transactions on Software Engineering*, 28(8), 782–796, 2002.

12. Chillarege, R., ODC for Process Measurement, Analysis and Control, *Proceedings, Fourth International Conference on Software Quality*, ASQC Software Division, McLean, VA, 1994.

13. CMMI Product Team, CMMI for Development, Version 1.2 (CMU/SEI-2006-TR-008), Pittsburgh, PA: Software Engineering Institute, Carnegie Mellon University, 2006.

14. Haider, S.W., Cangussu, J.W., Cooper, K., and Dantu, R., Estimation of Defects Based on Defect Decay Model: ED^{3}M, *IEEE TSE—Transactions on Software Engineering*, 34(3), 336–356, 2008.

15. Harel, A., Kenett, R.S., and Ruggeri, F., Modeling Web Usability Diagnostics on the Basis of Usage Statistics, in *Statistical Methods in eCommerce Research*, W. Jank and G. Shmueli, Editors, New York: Wiley, 2008.

16. Jeske D.R. and Zhang, X., Some Successful Approaches to Software Reliability Modeling in Industry, *Journal of Systems and Software*, 74(1):85–99, 2005.

17. Kenett, R.S. and Koenig S., A Process Management Approach to Software Quality Assurance, *Quality Progress*, 1988, 66–70.

18. Kenett, R.S. and Pollak, M., Data-Analytic Aspects of the Shiryayev-Roberts Control Chart: Surveillance of a Non-homogeneous Poisson Process, *Journal of Applied Statistics*, 23(1), 125–127, 1996.

19. Kenett, R.S. and Pollak, M., A Semi-Parametric Approach to Testing for Reliability Growth, with Application to Software Systems, *IEEE Transactions on Reliability*, 1986, 304–311.

20. Kenett, R.S., Assessing Software Development and Inspection Errors, *Quality Progress*, p. 109–112, October 1994, with correction in issue of February 1995.

21. Kenett, R.S., Software Failure Data Analysis, in *Encyclopedia of Statistics in Quality and Reliability*, Ruggeri, F., Kenett, R.S., and Faltin, F., Editors in Chief, New York: Wiley, 2007.

22. Kenett, R.S., Making Sense Out of Two Pareto Charts, *Quality Progress*, 1994, 71–73.

23. Kenett, R.S., Managing a Continuous Improvement of the Software Development Process, *Proceedings of the 8th IMPRO Annual Conference of the Juran Institute*, Atlanta, GA, 1989.

24. Kenett, R.S., Understanding the Software Process, in *Total Quality Management for Software*, G. Schulmeyer and J. McManus, Editors, New York: Van Nostrand Reinhold, 1992.

25. Kenett, R.S. and Baker, E., *Software Process Quality: Management and Control*, New York: Marcel Dekker, Inc., 1999.

26. Kenett, R.S. and Zacks, S., *Modern Industrial Statistics: Design and Control of Quality and Reliability*, San Francisco: Duxbury Press, 1998 (Spanish edition, 2000; 2nd edition, 2003; Chinese edition, 2004.)

27. Kenett, R.S., Harel, A., and Ruggeri, F., Controlling the Usability of Web Services, *International Journal of Software Engineering and Knowledge Engineering*, World Scientific Publishing Company, Vol. 19, No. 5, pp. 627–651, 2009.

28. Khoshgoftaar, T.M., Allen, E.B., Jones, W.D., and Hudepohl, J.P., Classification-Tree Models of Software-Quality over Multiple Releases, *IEEE Transactions on Reliability*, 49(1), 4–11, 2000.

29. Knuth, D., The Errors of TEX, Report No. STAN-CS-88-1223, Department of Computer Science, Stanford University, Stanford, CA, 1988.

30. Larman, C., *Agile and Iterative Development: A Manager's Guide,* Boston, MA: Addison-Wesley, 2004.

31. Lyu, M. (Ed.), *Handbook of Software Reliability Engineering*, New York: McGraw-Hill, 1996.

32. Miller, S., Decarlo, R., and Matrhur A., Quantitative Modeling for Incremental Software Process Control, in *Proceedings of the Computer Software and Applications Conference (COMPSAC'08)*, 2008, p. 830–835, 2008.

33. Miller, S., DeCarlo, R., Mathur, A., and Cangussu, J.W., A Control Theoretic Approach to the Management of the Software System Test Phase, *Journal of System and Software(JSS)*, 79(11), 1486–1503, 2006.

34. MIL-STD-1679 (Navy), Military Standard Weapon System Software Development, Department of Defense, Washington, D.C. 20360, 1978.

35. Pham, L. and Pham, H., Software Reliability Models with Time-Dependent Hazard Function Based on Bayesian Approach, *IEEE Transactions on Systems Man and Cybernetics Part A—Systems and Humans*, 30, 25–35, 2000.

36. Pittman, M., Lessons Learned in Managing Object-Oriented Development, *IEEE Software,* 1992, p. 43–53.

37. Takeuchi, H. and Nonaka, I., The New New Product Development Game, *Harvard Business Review*, January–February 1986.

38. Tartalja, I. and Milutinovic, V., Classifying Software-Based Cache Coherence Solutions, *IEEE Software*, 1997, p. 90–101.

39. Thwin, M.M.T. and Quah, T.S., Application of Neural Networks for Software Quality Prediction Using Object-Oriented Metrics, *Journal of Systems and Software*, 76, 147–156, 2005.

40. Whittaker, J., Rekab, K., and Thomason, M.G., A Markov Chain Model for Predicting the Reliability of Multi-build Software, *Information and Software Technology*, 42, 889–894, 2000.

41. Wiener, N., *Cybernetics: or Control and Communication in the Animal and the Machine,* New York: John Wiley & Sons, 1948.

Appendix: Software Reliability Models*

The standard definition of software reliability is the probability of execution without failure for some specified interval, called the mission time (see [4, 21, 29, 32, 49]). This definition is compatible with that used for hardware reliability, although the failure mechanisms in software components may differ significantly. Software reliability is applicable both as a tool complementing development testing, in which faults are found and removed, and for certification testing, when a software product is either accepted or rejected. When software is in operation, failure rates can be computed by computing the number of failures, say, per hours of operation. The predicted failure rate corresponding to the steady-state behavior of the software is usually a key indicator of great interest. The predicted failure rate may be regarded as high when compared to other systems. We should keep in mind, however, that the specific weight of the failures indicating severity of impact is not accounted for within the failure category being tracked. In the bug tracking database, all failures within one category are equal. Actual interpretation of the predicted failure rates, accounting for operational profile and specific failure impact, is therefore quite complex. Predicted failure rates should therefore be considered in management decisions at an aggregated level. For instance, the decision to promote software from system test status to acceptance test status can be based on a comparison of predicted failure rates to actual, as illustrated below. As early as 1978, the now-obsolete MIL-STD-1679(Navy) [30] document specified that one error per 70,000 machine instruction words, which degrades performance with a reasonable alternative work-around solution, is allowed; and one error causing inconvenience and annoyance is allowed for every 35,000 machine instruction words. However, no errors that cause a program to stop or degrade performance without alternative work-around solutions are allowed.

Kanoun, Kaaniche, and Laprie [18] analyzed four systems (telephony, defense, interface, and management), and derived average failure rates over ten testing periods and predicted failure rates. Their results are presented in Table 6A.1.

As expected in reliability growth processes, in all systems, the average failure rate over ten test periods is higher than the predicted failure rates. However, the differences in the defense and management systems are smaller, indicating that these systems have almost reached steady state, and no significant reliability improvements should be expected. The telephony and interface systems, however, are still evolving. A decision to promote the defense and interface systems to acceptance test status could be made, based on the data.

These reliability specifications should be verified with a specified level of confidence. Interpreting failure data is also used to determine if a system can be moved from development to beta testing with selected customers, and than from beta

---

* This section requires proper mathematical background and can be skipped without loss of continuity. It is an expansion of Section 6.6, and prepared as a stand-alone document providing a broad review of software reliability models.

Table 6A.1   Typical Failure Rates, by System Type

System	Predicted Failure Rate	Average Failure Rate over 10 Test Periods
Telephony	1.2×10^{-6}/hour	1.0×10^{-5}/hour
Defense	1.4×10^{-5}/hour	1.6×10^{-5}/hour
Interface	2.9×10^{-5}/hour	3.7×10^{-5}/hour
Management	8.5×10^{-6}/hour	2.0×10^{-5}/hour

testing to official shipping. Both applications to development testing and certification testing rely on mathematical models for tracking and predicting software reliability. Many system and software reliability models have been suggested in the literature ([1, 7, 10, 11, 14, 21, 34, 35, 43, 46, 48, 51], among others). Moreover, lessons learned from object-oriented development are used to better design Web Services and Service-Oriented Architecture systems for both better reliability and usability (see [2, 12, 13, 22, 38, 42]). In this appendix we review the classical reliability models that apply to software and overall system reliability.

6A.1 Markov Models

The Jelinski–Moranda model [15, 31] was the first published Markov model found to have profound influences on software reliability modeling. The main assumptions of this model include the following:

- The number of initial faults is an unknown but fixed constant.
- A detected fault is removed immediately and no new faults are introduced.
- Times between failures are independent, exponentially distributed random quantities.
- Each remaining software fault contributes the same amount to the software failure intensity.

Denote by N_0 the number of initial faults in the software before the testing starts; then the initial failure intensity is $N_0 \phi$, where ϕ is a proportionality constant denoting the failure intensity contributed by each fault. Denote by T_i, $i = 1, 2, \cdots, N_0$, the time between $(i-1)$:th and i:th failures, then T_i's are independent, exponentially distributed random variables with parameter:

$$\lambda_i = \phi[N_0 - (i-1)], \quad i = 1, 2, \ldots N_0$$

Many modified versions of the Jelinski-Moranda model have been studied in the literature. Schick and Wolverton [40] proposed a model assuming that times between failures are not exponential but follow a Rayleigh distribution. Shanthikumar [41] generalized the Jelinski-Moranda model using a general time-dependent transition probability function. Xie [52] developed a general decreasing failure intensity model allowing for the possibility of different fault sizes. Whittaker et al. [45] considered a model that allows use of prior testing data to cover the real-world scenario in which the release build is constructed only after a succession of repairs to buggy pre-release builds. Boland and Singh [3] consider a birth-process approach to a related software reliability model, and Lipow considers data aggregated over various time intervals [25]. The Lipow model assumes that:

- The rate of error detection is proportional to the current error content of a program.
- All errors are equally likely to occur and are independent of each other.
- Each error is of the same order of severity as any other error.
- The error rate remains constant over the testing interval.
- The software operates in a similar manner as the anticipated operational usage.
- During a testing interval i, f_i errors are discovered but only ni errors are corrected in that time frame.

When software is in operation, failure rates can be computed by computing the number of failures, say, per hours of operation. The predicted failure rate corresponding to the steady-state behavior of the software is usually a key indicator of great interest. The predicted failure rate may be regarded as high when compared to other systems. We should keep in mind, however, that the specific weight of the failures indicating severity of impact is not accounted for within the failure category being tracked. In the bug tracking database, all failures within one category are equal. Actual micro-interpretation of the predicted failure rates, accounting for operational profile and specific failure impact, is therefore quite complex. Predicted failure rates should therefore be considered in management decisions at a macro, aggregated level. For instance, the decision to promote software from system test status to acceptance test status can be based on a comparison of predicted failure rates to actual, as illustrated below. Because the error rate remains constant during each of the M testing periods (the "error rate remains constant over the testing interval" assumption from above), the failure rate during the i-th testing period is:

$$Z(t) = \varphi[N - F_i], \quad t_{i-1} \le t \le t_i$$

where φ is the proportionality constant, N is again the total number of errors initially present in the program,

$$F_{i-1} = \sum_{j=1}^{i-1} n_j$$

is the total number of errors corrected up through the (i − 1)st testing intervals, and t_i is the time measured in either CPU or wall clock time at the end of the i-th testing interval, $x_i = t_i - t_{i-1}$. The t_i values are fixed and thus are not random variables as in the Jelinski-Moranda model. Taking the number of failures, f_i, in the i-th interval to be a Poisson random variable with mean $Z(t_i)x_i$, the likelihood is:

$$L(f_1,...f_M) = \prod_{i=1}^{M} \frac{[\phi[N - F_{i-1}]x_i]^{f_i} \exp\{-\phi[N - F_{i-1}]x_i\}}{f_i!}$$

Taking the partial derivatives of L(f) with respect to φ and N, and setting the resulting equations to zero, we derive the following equations satisfied by the maximum likelihood estimators $\hat{\phi}$ and $\hat{N}$ of φ and N:

$$\hat{\phi} = \frac{F_M / A}{\hat{N} + 1 - B / A} \quad \text{and} \quad \frac{F_M}{\hat{N} + 1 - B / A} = \sum_{i=1}^{M} \frac{f_i}{\hat{N} - F_{i-1}}$$

where:

$$F_M = \sum_{i=1}^{M} f_i = \text{Total number of errors found in the M periods of testing}$$

$$B = \sum_{i=1}^{M} (F_{i-1} + 1)x_i$$

$$A = \sum_{i=1}^{M} x_i = \text{Total length of the testing period}$$

From these estimates, the maximum likelihood estimate of the mean time until the next failure (MTTF), given the information accumulated in the M testing periods, is:

$$\frac{1}{\hat{\phi}(\hat{N} - F_M)}$$

6A.1.1 A Case Study of the Lipow Model

To demonstrate the application of Lipow's extension to the Jelinski and Moranda de-eutrophication model, we can quote some data collected at Bell Laboratories

Table 6A.2   Typical Failure Rates, by System Type

Test Period	Observed Failures	Predicted Failures
1	72	74
2	70	67
3	57	60
4	50	54
5	—	49
6	—	44
7	—	40

during the development of the No. 5 Electronic Switching System. The data is presented in Table 6A.2 and consists of failures observed during four consecutive testing periods of equal length and intensity:

In operational software, failure rates can be computed by computing the number of failures, say, per hours of operation. A key indicator of interest is the predicted failure rate corresponding to the steady-state behavior of the software. The predicted failure rate can be regarded as high or low, or "as expected," when compared to other systems. Severity of impact is typically assigned a generic interpretation for a class of failures and is not assessed for individual failures. In the bug tracking database, all failures within one category are therefore considered as having equal impact. Predicted failure rates should therefore be considered in macro, aggregated level management decisions. For instance, the decision to promote software from system test status to acceptance test status can be based on a comparison of predicted failure rates in severity classes 1 through 3, to actual, as illustrated.

The column of predicted failures in test periods 5, 6, and 7 is computed from the Lipow model with the above estimates for φ and N. Comparing observed and predicted values over the first four test periods can help us evaluate how well the Lipow model fits the data. The Chi-square goodness of fit test statistic is 0.63, indicating extremely good fit (at a significance level better than 0.001). We can therefore provide reliable predictions for the number of failures expected at test periods 5 through 7, provided the test effort continues at the same intensity and that test periods are of the same length as before.

The Jelinski–Moranda model and the various extensions to this model are classified as time domain models. They rely on a physical modeling of the appearance and fixing of software failures. The different sets of assumptions are translated into differences in mathematical formulations. We expand on various alternatives proposed in the literature in the next section.

6A.2 NHPP Models

For nonindependent data, a nonhomogeneous Poisson process (NHPP) can be assumed. Denote by $N(t)$ the number of observed failures until time t; the main assumptions of this type of models are:

$$N(0) = 0$$

- $\{N(t), t \geq 0\}$ has independent increments
- At time t, $N(t)$ follows a Poisson distribution with parameter $m(t)$, that is,

$$P\{N(t) = n\} = \frac{[m(t)]^n}{n!} \exp[-m(t)], \quad n = 0,1,2,\ldots$$

where $m(t)$ is called the *mean value function* of the NHPP, which describes the expected cumulative number of experienced failures in time interval $(0,t]$.

The *failure intensity function* $\lambda(t)$ is defined as:

$$\lambda(t) \equiv \lim_{\Delta t \to 0_+} \frac{P(N(t + \Delta t) - N(t) > 0\}}{\Delta t} = \frac{dm(t)}{dt}, \quad t \geq 0$$

Generally, by using a different mean value function of $m(t)$, we get different NHPP models (Figure 6A.1).

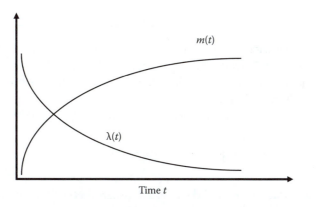

Figure 6A.1   An example of failure intensity function $\lambda(t)$ and mean value function $m(t)$.

6A.2.1 The Goel–Okomoto Model

The most representative NHPP model was proposed by Goel and Okumoto [8]. In fact, many other NHPP models are a modification or generalization of this model. The mean value function of the Goel–Okumoto model is:

$$m(t) = a\left[1 - \exp(-bt)\right], \quad a > 0, b > 0$$

where a is the expected number of faults to be eventually detected, and b is the failure occurrence rate per fault. The failure intensity function of the Goel–Okumoto model is:

$$\lambda(t) = ab \exp(-bt)$$

6A.2.2 The Musa–Okomoto Logarithmic Model

Musa and Okumoto [33] developed a logarithmic Poisson execution time model. The mean value function is:

$$m(t) = \frac{1}{\varphi} \ln(\lambda_0 \varphi t + 1), \quad \varphi > 0, \lambda_0 > 0$$

where λ_0 is the initial failure intensity, and ϕ is the failure intensity decay parameter. Because this model allows an infinite number of failures observed, it is also called an infinite failure model.

6A.2.3 The Yamada S-Shaped NHPP Models

Based on experience, it is sometimes observed that the curve of the cumulative number of faults is S-shaped. Several different S-shaped NHPP models have been proposed in the existing literature. Among them, the most interesting ones are the delayed S-shaped NHPP model (Yamada, [54]) and the inflected S-shaped NHPP model (Ohba, [36]). The mean value function of the delayed S-shaped NHPP model is:

$$m(t) = a\left[1 - (1 + bt)\exp(-bt)\right], \quad a > 0, b > 0$$

and the mean value function of the infected S-shaped NHPP model is:

$$m(t) = \frac{a\left[1 - \exp(-bt)\right]}{\left[1 + c\exp(-bt)\right]}, \quad a > 0, b > 0, c > 0$$

6A.2.4 The Log-Power Model

An interesting model called the log-power model was proposed by Xie and Zhao [50]. It has the following mean value function:

$$m(t) = a \ln^b (1+t), \quad a > 0, b > 0$$

This model is a modification of the traditional Duane [25] model for a general repairable system. A useful property is that if we take the logarithm on both sides of the mean value function, we have:

$$\ln m(t) = \ln a + b \ln \ln(1+t)$$

Hence, a graphical procedure is established. If we plot the observed cumulative number of failures versus $(t + 1)$, the plot should tend to be a straight line on a log-log scale. This can be used to easily estimate the model parameters and, more importantly, to validate the model.

6A.2.5 Other NHPP Extensions

NHPP, a very favorable alternative in software reliability modeling, has continuously received a lot of attention. Yamada et al. [53] incorporated the concept of test effort into the Goel–Okumoto model for a better description of the failure phenomenon. Later, Xie et al. [47] incorporated the concept of a learning factor into the same model. Gokhale and Trivedi [9] proposed an enhanced NHPP model that incorporates explicitly the time-varying test-coverage function in its analytical formulation. Teng and Pham [44] considered a model for predicting software reliability by including a random environmental factor.

6A.3 Bayesian Models

For this type of model, a Bayesian assumption is made or a Bayesian inference is adopted. The well-known model by Littlewood and Verrall [27] is the first Bayesian software reliability model assuming that times between failures are exponentially distributed with a parameter that is treated as a random variable, which is assumed to have a Gamma prior distribution. There are also many papers dealing with Bayesian treatment of the Jelinski–Moranda model; see [6, 24, 26, 28]. The most important feature of using a Bayesian model is that prior information can be incorporated into a parameter estimation procedure. There is also abundant research on the Bayesian inference for NHPP software reliability models. Kuo et al. [23] investigated the Bayesian inference for an NHPP with an *S*-shaped mean value function. Cid and Achcar [5] presented Bayesian analysis for NHPP in software reliability models assuming non-monotonic intensity functions. Pham and Pham [37] proposed a Bayesian model, for which times between failures follow Weibull distributions with stochastically decreasing ordering on the hazard functions of successive failure time intervals. Rees et al. [39] used Bayesian graphical models to model the uncertainties involved in the testing process.

6A.4 Estimation for Model Parameters and Semi-Parametric Control

In each software reliability model, there always exist some parameters that must be estimated based on historical data. Maximum likelihood estimation (MLE), due to its many good mathematical properties, is a commonly adopted method for parameter estimation [17]. This is discussed here.

6A.4.1 Point Estimation for NHPP Model Parameters

In practice, failure data can arise in two different types: numbers of failures or failure times. For the former case, denote by n_i the number of failures observed in the time interval $(s_{i-1}, s_i]$, where $0 \equiv s_0 < s_1 < \cdots < s_k$ and $s_i \, (i > 0)$ is the prescribed time point in the software testing process. Then the likelihood function for an NHPP model with mean value function $m(t)$ is:

$$L(n_1, n_2, \cdots, n_k) = \prod_{i=1}^{k} \frac{\left[m(s_i) - m(s_{i-1}) \right]^{n_i} \exp\left\{ -\left[m(s_i) - m(s_{i-1}) \right] \right\}}{n_i!}$$

The parameters in $m(t)$ can be estimated by maximizing the likelihood function given above. Usually, numerical procedures must be used.

For the latter case, denote by $T_i, \; i = 1, 2, \cdots, k$, the observed k failure times in a software testing process. Then the likelihood function is:

$$L(T_1, T_2, \cdots, T_k) = \exp\left[-m(T_k) \right] \prod_{i=1}^{k} \lambda(T_i)$$

Similarly, the NHPP model parameters can be estimated by maximizing the likelihood function given above.

6A.4.2 A General Approach for Parameter Estimation

To find asymptotic confidence intervals for the k model parameters, the derivation of the *Fisher information matrix* is needed, which is given by:

$$
I(\theta_1,\cdots,\theta_k) \equiv
\begin{bmatrix}
-E\left[\dfrac{\partial^2 \ln L}{\partial \theta_1^2}\right] & -E\left[\dfrac{\partial^2 \ln L}{\partial \theta_2 \partial \theta_1}\right] & \cdots & -E\left[\dfrac{\partial^2 \ln L}{\partial \theta_k \partial \theta_1}\right] \\
-E\left[\dfrac{\partial^2 \ln L}{\partial \theta_1 \partial \theta_2}\right] & -E\left[\dfrac{\partial^2 \ln L}{\partial \theta_2^2}\right] & \cdots & -E\left[\dfrac{\partial^2 \ln L}{\partial \theta_k \partial \theta_2}\right] \\
\cdots & \cdots & \cdots & \cdots \\
-E\left[\dfrac{\partial^2 \ln L}{\partial \theta_1 \partial \theta_k}\right] & -E\left[\dfrac{\partial^2 \ln L}{\partial \theta_2 \partial \theta_k}\right] & \cdots & -E\left[\dfrac{\partial^2 \ln L}{\partial \theta_k^2}\right]
\end{bmatrix}
$$

From the asymptotic theory of MLE, when n approaches infinity, $[\hat{\theta}_1,\cdots,\hat{\theta}_k]$ converges in distribution to k-variate normal distribution with mean $[\theta_1,\cdots,\theta_k]$ and covariance matrix I^{-1}. That is, the asymptotic covariance matrix of the MLEs is:

$$
V_e \equiv
\begin{bmatrix}
Var\left(\hat{\theta}_1\right) & Cov\left(\hat{\theta}_1,\hat{\theta}_2\right) & \cdots & Cov\left(\hat{\theta}_1,\hat{\theta}_k\right) \\
Cov\left(\hat{\theta}_2,\hat{\theta}_1\right) & Var\left(\hat{\theta}_2\right) & \cdots & Cov\left(\hat{\theta}_2,\hat{\theta}_k\right) \\
\cdots & \cdots & \cdots & \cdots \\
Cov\left(\hat{\theta}_k,\hat{\theta}_1\right) & Cov\left(\hat{\theta}_k,\hat{\theta}_2\right) & \cdots & Var\left(\hat{\theta}_k\right)
\end{bmatrix}
= I^{-1}
$$

Therefore, the asymptotic $100(1-\_)\%$ confidence interval for $\hat{\theta}_i$ is:

$$
\left(\hat{\theta}_i - z_{\alpha/2}\sqrt{Var(\hat{\theta}_i)},\quad \hat{\theta}_i + z_{\alpha/2}\sqrt{Var(\hat{\theta}_i)}\right),\quad i=1,\ldots k
$$

where $z_{\alpha/2}$ is the $1-\alpha/2$ percentile of the standard normal distribution, and the quantity $Var(\hat{\theta}_i)$ can be obtained from the covariance matrix given above.

Because the true values of θ_i are unknown, the *observed information matrix* is often used:

$$
\hat{I}(\hat{\theta}_1,\cdots,\hat{\theta}_k) \equiv
\begin{bmatrix}
-\dfrac{\partial^2 \ln L}{\partial \theta_1^2} & -\dfrac{\partial^2 \ln L}{\partial \theta_2 \partial \theta_1} & \cdots & -\dfrac{\partial^2 \ln L}{\partial \theta_k \partial \theta_1} \\
-\dfrac{\partial^2 \ln L}{\partial \theta_1 \partial \theta_2} & -\dfrac{\partial^2 \ln L}{\partial \theta_2^2} & \cdots & -\dfrac{\partial^2 \ln L}{\partial \theta_k \partial \theta_2} \\
\cdots & \cdots & \cdots & \cdots \\
-\dfrac{\partial^2 \ln L}{\partial \theta_1 \partial \theta_k} & -\dfrac{\partial^2 \ln L}{\partial \theta_2 \partial \theta_k} & \cdots & -\dfrac{\partial^2 \ln L}{\partial \theta_k^2}
\end{bmatrix}_{\substack{\theta_1=\hat{\theta}_1 \\ \cdots \\ \theta_k=\hat{\theta}_k}}
$$

and the confidence interval for $\hat{\theta}_i$ can be calculated.

If $\Phi \equiv g(\theta_1, \cdots, \theta_k)$, where $g(\cdot)$ is a continuous function, then as n converges to infinity, $\hat{\Phi}$ converges in distribution to normal distribution with mean Φ and variance

$$Var(\hat{\Phi}) = \sum_{i=1}^{k} \sum_{j=1}^{k} \frac{\partial g}{\partial \theta_i} \cdot \frac{\partial g}{\partial \theta_j} \cdot v_{ij}$$

where v_{ij} is the element of the i-th row and j-th column of the matrix above. The asymptotic $100(1 - \_)\%$ confidence interval for $\hat{\Phi}$ is:

$$\left(\hat{\Phi} - z_{\alpha/2} \sqrt{Var(\hat{\Phi})}, \quad \hat{\Phi} + z_{\alpha/2} \sqrt{Var(\hat{\Phi})} \right)$$

The above result can be used to obtain the confidence interval for some useful quantities such as failure intensity or reliability.

6A4.2.1 An Example

Assume that there are two model parameters, a and b, which is quite common in most software reliability models. The Fisher information matrix is given by:

$$I \equiv \begin{bmatrix} -E\left[\dfrac{\partial^2 \ln L}{\partial a^2}\right] & -E\left[\dfrac{\partial^2 \ln L}{\partial a \partial b}\right] \\ -E\left[\dfrac{\partial^2 \ln L}{\partial a \partial b}\right] & -E\left[\dfrac{\partial^2 \ln L}{\partial b^2}\right] \end{bmatrix}$$

The asymptotic covariance matrix of the MLEs is the inverse of this matrix:

$$V_e = I^{-1} = \begin{bmatrix} Var(\hat{a}) & Cov(\hat{a}, \hat{b}) \\ Cov(\hat{a}, \hat{b}) & Var(\hat{b}) \end{bmatrix}$$

For illustrative purposes, we take the data cited in Table 1 in Ref. [56]. The observed information matrix is:

$$\hat{I} \equiv \begin{bmatrix} -\dfrac{\partial^2 \ln L}{\partial a^2} & -\dfrac{\partial^2 \ln L}{\partial a \partial b} \\ -\dfrac{\partial^2 \ln L}{\partial a \partial b} & -\dfrac{\partial^2 \ln L}{\partial b^2} \end{bmatrix}_{\substack{a=\hat{a} \\ b=\hat{b}}} = \begin{bmatrix} 0.0067 & 1.1095 \\ 1.1095 & 4801.2 \end{bmatrix}$$

We can then obtain the asymptotic variance and covariance of the MLE as follows:

$$Var(\hat{a}) = 154.85, \ Var(\hat{b}) = 2.17 \times 10^{-4}$$

$$Cov(\hat{a}, \hat{b}) = -0.0358$$

The 95% confidence intervals on parameters a and b are (117.93, 166.71) and (0.0957, 0.1535), respectively. Plugging these numbers in

$$m(t) = a \ln^b (1+t), \quad a > 0, \ b > 0$$

gives us a range of estimates of the software reliability growth performance. With such models we can determine if and when the system is ready for deployment in the market.

In general, a software development process consists of the following four major phases: specification, design, coding, and testing. The testing phase is the most costly and time-consuming phase. It is thus very important for management to spend the limited testing resource efficiently. For the optimal testing-resource allocation problem, the following assumptions are made:

■ A software system is composed of n independent modules that are developed and tested independently during the unit testing phase.
■ In the unit testing phase, each software module is subject to failures at random times caused by faults remaining in the software module.
■ The failure observation process of software module i is modeled by an NHPP with mean value function $m_i(t)$, or failure intensity function $\lambda_i(t) \equiv dm_i(t)/dt$.

A total amount of testing-time T is available for the entire software system, which consists of a few modules. Testing time should be allocated to each software module in such a way that the reliability of the system after the unit testing phase will be maximized. For more on such optimization problems, including optimal time to release a system, see [14, 47, 48].

6A4.3 The Shiryayev–Roberts Semi-Parametric Procedure

Kenett and Pollak [19, 20] propose the application of the Shiryayev–Roberts sequential change control procedure to determine if a system or software under test has reached a required level of reliability. This approach does not require a specific model, such as those listed above, and has therefore been labeled "semi-parametric." The Shiryayev–Roberts procedure is based on a statistic that is a sum of likelihood ratios. If there is no change, it is assumed that the observations follow a known distribution whose density is denoted as $f_{v=\infty}$; and if there is a change at time $v = k$, then the density of the observations is denoted as $f_{v=k}$. Denoting the likelihood ratio of the observations $X_1, X_2, \ldots, X_n$ when $v = k$, by

$$\Lambda_{n,k} = \frac{f_{v=k}(X_1, X_2, \ldots X_n)}{f_{v=\infty}(X_1, X_2, \ldots X_n)}$$

where the observations may be dependent, the Shiryayev–Roberts surveillance statistic is:

$$R_n = \sum_{k=1}^{n} \Lambda_{n,k}$$

The scheme calls for releasing the system for distribution at

$$N_A = \min\{n \,|\, R_n \geq A\}$$

where A is chosen so that the average number of observations until false alarm (premature release) is equal to (or larger than) a prespecified constant B. Under fairly general conditions, there exists a constant $c > 1$ such that:

$$\lim_{A \to \infty} E_{v=\infty}(N_A / A) = c$$

where $E(\cdot)$ stands for the expected value so that setting $A = B/c$ yields a scheme that approximately has B as its average number of observations until a possible premature release of the software.

If $f_\theta(x)$ represents an exponential distribution with parameter θ, where $\theta = w_0$ is a prespecified level of failure rate where the system is deemed unreliable if its failure rate exceeds it, and $\theta = w_1$ is an acceptable failure rate ($w_1 < w_0$), then

$$R_n = \sum_{k=1}^{n} \Lambda_{n,k} = \sum_{k=1}^{n} \left(\frac{w_1}{w_0}\right)^{n-k+1} e^{(w_0-w_1)\sum_{i=k}^{n} X_i}$$

In this case, the computation is recursive and

$$R_{n+1} = \left(\frac{w_1}{w_0}\right) e^{(w_0-w_1)X_{n+1}} (1 + R_n)$$

Tracking R_n, and comparing it to A, provides an easily implemented procedure for determining when the software is ready for release. The procedure has been applied to tracking data on CPU execution time between failures of a telecommunication system [20] and computer crashes in a large computer center [19].

References

1. Almering, V., van Genuchten, M., Cloudt, G., and Sonnemans, P., Using Software Reliability Growth Models in Practice, *IEEE Software*, 24(6), 82–88, 2007.
2. Bai, X. and Kenett, R.S., Risk-Based Adaptive Group Testing of Web Services, *Proceedings of the 33rd Annual IEEE International Computer Software and Applications Conference, COMPSAC 2009*, Seattle, WA, 2009.
3. Boland, P.J. and Singh, H., A Birth-Process Approach to Moranda's Geometric Software-Reliability Model, *IEEE Transactions on Reliability*, 52(2), 168–174, 2003.
4. Brown, J.R. and Lipow, M., Testing for Software Reliability, *Proceedings of the International Conference on Reliable Software, IEEE*, Los Angeles, CA, 1975, p. 518–527.
5. Cid, J.E.R. and Achcar, J.A., Bayesian Inference for Nonhomogeneous Poisson Processes in Software Reliability Models Assuming Nonmonotonic Intensity Functions, *Computational Statistics & Data Analysis*, 32, 147–159, 1999.
6. Csenki, A., Bayes Predictive Analysis of a Fundamental Software-Reliability Model, *IEEE Transactions on Reliability*, R-39, 177–183, 1990.
7. Dai, Y.S., Xie, M., and Poh, K.L., Modeling and Analysis of Correlated Software Failures of Multiple Types, *IEEE Transactions on Reliability*, R-54(1), 100–106, 2005.
8. Goel, A.L. and Okumoto, K., Time Dependent Error-Detection Rate Model for Software Reliability and Other Performance Measures, *IEEE Transactions on Reliability*, R-28, 206–211, 1979.
9. Gokhale, S.S. and Trivedi, K.S., A Time/Structure Based Software Reliability Model, *Annals of Software Engineering*, 8, 85–121, 1999.
10. Goseva-Popstojanova, K. and Trivedi, K.S., Failure Correlation in Software Reliability Models, *IEEE Transactions on Reliability*, 49, 37–48, 2000.
11. Haider, S.W., Cangussu, J.W., Cooper, K., and Dantu, R., Estimation of Defects Based on Defect Decay Model: ED^{3}M, *IEEE TSE—Transaction on Software Engineering*, 34(3), 336–356, 2008.
12. Harel, A., Kenett R.S., and Ruggeri, F., Decision Support for User Interface Design: Usability Diagnosis by Time Analysis of the User Activity, *Proceedings of the 32nd Annual IEEE International Computer Software and Applications Conference, COMPSAC 2008*, Turku, Finland, 2008.
13. Harel, A., Kenett, R.S., and Ruggeri, F., Modeling Web Usability Diagnostics on the Basis of Usage Statistics, in *Statistical Methods in eCommerce Research*, W. Jank and G. Shmueli, Editors, New York: Wiley, 2008.
14. Huang, C.Y. and Lyu, M.R., Optimal Release Time for Software Systems Considering Cost, Testing-Effort, and Test Efficiency, *IEEE Transactions on Reliability*, 54(4), 583–591, 2005.

15. Jelinski, Z. and Moranda, P.B., Software Reliability Research, in *Statistical Computer Performance Evaluation*, Freiberger, W., Editor, New York: Academic Press, 1972, p. 465–497.

16. Jeske D.R. and Zhang, X., Some Successful Approaches to Software Reliability Modeling in Industry, *Journal of Systems and Software*, 74(1), 85–99, 2005.

17. Jeske, D.R. and Pham, H., On the Maximum Likelihood Estimates for the Goel-Okumoto Software Reliability Model, *American Statistician*, 55(3), 219–222, 2001.

18. Kanoun, K., Kaaniche, M., and Laprie, J.-C., Qualitative and Quantitative Reliability Assessment, *IEEE Software*, 1997, p. 77–87.

19. Kenett, R.S. and Pollak, M., Data-Analytic Aspects of the Shiryayev-Roberts Control Chart: Surveillance of a Non-homogeneous Poisson Process, *Journal of Applied Statistics*, 23(1), 125–127, 1996.

20. Kenett, R.S. and Pollak, M., A Semi-Parametric Approach to Testing for Reliability Growth, with Application to Software Systems, *IEEE Transactions on Reliability*, 1986, p. 304–311.

21. Kenett, R.S., Software Failure Data Analysis, in *Encyclopedia of Statistics in Quality and Reliability*, Ruggeri, F., Kenett, R. S. and Faltin, F., Editors in Chief, New York: Wiley, 2007.

22. Kenett, R.S., Harel, A., and Ruggeri, F., Controlling the Usability of Web Services, *International Journal of Software Engineering and Knowledge Engineering*, World Scientific Publishing Company, Vol. 19, No. 5, pp. 627–651, 2009.

23. Kuo, L., Lee, J.C., Choi, K., and Yang, T.Y., Bayes Inference for S-Shaped Software-Reliability Growth Models, *IEEE Transactions on Reliability*, 46, 76–81, 1997.

24. Langberg, N.A. and Singpurwalla, N.D., A Unification of Some Software Reliability Models, *SIAM Journal on Scientific and Statistical Computing*, 6, 781–790, 1985.

25. Lipow, M., Models for Software Reliability, Proceedings of the Winter Meeting of the Aerospace Division of the American Society for Mechanical Engineers, 78-WA/Aero-18, 1978 p. 1–11.

26. Littlewood, B. and Sofer, A., A Bayesian Modification to the Jelinski-Moranda Software Reliability Growth Model, *Software Engineering Journal*, 2, 30–41, 1987.

27. Littlewood, B. and Verall, B., A Bayesian Reliability Growth Model for Computer Software, *The Journal of the Royal Statistical Society, Series C*, 22(3), 332–346, 1973.

28. Littlewood, B. and Verral, J.L., A Bayesian Reliability Growth Model for Computer Software, *Applied Statistics*, 22, 332–346, 1973.

29. Lyu, M., Ed., *Handbook of Software Reliability Engineering*, New York: McGraw-Hill, 1996.

30. MIL-STD-1679(Navy), Military Standard Weapon System Software Development, Department of Defense, Washington, D.C. 20360, 1978.

31. Moranda, P. and Jelinski, Z., Final Report on Software Reliability Study, McDonnell Douglas Astronautics Company, MDC Report No. 63921, 1972.

32. Musa, J., Iannino, A., and Okumuto, K., *Software Reliability Measurement, Prediction, Application*, New York: McGraw-Hill, 1987.

33. Musa, J.D. and Okumoto, K., A Logarithmic Poisson Execution Time Model for Software Reliability Measurement, *Proceedings of the 7th International Conference on Software Engineering*, 1984, 230–238.

34. Nelson, E., Estimating Software Reliability from Test Data, *Microelectronics and Reliability*, 17, 67–73, 1978.

35. Neufelder, A., *Ensuring Software Reliability*, New York: Marcel Dekker, 1993.

36. Ohba, M., Software reliability analysis models, *IBM Journal of Research and Development*, 28, 428–443, 1984.
37. Pham, L. and Pham, H., Software Reliability Models with Time-Dependent Hazard Function Based on Bayesian Approach, *IEEE Transactions on Systems Man and Cybernetics. Part A—Systems and Humans*, 30, 25–35, 2000.
38. Pittman, M., Lessons Learned in Managing Object-Oriented Development, *IEEE Software*, 1992, 43–53, 1992.
39. Rees, K., Coolen, F., Goldstein, M., and Wooff, D., Managing the Uncertainties of Software Testing: A Bayesian Approach, *Quality and Reliability Engineering International*, 17, 191–203, 2001.
40. Schick, G.J. and Wolverton, R.W., An Analysis of Competing Software Reliability Models, *IEEE Transactions on Software Engineering*, SE-4, 104–120, 1978.
41. Shanthikumar, J.G., A General Software Reliability Model for Performance Prediction, *Microelectronics and Reliability*, 21, 671–682, 1981.
42. Takeuchi, H. and Nonaka, I., The New New Product Development Game, *Harvard Business Review*, January–February 1986.
43. Tartalja, I. and Milutinovic, V., Classifying Software-Based Cache Coherence Solutions, *IEEE Software*, 1997, p. 90–101.
44. Teng, X.L. and Pham, H., A New Methodology for Predicting Software Reliability in the Random Field Environments, *IEEE Transactions on Reliability*, 55(3), 458–468, 2006.
45. Whittaker, J., Rekab, K., and Thomason, M.G., A Markov Chain Model for Predicting the Reliability of Multi-build Software, *Information and Software Technology*, 42, 889–894, 2000.
46. Wood, A., Predicting Software Reliability, *IEEE Software*, 1996, p. 69–77.
47. Xia, G., Zeephongsekul, P., and Kumar, S., Optimal Software Release Policies for Models Incorporating Learning in Testing, *Asia Pacific Journal of Operational Research*, 9, 221–234, 1992.
48. Xie, M. and Yang, B., Optimal Testing-Time Allocation for Modular Systems, *International Journal of Quality & Reliability Management*, 18, 854–863, 2001.
49. Xie, M. and Yang, B., Software Reliability Modeling and Analysis, in *Encyclopedia of Statistics in Quality and Reliability*, Ruggeri, F., Kenett, R.S., and Faltin, F., Editors in Chief, New York: Wiley, 2007.
50. Xie, M. and Zhao, M., On Some Reliability Growth Models with Graphical Interpretations, *Microelectronics and Reliability*, 33, 149–167, 1993.
51. Xie, M., Dai, Y.S., and Poh, K.L., *Computing Systems Reliability*, Boston: Kluwer Academic, 2004.
52. Xie, M., *Software Reliability Modeling*, Singapore: World Scientific Publisher, 1991.
53. Yamada, S., Hishitani, J., and Osaki, S., Software Reliability Growth with a Weibull Test-Effort: A Model Application, *IEEE Transactions on Reliability*, 42, 100–105, 1993.
54. Yamada, S., Ohba, M., and Osaki, S., S-Shaped Software Reliability Growth Models and Their Applications, *IEEE Transactions on Reliability*, R-33, 289–292, 1984.
55. Yang, B. and Xie, M., A Study of Operational and Testing Reliability in Software Reliability Analysis, *Reliability Engineering & System Safety*, 70, 323–329, 2000.
56. Zhang, X.M. and Pham, H., A Software Cost Model With Error Removal Times and Risk Costs, *International Journal of Systems Science*, 29, 435–442, 1998.

Chapter 7

System and Software Beta and Usability Programs

Synopsis

This chapter deals with the design and implementation of beta testing, usability analysis, and readiness assessments. A key objective of these efforts is to establish the level of readiness of a system or software product, prior to official release. A development organization can follow a very good process faithfully and still field a system that doesn't quite meet the user's needs. When this results, it's usually because of an inability of the user to adequately articulate the real requirements for the system. Even users who generally do a good job of articulating requirements will sometimes fall prey to IKIWISI (I'll know it when I see it) because of the complexity or uniqueness of the new system. Beta testing and usability analysis are, in a way, the final proof of whether the process has worked properly.

The beta process consists of three main phases: (1) planning, (2) monitoring, and (3) data analysis. The analysis of beta testing data is the prime source that information management relies on to decide on the release of a product. Monitoring the post-introduction phase (after release) is also obviously very important. We discuss monitoring this phase with a focus on usability. This chapter also discusses basic elements of risk management as a methodology used to assess the readiness status of a system or software product. We begin the chapter with a section on the planning of a beta test.

7.1 Planning Beta Programs

7.1.1 Introduction to Beta Planning

Planning a beta test is planning an experiment. The purpose of the beta test is to test the product in a customer environment and learn about the maturity level of the product or software version. A beta test consists of three main phases—(1) planning, (2) monitoring during beta execution, and (3) analysis—in order to present recommendations and results to the Readiness Review Board. In this section we focus on the planning phase.

In planning a beta test, one should consider the following questions:

1. How do you determine potential customers or users of the product under test?
2. How do you segment/classify these customers?
3. How do you determine who will be asked to participate in the beta test?
4. How do you set expectations with these potential participants?
5. What is the acceptance rate (customers who agreed to participate)/(customers approached)?
6. What is the compliance rate (customers who comply with the beta procedures)/(participating customers)?
7. How do you determine for how long to run the beta test?
8. Did you plan the beta test logistics for equipment supply and monitoring of the beta test?
9. How do you plan the reporting processes, both internal (to engineering, marketing, sales, quality, and R&D) and external (to the beta site customer, to other participating customers, to others)?
10. How do you determine the beta test objectives?
11. How do you set the beta test success criteria?
12. How do you design the beta test survey questionnaire?
13. How do you analyze beta test data from the survey and tracking systems?
14. Who will assess results at the end of a beta test?
15. Did you determine risks of the beta test and the product, their frequency and impact?
16. How do you present results to the Readiness Review Board?

Statistically designed experiments have been used to accelerate learning since their introduction by R.A. Fisher in the first half of the twentieth century. Fisher's 1935 book on *The Design of Experiments* [8] revolutionized agricultural research by providing methods, now used in all disciplines all over the world, for evaluating the results of small sample experiments. One of Fisher's major contributions was the development of factorial experiments, which can simultaneously study several factors. Such experiments ran completely counter to the common wisdom of isolating one factor at a time, with all other factors held constant. Quoting Fisher [9]:

No aphorism is more frequently repeated in connection with field trials, than that we must ask Nature few questions, or, ideally, one question, at a time. The writer is convinced that this view is wholly mistaken. Nature, he suggests, will best respond to a logical and carefully thought out questionnaire. A factorial design allows the effect of several factors and interactions between them, to be determined with the same number of trials as are necessary to determine any one of the effects by itself with the same degree of accuracy.

In this section we show how to consider beta tests as experiments designed to effectively capture data for deciding on product readiness. For a modern comprehensive treatment of experimental design, see [22, 23].

The first step in planning any experiment is to clearly define its goals. The next step is to determine how to assess performance. The third step is to list the controllable factors that may affect performance. It is very useful to carry out these steps in a "brainstorming" session, bringing together individuals who are familiar with different aspects of the process under study. Typically the list of potential factors will be long and often there will be disagreements about which factors are most important. The experimental team needs to decide which factors will be systematically varied in an experiment, and how to treat the ones that will be left out. For each factor that is included in an experiment, the team must decide on its experimental range. Some factors may be difficult or impossible to control during actual production or use, but can be controlled for the purpose of an experiment. Such factors are called noise factors, and it can be beneficial to include them in the experiment. In general, for an experiment to be considered statistically designed, the nine issues in Table 7.1 need to be addressed. Some of these issues, such as points 6 and 7 on determination of the experimental array and number of replications, are purely statistical in nature. Some other points deal with basic problem-solving and management aspects such as the statement of the problem (point 1) and budget control (point 9).

For example, consider a beta test of a product where we would like to characterize the effect of workload, type of customer, and technological sophistication of the customer on the way our system is used. After segmenting customers that are potential participants in the beta test, we identify three control factors that characterize customers:

1. *Workload*: high workload (HW) versus low workload (LW).
2. *Size*: small and medium enterprises (SME) versus large companies (LC).
3. *Usage*: basic (B) versus sophisticated (S).

A Full Factorial Experiment consists of all factor level combinations. In our example this consists of the eight groups of customers characterized in Table 7.2.

In designing such a beta test we can learn, with information from the eight groups of participating customers, about the effect of workload, company size,

Table 7.1   Issues for Consideration in Statistically Designed Experiments

	Issue	*Description*
1	Problem Definition	Plain language description of the problem, how it was identified, who is in charge of the process/product under consideration.
2	Response Variable(s)	What will be measured and with what device? How will the data be collected?
3	Control Factors	What factors can be controlled by the operator or by design decisions?
4	Factor Levels	What are the current factor levels and what are reasonable alternatives?
5	Noise Factors	What factors cannot be controlled during regular operations but could be controlled in an experiment, and what factors simply cannot be controlled but could be observed and recorded?
6	Experimental Array	What main effects and interactions are of interest? What is the total practically possible number of runs? What is the detection power required against what alternatives? What are the experimental runs?
7	Number of Replications and Order of Experimentation	What is the practically possible number of replications per experimental run? In what order the experimental runs will be conducted?
8	Data Analysis	How will the data be analyzed? Who will write the final report?
9	Budget and Project Control	Who is in charge of the project? What the allocated budget is? What is the timetable? What resources are needed?

and usage sophistication on the product performance, and possible interactions between these factors. This is simply achieved by comparing the results achieved in the different levels of the factors. For example, comparing results in groups 1 through 4 with results in groups 5 through 8 allows us to determine the impact of usage sophistication on the reliability of the system. At the limit, we can have one customer per group. This, however, will not provide us with information on possible variability in responses within a specific group of customers.

Table 7.2 Full Factorial Beta Test Experiment

Group	Workload	Size	Usage
1	HW	SME	B
2	LW	SME	B
3	HW	LC	B
4	LW	LC	B
5	HW	SME	S
6	LW	SME	S
7	HW	LC	S
8	LW	LC	S

The statistical theory of experimental design allows us to consider smaller experiments without losing critical information. A Fractional Factorial Experiment consists of a statistically designed subset of all factor level combinations. An example of such an experiment with four groups of customers is presented in Table 7.3.

By implementing the beta test in groups 2, 3, 5, and 8 only, we still maintain the ability to compare evenly the effect of the three factors at the two levels under consideration. For example, we can learn about the impact of workload on product reliability by comparing groups 2 and 8 with groups 3 and 5. In fact, we are able to do learn from such a fractional factorial beta test design almost as much as from a full factorial experiment, for half the budget. What is lost by running the smaller

Table 7.3 Fractional Factorial Beta Test Experiment

Group	Workload	Size	Usage
2	LW	SME	B
3	HW	LC	B
5	HW	SME	S
8	LW	LC	S

beta test is the ability to measure high-level interactions. We also need to account for some confounding effects, which requires careful analysis. For more on statistically designed experiments, see [23].

Full factorial and fractional factorial designs are used in development, beta tests, and general marketing applications. Suppose your development organization designed two types of service contracts to complement the product under beta test. We call them *Standard* and *Platinum*. We also consider two levels of annual fees for these types of service. The cost levels are $4000 and $7000, respectively, per year for an annual service contract. If cost is the only consideration, then the choice is clear: the lower-priced service is preferable. What if the only consideration in buying an annual service agreement is regular (within 1 day) or fast service (within 4 hours)? If service response time is the only consideration, then you would probably prefer a fast service with a promised response time of technicians within 4 hours. Finally, suppose you can take either a service contract with access to the technician's hotline only between 9:00 a.m. and 5:00 p.m., or buy a service level with full-time accessibility, 24 hours a day, 7 days a week? Virtually everyone would prefer the full-time accessibility. In a real purchase situation, however, customers do not make choices based on a single attribute such as price, response time, or access to technicians. Rather, customers examine a range of features or attributes and then make judgments or trade-offs to determine their final purchase choice. Marketing experts call the application of factorial designs to these types of problems "conjoint analysis." Conjoint analysis examines feature trade-offs to determine the combination of attributes that will be most satisfying to the customer. In other words, by using conjoint analysis, a company can determine the optimal features for its product or service. In addition, conjoint analysis will identify the best advertising message by identifying the features that are most important in product choice. Conjoint analysis therefore presents choice alternatives between products/services defined by sets of attributes. This is illustrated by the following choice: Would you prefer a service agreement with 1-day service that costs $4000 and gives you access to the service technicians only during working hours, or a service agreement that costs $7000, has 24/7 coverage, and takes a maximum of 4 hours for the technician to arrive? Extending this, there are potentially eight types of service (see Table 7.4).

Given the above alternatives, service type 4 is very likely the most preferred choice while service type 5 is probably the least preferred product. How the other choices are ranked is determined by what is important to that individual. Conjoint analysis can be used to determine the relative importance of each attribute, attribute level, and combinations of attributes. If the most preferable product is not feasible for some reason, then the conjoint analysis will identify the *next* most preferred alternative. If you have other information on customers, such as background demographics, you might be able to identify market segments for which distinct products may be appealing. In evaluating products, consumers will always make trade-offs. For more on conjoint analysis, see Green et al. [10, 11]; Gustaffson et al. [12]; and Rossi et al. [26].

Table 7.4 Factorial Design Used in Conjoint Analysis

Choice	Access to Technicians	Price	Maximum Response Time
1	24/7	$7000	1 day
2	24/7	$7000	4 hours
3	24/7	$4000	1 day
4	24/7	$4000	4 hours
5	9:00–17:00	$7000	1 day
6	9:00–17:00	$7000	4 hours
7	9:00–17:00	$4000	1 day
8	9:00–17:00	$4000	4 hours

Once the customers we would like to have in a specific beta test are chosen, we need to communicate to them our intention and properly set up expectations. In the next subsection we present a generic set of Standard Operating Procedures (SOPs) for conducting beta tests that cover all aspects of organizing and conducting a beta test.

7.1.2 Beta Test Standard Operating Procedure

General: Every new product or a major change in a current product requires a beta test. This applies to systems and major software releases. The beta test is the final test in a sequence of tests, and it is held at a user's site in order to have a bias-free, objective judgment of the system's performance in its targeted work environment. The beta test results provide information regarding the system/change maturity and its readiness.

Purpose: The purpose of a Beta Test Standard Operating Procedure is to define and assign all activities and responsibilities regarding the beta process. Each activity that is related to the beta process should be defined, and assigned to a single predefined owner who is responsible for his/her part in the process. In some cases, where specific activities may not be relevant, the beta test procedure can be tailored by the project manager. The revised procedure needs to be approved by the Readiness Review Board (RRB).

Method:

- Define all activities related to the process.
- Assign ownership for each activity.

- Monitor all activities by periodic meetings.
- Define a progress-tracking mechanism.

Concept and Ownership: The procedure is divided into activities and subactivities. Each activity's time frame is defined, as well as its owner. The owner has responsibility to define his/her subactivities schedules and obtain the commitment to complete them. The general owner of the beta process is the project manager.

Initiation: The implementation of this procedure occurs immediately after approval by QA.

Distribution: Marketing, QA, R&D, Product Engineering, Customer Support, Manufacturing, Logistics.

The following are descriptions of the various beta test activities, at each stage of a system development project, and a definition of the role responsible for this activity:

A. *Concept Phase:* Definition of the product success criteria is needed at the beginning of the product life in order to evaluate the design and performance relative to the customers' needs.
   Deliverables:
   - General specification and beta test success criteria—to be validated by marketing

B. *R&D Test Phase*

B1. *Beta Requirements—Project Manager:* Before beginning the formal beta test, a dry run of the beta test is performed in-house, and is referred to as an R&D test. During the lab test phase, the project manger has to define the following activities and schedule. The checklist below must be completed before the beginning of the alpha test.
   Deliverables:
   - Beta package (description of the system configuration and the participating beta test customers)
   - Test requirements
   - Test success criteria
   - Timetables for activities (e.g., a Gantt chart defining when each activity is going to take place)
   - Beta customers characterization
   - Number of units and geographical preferences allocated to the beta test
   - Manufacturing plan for beta systems
   - Monitoring process and report formats

B2. *Customer Support Plan—Customer Support:* Customer support is responsible for planning the beta test, preparing the support documentation at site level, and training the customers. The customer support planning activity should be complete at the end of the lab test phase.
   Deliverables:
   - Customer support plan for beta program

- Installation plan
- Site support documentation definition
- Definition of support needed from R&D during the beta test
- Training courses plan
- Spare parts requirements

B3. *Update Meetings (ongoing)—Project Manager:* Coordination and update meetings should be held by the project manager, to keep all participants updated regarding the project progress and get feedback.

Deliverables:
- Periodic update meetings—project manager
- Data transfer to the field—product launch manager

C. *Alpha Phase*

C1. *System Production—Operations:* The purpose of this activity is to produce the beta systems and prepare the systems for shipment. The performer of this activity is operations. The activity should be completed by the end of the alpha phase. In cases of software or options that do not require manufacturing activities, the performer will be the project manager.

Deliverables:
- Bill of Materials (BOM) ready—engineering
- Beta systems production—operations
- Acceptance test procedures (ATP) preparation—operations
- Packaging—operations
- Spare parts production—operations

C2. *Training and Documentation—Customer Support:* Prior to the system shipment to customers for beta testing, all documentation should be ready, service engineers should be trained to operate the new system/feature, customer training packages should be ready, spare parts should be in place, and the beta test site must be ready for the test. These activities take place during the alpha phase, and the owner is customer support.

Deliverables:
- Product's service policy
- Assign owners for test running in the field
- Site preparation guide
- Installation guide
- Operator manual
- Service manual
- Letter to the customer
- Feedback form and beta test survey questionnaire
- Other documents, if needed
- Spare parts catalog
- Training course for customers
- Other training, if needed
- Spare parts distribution

C3. *Beta Agreement—Marketing :* The activity of identifying proper customers for the beta test and signing agreements with them is led by marketing.
Deliverables:
- Identify potential customers that meet requirements
- Preparation of a beta agreement
- Beta agreement signed
- Pricing for beta

C4. *Shipments and Logistics—Sales Administration:* Before the beta test starts, all systems should arrive at their destination, as well as all supporting materials like spare parts, documentation, manuals, etc.
Deliverables:
- System shipment
- Spare parts shipment
- Other supporting material like documentation

D. *Beta Test Phase:* During the beta test phase, the activities listed below should take place. These activities include the installation of system/software change, running the tests, and feedback reports.

D1. *Beta Installation and Support—R&D:* Customer support is the leader of the following activities, which include installing the system and getting them ready to run the tests.
Deliverables:
- Initial installation of the system or software or upgrades to them at the customer's site
- Installation reports
- Running the tests
- Routine service

D2. *Reports—Customer Support:* During the test, periodic and final reports are needed in order to evaluate performance. The data should be processed under the format supplied by the project manager.
Deliverables:
- Periodic reports
- Summary report

D3. *Beta Results Analysis—R&D:* After data analysis, the project manager should draw conclusions and present his final recommendation to the Readiness Review Board (RRB) for final approval, and get a "green light" to start production.
Deliverables:
- Conclusions and recommendation
- RRB approval

E. *Final Product Approval—RRB—Project Manager:* Approve the product at the RRB and release the product.
Deliverables:
- RRB decision

In the next section we discuss the tools and procedures used to track and assess the outcomes of a beta test.

7.2 Beta Test Data Collection and Analysis

7.2.1 Beta Survey Design

A main goal of a beta test is to gather information on the new or updated product that represents actual customer experience. Some of the information is available in tracking and remote diagnostics monitoring systems. More information is collected in customer relationship management (CRM) systems documenting interactions between the customers participating in the beta test and technicians or other support personnel. In addition to these, one can use standard survey techniques such as Internet-based or paper surveys to gather impressions and perceptions. In many cases these are self-reports where representatives of the customer participating in the beta test are asked to complete a questionnaire. Self-reports, however, are a potentially unsafe source of data, and changes in question wording, question format, or question context can result in changes in the obtained results, as a few examples from social science research illustrate:

- When asked what they consider "the most important thing for children to prepare them for life," 61.5% of a representative sample chose the alternative "To think for themselves" when this alternative was offered on a list. Yet, only 4.6% volunteered an answer that could be assigned to this category when no list was presented.
- When asked how successful they have been in life, 34% of a representative sample reported high success when the numeric values of the rating scale ranged from –5 to +5, whereas only 13% did so when the numeric values ranged from 0 to 10.
- When asked how often they experience a variety of physical symptoms, 62% of a sample of psychosomatic patients reported symptom frequencies of more than twice a month when the response scale ranged from "twice a month or less" to "several times a day." Yet, only 39% reported frequencies of more than twice a month when the scale ranged from "never" to "more than twice a month."

The underlying cognitive and communicative processes causing such effects are systematic and increasingly well understood. Since the early 1980s, psychologists and survey methodologists have developed an interdisciplinary field of research devoted to understanding the nature of self-reports and to improving the quality of data collection. Research in this field has addressed a wide range of topics: How do respondents make sense of the questions asked of them? What is the role of autobiographical memory in retrospective reports of behaviors, and how can we increase the accuracy of these reports? What are the judgmental processes underlying the emergence of context effects in attitude measurement? Do the processes underlying self-reports of behaviors and attitudes change across the adult life span? Which techniques can we use to determine if a question "works" as intended? This area of research is typically referred to as "cognitive aspects of survey methodology"; the cognitive and communicative processes investigated apply to the question–answer process in all standardized research situations. For more on such aspects of survey design and a comprehensive bibliography, see [18].

Not surprisingly, the first task that respondents face is to understand the question asked of them. The key issue is whether the respondent's understanding of the question matches what the survey designer had in mind: Is the attitude object, or the behavior, that the respondent identifies as the referent of the question the one that the researcher intended? Does the respondent's understanding tap the same facet of the issue and the same evaluative dimension?

From a psychological point of view, question comprehension reflects the operation of two intertwined processes of semantic and contextual understanding. Comprehending the literal meaning of a sentence involves the identification of words, the recall of lexical information from semantic memory, and the construction of a meaning of the utterance, which is constrained by its context. Not surprisingly, textbooks urge researchers to write simple questions and to avoid unfamiliar or ambiguous terms. However, understanding the words is not sufficient to answer a question. For example, if respondents are asked, "How many cases of part replacements did you experience during the beta test?", they are likely to understand the meaning of the words and might even go to some logbook to check the records. Yet, they still need to determine what kinds of parts are involved. Should they report, for example, that they took parts of the system under beta test because of some shortage in similar parts in another working system or not? Hence, understanding a question in a way that allows an appropriate answer requires not only an understanding of the literal meaning of the question, but also the inferences about the questioner's intention to determine the pragmatic meaning of the question. Suppose that respondents are asked how frequently they made complaints to the beta test call center service about the new system. To provide an informative answer, respondents have to determine what is meant by "complaint." Does this term refer to major or to minor annoyances? To identify the intended meaning of the question, they may consult the response alternatives provided by the researcher. If the response alternatives present low-frequency categories, for example, ranging from "less than once a year" to "more than once a month," respondents may conclude that the researcher has relatively rare events in mind. Hence, the question cannot refer to minor complaints that are likely to occur more often, so the researcher is probably interested in more severe complaints.

In specifying a scale (in %) of:

Very Low	Low	Average	High	Very High
5	25	50	75	95

as possible responses to the question: "In how many problems did you use the diagnostics tools?", we imply that we are interested in all types of problems, minor or major. Should the scale have been from 1% very low to 5% very high, the implication would have been that the question addresses only severe problems, not minor cosmetic annoyances. Figure 7.1 represents a beta test survey questionnaire appropriate for a fictitious system combining software with electronic and mechanical parts that we call *System 1950 Update*.

All customers participating in the *System 1950 Update* beta test received (1 week before it was launched) a letter announcing the survey. The survey itself was conducted over the Internet. All customers participating in the beta test knew about such a survey from the very beginning, including the fact that they were expected to complete it, and guidance on who should do that. Typically, we will look for inputs from various people at the customer's site. This could include decision makers who make financial and technical decisions, operators, technicians, sales and marketing experts who have to promote services offered by the new system to the customer's customers, etc. In the next subsection we discuss how to analyze data collected through such questionnaires.

7.2.2 Beta Test Survey Data Analysis

Survey analysis typically involves a descriptive study of demographic parameters and satisfaction levels from various items. In this subsection we present several powerful nonstandard statistical methods for survey data analysis. Standard survey data analysis uses bar charts and averages. Once questionnaires are returned, an analysis of possible bias in the returned questionnaires is necessary. Such bias can be evaluated using the M-test presented in Chapter 6 (Section 6.2.6) to compare the distribution of bug severity reports across system releases. Here we use the M-test to determine if the responses received from the survey are representative of the group of customers we decided to approach in the beta test. In this context, the M-test consists of comparing the number of expected returns, by demographic strata, conditioned on the total number of returns. With this approach we can determine if there are groups with under(over)-representation in the responses. If bias is identified, a follow-up analysis is carried out to determine if there are significant differences in satisfaction levels between the various groups. If such differences are demonstrated, weighted expansion estimators are used to determine overall satisfaction levels; if not, unweighted estimators are computed (see [23]). For example, consider a beta test performed at eight customer sites where customers have been chosen to represent the four groups specified in Table 7.3, two customers per group. At each

System 1950 Update Beta Customer Survey

Dear customer,

As part of the Beta test that your company is taking part of I would like to ask you some questions regarding the System 1950 Update. We value your feedback and intend to use it as the basis for future improvements of the hardware and software. Please answer the following questions, which should take an estimated 30 minutes of your time.

Before starting with the questions let me explain that on this survey you will be asked about the various new features of the System 1950 Update. The questionnaire's objective is to hear your opinion regarding the improvements made in the System 1950 Update compared to the System 1950. The questionnaire will focus on your level of satisfaction from the improvement in your productivity, print quality and profitability generated by the new features.

In order to answer these questions please state the number which best reflects how satisfied you feel about the subject asked, on a scale from 1 to 5 when:

5 means Very high satisfaction level
4 means High satisfaction level
3 means Average satisfaction level
2 means Low satisfaction level
1 means Very Low satisfaction level

And 0 means Not relevant- you did not use the feature

Thank you in advance for your cooperation

John Doe
Technical Marketing
Systems Inc.

#	Questions	Responses					
1	What is your company name?						
2	What is your name?						
3	What is your title? Technical, Supervisor, plant manager, other (please specify)	1. Operator	2. Supervisor	3. Manager	4. Other:		
	Site Preparation	Very Low	Low	Average	High	Very High	Not relevant
4	Your satisfaction level with the site preparation instructions	1	2	3	4	5	0
5	Your satisfaction level with the site preparation timing	1	2	3	4	5	0
6	Your satisfaction level with the site preparation requirements	1	2	3	4	5	0
	Installation	Very Low	Low	Average	High	Very High	Not relevant
7	Your satisfaction level with the shipment timing	1	2	3	4	5	0
8	Your satisfaction level with the installation length	1	2	3	4	5	0
9	Your satisfaction level with the cleanliness of equipment after installation	1	2	3	4	5	0
	Training	Very Low	Low	Average	High	Very High	Not relevant
10	Your satisfaction level with the operator training	1	2	3	4	5	0
11	Your satisfaction level with the technical training	1	2	3	4	5	0
12	Your satisfaction level with the on site training	1	2	3	4	5	0
	Production ramp up	Very Low	Low	Average	High	Very High	Not relevant
13	What was the expected usage of system 1950 Update	1	2	3	4	5	0
14	Actual usage of system 1950 Update during beta test	1	2	3	4	5	0
15	What is the billable usage during the beta test? (%)	5	25	50	75	95	0
	Diagnostics	Very Low	Low	Average	High	Very High	Not relevant
16	In how many problems did you use the diagnostics tools (%)	5	25	50	75	95	0
17	In how many of these cases did the diagnostics tool help resolve the problem? (%)	5	25	50	75	95	0
	Reliability and maintenance	Very Low	Low	Average	High	Very High	Not relevant
18	How many cases of part replacements did you experience during the beta test?	1	5	10	15	20	0
19	Your satisfaction level with component A replacement procedure	1	2	3	4	5	0
20	Your satisfaction level with component B replacement procedure	1	2	3	4	5	0
21	Your satisfaction level with component C replacement procedure	1	2	3	4	5	0
22	Your satisfaction level of other replacements and handling procedures	1	2	3	4	5	0
23	Comments:						
	Overall productivity	Very Low	Low	Average	High	Very High	Not relevant
24	Rate the overall productivity od System 1950 Update	1	2	3	4	5	0
25	Rate the failure rate in the System 1950 Update	1	2	3	4	5	0
26	Rate the improvement in supplies that can be used in the System 1950 Update	1	2	3	4	5	0
27	Rate the improvement in the System 1950 Update utilization level	1	2	3	4	5	0
28	Rate the performance of system user interface	1	2	3	4	5	0
29	Rate the ease-of-use when fixing a technical problem	1	2	3	4	5	0
	Software	Very Low	Low	Average	High	Very High	Not relevant
30	The number of restarts needed when using the System 1950 Update was	5	25	50	75	95	0
31	Comments on main software restart cause:						
32	Rate the user inteface connectivity	1	2	3	4	5	0
33	Comments on user interface connectivity:						
	Overall satisfaction level	Very Low	Low	Average	High	Very High	Not relevant
34	Your satisfaction level of improvements in the quality of the System 1950 Update	1	2	3	4	5	0
35	Your satisfaction of improvements in the ease-of-use of the System 1950 Update	1	2	3	4	5	0
36	**What is your overall satisfaction level from the System 1950 Update?**	1	2	3	4	5	0
37	If you were in the market to buy a system, with what likelihood would you purchase the System 1950 Update?	1	2	3	4	5	0
38	**General comments on the System 1950 Update:**						

Figure 7.1 Sample beta test questionnaire.

customer site, three roles of respondents have been identified: (1) decision makers (D), (2) technical experts (T), and (3) operators (O). At each customer site, three respondents of each type have been asked to complete the survey, for a total of 72 questionnaires to be completed. Table 7.5 presents the number of questionnaires sent out (72), the number of responses (61), and the M-test Z scores.

Table 7.5 shows Z scores computed with the M-test with respect to both the groups (2, 3, 5, and 8 defined in Table 7.3) and the respondents' roles (D, T, and O). Because all Z scores are below the critical value, we can conclude that the 61 returned questionnaires are representative of the customer groups and respondents' types so that we can go ahead and perform the data analysis without the need to weight the results. Once we determined that the responses represent the intended populations, we can proceed with the data analysis of the raw data collected in the survey.

Take, for example, question 36 on the overall satisfaction from *System 1950 Update*. We received 61 responses. Figure 7.2 presents the results using a basic bar chart.

We see that 24 respondents rated the system "5"—Very High Satisfaction (39%), and 5 respondents rated the system as "1" or "2" (8%). The 24 very satisfied respondents were: 7 decision makers, 8 technical experts, and 9 operators, out of 18, 22, and 21 respondents, respectively. These 24 respondents can also be split by customer groups: 7 came from group 2, 4 from group 3, 6 from group 5, and 7 from group 8 out of 16, 15, 18, and 12 respondents, respectively.

A powerful technique for questionnaire analysis involves control charts. This nonstandard application of control charts to the analysis of satisfaction surveys data is described in [18, 23]. In Chapter 4 (Section 4.5.1) and Chapter 6 (Section 6.2.5), control charts were used to identify significant changes over time. For analyzing survey data, we use a control chart designed for comparing proportions—the p-chart. In this type of chart, we analyze observations of a binary event that follow a binomial distribution. In a beta test survey, we want to analyze, for example, the proportion of responses that correspond to the top evaluation rating, a "5" on a 1 to 5 scale. (Similar analysis can be done for the bottom part of the scale, a "1" or "2.") These responses correspond to very satisfied customers that give the top rating to questions on satisfaction levels from specific product features and dimensions of service, and represent a measure of success of the product under test. A customer responding "5" to a question on overall satisfaction is typically a loyal customer. A response of "4" indicates mere satisfaction and implies significantly reduced loyalty levels.

Figure 7.3 presents proportions of "5" to the 16 questions rating satisfaction levels. The average proportion is 50.28%. The formulae for computing control limits for a proportion derived from n_i responses, where the average proportion of "5" over all questions is 50.28%, are:

The varying number of responses, per question, is reflected by varying control limits. The more responses, the narrower the control limits. The control limits delineate three regions: significantly above average, as expected by the grand average, and significantly below average. Points above or below the control limits indicate proportions significantly higher, or lower, than the grand average. The

Table 7.5   Distributed and Actual Number of Returned Questionnaires

Group	Workload	Size	Usage	D	D Actual	T	T Actual	O	O Actual	Z
2	LW	SME	B	6	5	6	5	6	6	0.22
3	HW	LC	B	6	3	6	6	6	6	−0.07
5	HW	SME	S	6	6	6	6	6	6	0.81
8	LW	LC	S	6	4	6	5	6	3	−0.96
				Z	−0.63		0.45		0.18	

Note: D = decision makers; T = technical experts; O = operators.

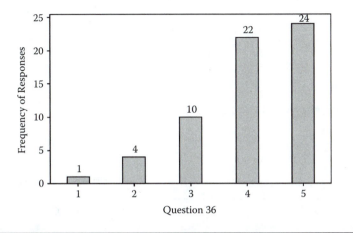

Figure 7.2 **Bar chart of responses to question 36 (overall satisfaction from system).**

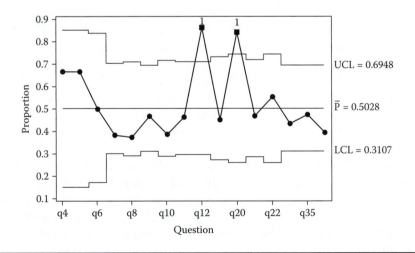

Figure 7.3 **Control chart for analyzing proportions of "5" in beta test questionnaire data.**

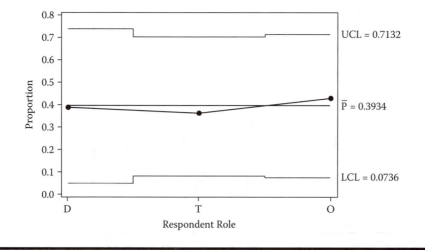

Figure 7.4 Control chart for analyzing proportions of "5" in overall satisfaction, by role.

probability of making such a statement when it should not be made (a Type I error) is about 1/370—a rather low risk. We can now use the p-chart to determine if there are differences between the various groups included in the beta test survey.

In the control chart in Figure 7.3, questions 12 ("Your satisfaction level with the on-site training") and question 20 ("Your satisfaction level with component B replacement procedure") have an unusually high proportion of "5"s, indicating outstanding strengths. All other questions provide responses consistent with the grand average—that is, not worth pointing out as unusual or significantly high. Figure 7.4 and Figure 7.5 present p-charts by percentage of "5" responses, by the respondent role, and by the customer group as defined in Table 7.3. We can see from these control charts that there was no significant difference in overall satisfaction level "5" between customer groups and respondents with various roles (decision makers (D), technical experts (T), and operators (O)).

The percentage of "5"s in the four groups presented in Figure 7.4 can be analyzed in more depth, because of the fractional factorial design that led to the choice of these groups. In fact, these groups were defined by three dimensions: workload, usage, and size (see Table 7.3).

Figure 7.6 presents a three-dimensional cube with the percents of "5" in overall satisfaction. We can see visually that lower workload improves satisfaction. The average results for high workload were 27% and 33%, and for low workload, 44% and 58%. Usage and company size behave in a more complicated way and seem to interact. The effect of usage depends on the company size, and vice versa.

Figure 7.7 presents main effects plots where results for each factor used to define the groups are shown at specific levels of the factors. The average results for high

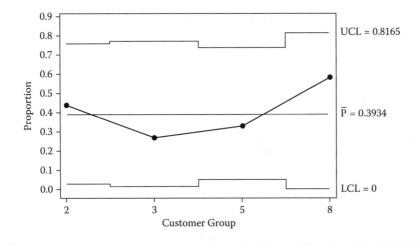

Figure 7.5   Control chart for analyzing proportions of "5" in overall satisfaction, by group.

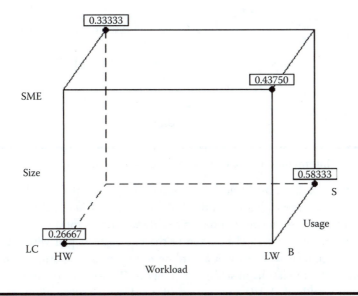

Figure 7.6   Cube plot of overall satisfaction top ratings.

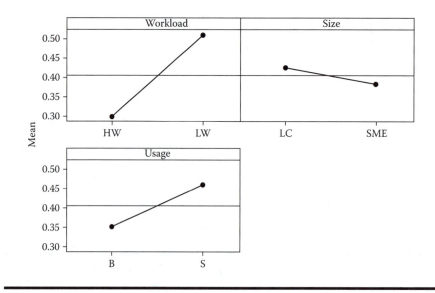

Figure 7.7 Main effects plot of overall satisfaction top ratings.

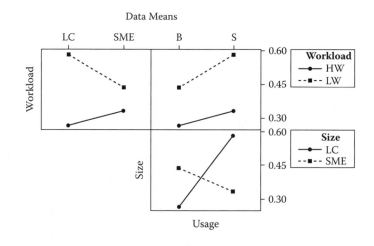

Figure 7.8 Interaction plot of overall satisfaction top ratings.

and low workloads are 30% and 51%, respectively. We see that installing the system at customer sites with low workload increases satisfaction by 21%. Finally, we present in Figure 7.8. the interaction plots of these three factors.

Interaction plots are an elaboration of the main effects plot with the average results at specific factor levels split according to the levels of another factor. Take, for example, the top left-hand side graph in Figure 7.7. Results for large companies (LC), high workload (HW) and low workload (LW) are 58% and 27%, respectively. Similarly, results for small and medium enterprises (SME), high workload

(HW) and low workload (LW) are 44% and 33%, respectively. Parallel lines indicate no interaction because changes in factor have a fixed effect, irrespective of the other factor. Company size and usage sophistication do show an interaction where sophisticated usage in large companies brings record highs in satisfaction of 58% and record lows (27%) for large companies with basic usage. All these conclusions have an obvious impact on the analysis of the beta test data and our understanding of how the system performs at various customer sites. For example, we can determine that the system will not perform adequately with large customers having only basic usage. Such customers should probably be offered a more basic system.

This subsection focused on designing, implementing and analyzing data from a beta test survey. In the next subsection we address the issue of analyzing operational data, such as information on usage being tracked by online monitoring systems.

7.2.3 Reliability Models

When manufacturers claim that their products are very reliable, they essentially mean that the products can function as required for a long period of time, when used as specified in the operating manual. In this subsection we show how to assess the reliability of a system under beta test, so that decision makers can determine if it is ready for release or reliability improvements are required. To discuss the reliability of a system we need a more formal definition of reliability. When an item stops functioning satisfactorily, it is said to have failed. The time-to-failure or lifetime of an item is intimately linked to its reliability, and this is a characteristic that will vary from system to system even if they are identical in design and structure. For example, say we have 100 identical-brand LCD displays that we plug into a test circuit, turn on simultaneously, and observe how long they last. We would not, of course, expect them all to fail at the same time. Their times to failure will differ. The random element of their failure times is captured by a random variable whose behavior can be modeled by a probability distribution. This is the basis of reliability theory (see Chapter 14 in [23]. Reliability analysis enables us to answer such questions as:

- What is the probability that a system will fail before a given time?
- What percentage of systems will last longer than a certain time?
- What is the expected lifetime of a component within a system?

In Chapter 6 we discussed software reliability models such as the Weibull distribution (Section 6.2.5) and more sophisticated models (Section 6.6 and the appendix to Chapter 6). In this subsection we discuss additional aspects of reliability analysis that can be used to analyze problem reports collected during a beta test. For more information on software failures data analysis, see [2–5, 7].

Let's begin with a simple plot of the timing of critical events such as a system failure during beta test. Figure 7.9 shows three possible scenarios of improvements,

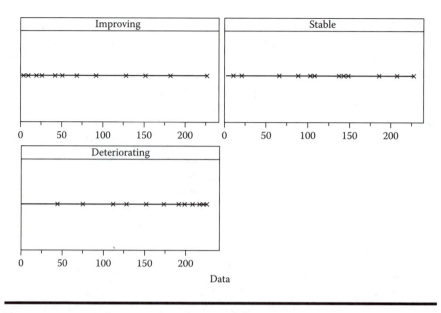

Figure 7.9 Event plot scenarios of system failures.

stability, and deterioration in such reports. Identifying the actual scenario is key to the ability to predict future behavior of the system and thus impact the decision of the RRB.

An additional tool we can use to analyze beta test data is the Duane plot (Figure 7.10). A Duane plot is a scatterplot of the cumulative number of failures at a particular time, divided by the time (cumulative failure rate), versus time. Duane plots are used to determine if a system under test is improving, deteriorating, or remaining stable. The fitted line on the Duane plot is the best fitted line when the assumption of the power-law process is valid, and the shape and scale are estimated using the least squares method.

A simple model for estimating system reliability has been proposed by Wall and Ferguson [29]. The model relies on the basic premise that the failure rate decreases with time. The relationship proposed for this phenomenon is:

$$C1 = C0\left(\frac{M1}{M0}\right)^a$$

where:

$C1$ = Future period cumulative number of errors

$C0$ = Previous period cumulative number of errors, a constant determined empirically

$M1$ = Units of test effort in future period (e.g., number of tests, testing weeks, CPU time)

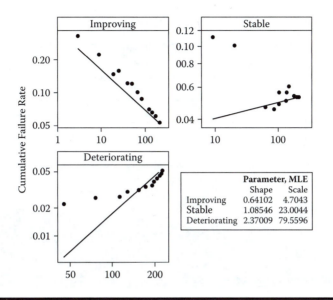

Figure 7.10   Duane plots under various scenarios of system failures.

$M0$ = Test units in previous period, a scalable constant

a = Growth index, a constant determined empirically by plotting the cumulative number of errors on a logarithmic scale or using the following formula:

$$a = \frac{\log\left(\dfrac{C1}{C0}\right)}{\log\left(\dfrac{M1}{M0}\right)}$$

The failure rate $Z(t)$ is derived from the following formula:

$$Z(t) = \frac{dC1}{dt} = aC0 \frac{d\left(\dfrac{M1}{M0}\right)}{dt}\left(\frac{M1}{M0}\right)^{a-1}$$

This model is based on several assumptions:

■ Software test domain is stable (i.e., no new features are introduced).
■ Software errors are independent.
■ Errors detected are corrected within the same test period.

Table 7.6 Predicted Failures Using the Pragmatic Model

Test Period	Observed Failures	Cumulative Failures	a	Predicted Failures
1	72	72		
2	70	142	0.8	125.36
3	57	199	0.8	196.41
4	50	249	0.8	250.50
5	47.17		0.8	297.66

Cumulative Failures

- Testing is incremental so that new tests are introduced in subsequent testing periods.

Revisiting the data presented in Table 6.11 (in Section 6.6) with an empirically derived estimate of 0.8 for *a*, we get an estimated number of failures for test period 5 of 48 failures and a predicted cumulative number of failures of 298 (Table 7.6).

In this section we discussed the design and analysis of beta test surveys and briefly covered the analysis of failure reports over time. For more details on survey design and reliability analysis, the reader is referred to [17, 18, 20, 23]. Beta testing can also be considered an exercise in risk management. The information from beta testing should provide management in general, and the Readiness Review Board in particular, with information necessary for identifying risks and allowing effective risk management. In the next section we provide an introduction to risk management with a specific application to beta testing.

7.3 Beta Test Risk and Usability Analysis

Risk occurs at all stages of a product's life cycle: initial concept formulation, product development, production, installation, beta testing, and operations and maintenance. The techniques for performing risk management are the same, no matter in which phase of the life cycle risk exists. In this section we discuss risk management as it pertains to beta testing.

7.3.1 Introduction to Risk Management

Risk management is the systematic process of identifying, analyzing, and responding to project risk. It is focused on identifying risk issues and adverse events with a significant impact on the success of a project or the launch of a product. Risk management consists of the following major activities:

- *Risk identification*: Determining which risks might affect the project and documenting their characteristics.
- *Risk analysis*: Performing a qualitative and quantitative analysis of risks and conditions to prioritize their effects on project objectives.
- *Risk response planning*: Developing procedures and techniques to mitigate identified risks.
- *Risk monitoring and control*: Monitoring residual risks, identifying new risks, and executing risk mitigation plans.

Project risk is an uncertain event or condition that, if it occurs, has a significant effect. A risk has a cause and, if it occurs, a consequence. For example, a cause may be requiring a signed legal contract for running a beta test or having limited personnel assigned to the beta test execution. The risk event is that the legal contract may take longer to sign than planned, or that the personnel supporting the beta test may not be adequate for the task. If either of these uncertain events occurs, there will be a consequence on the beta test's cost, schedule, or quality. Known risks are those that have been identified and analyzed, with adequate mitigation plans prepared for them. Unknown risks cannot be managed, although project managers may address them by applying a general contingency based on past experience with similar projects.

Organizations perceive risk as it relates to threats to project success. Risks that are threats to the beta test may be accepted if they are in balance with the reward that may be gained by taking the risk. For example, adopting a fast-track schedule that may be overrun is a risk taken to achieve an earlier completion date.

In the remainder of this section we refer to a tool called Technical Risk Identification and Mitigation (TRIMS), which was developed for suppliers of the U.S. Department of Defense (DoD) and provides a comprehensive software environment for conducting risk management [28]. In the next subsections we present an application of TRIMS to beta testing.

7.3.2 Risk Identification

Risk identification involves determining which risks might affect the proper performance of the product or system when deployed. Risk identification is an iterative process. The first iteration may be performed by marketing, engineering, service support, and the beta management team during planning of the beta test. A second iteration might be conducted during beta monitoring, and a final iteration following the beta test completion, in preparation for the presentation to the Readiness Review Board.

Often, simple and effective risk responses can be developed and even implemented as soon as the risk is identified. Risks that may affect the project for better or worse can be identified and organized into risk categories. Risk categories should be well defined and should reflect common sources of risk such as:

■ Technical, quality, or performance risks such as unrealistic performance goals
■ Project-management risks such as poor allocation of time and resources
■ Organizational risks such as cost, time, and scope objectives that are internally inconsistent
■ External risks such as changing owner priorities
■ Operational risks such as IT infrastructure failures or inadequate logistics

Risk identification efforts mostly rely on interviews, team discussions, and lessons learned from previous beta tests. Once potential risks have been identified, analysis of risks is initiated. In Section 7.1.1 we listed several questions that are used to identify typical risks of a beta test project. Such questions are related to the three separate phases beta testing: planning, follow-up, and analysis. Figure 7.11 shows the basic elements of a software beta testing project. Each element is evaluated by its current status. Elements partially or not completed, therefore, present risks to the beta test and are typically colored in yellow or red, respectively, depending on the severity of the risk, red representing the most severe risk.

7.3.3 Risk Analysis

Risk analysis is the process of assessing the impact and likelihood of identified risks. Qualitative risk analysis is an approach based on expert judgment to determine the importance of addressing specific risks and guiding risk responses. Figure 7.12 provides a sample list of questions used to assess risks in a beta testing project. The figure is a screenshot from the TRIMS tool applied to beta testing. The responses to these questions, when entered in TRIMS, generate the risk level colors. Obviously, specific projects need to tailor and adapt generic checklists of questions.

Risk probability and risk consequences may be described in qualitative terms such as very high, high, moderate, low, and very low. Risk probability is the likelihood that a risk will occur. Risk consequences (sometimes referred to as risk impact) are the effect on project objectives if the risk event occurs. These two dimensions of risk are applied to specific risk events. Analysis of risks using probability and consequences helps identify those risks that should be managed aggressively. At any given time, a matrix may be constructed that assigns risk.

Ratings are assigned (very low, low, moderate, high, and very high) to risks or conditions based on combining probability and impact scales. Risks with high

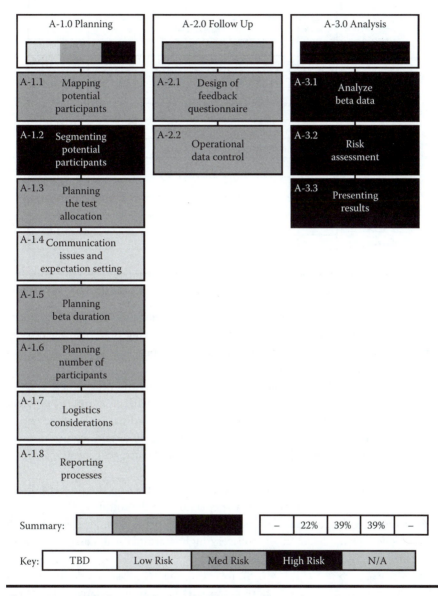

Figure 7.11   Risk elements in a typical beta testing project.

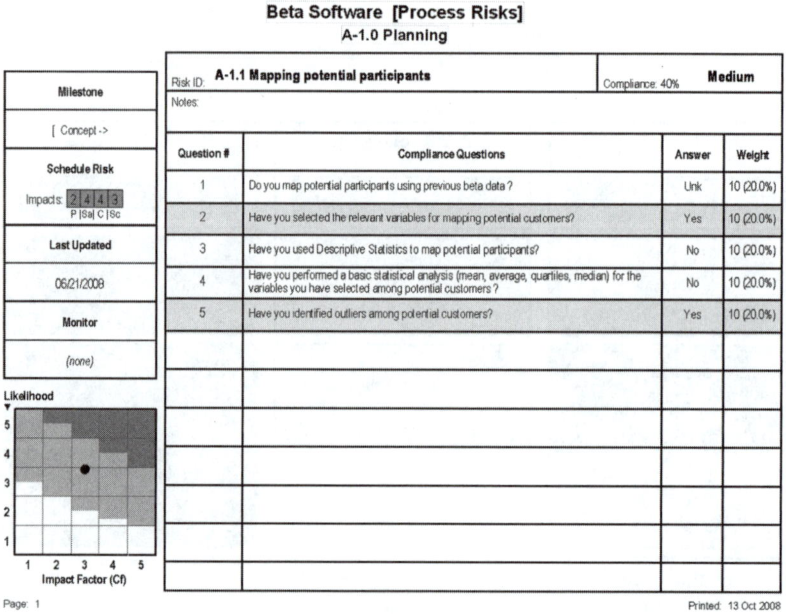

Beta Software [Process Risks]
A-1.0 Planning

Milestone	Risk ID:	A-1.1 Mapping potential participants		Compliance: 40%	**Medium**
	Notes:				
[Concept ->					

Schedule Risk	Question #	Compliance Questions	Answer	Weight
Impacts: 2 4 4 3	1	Do you map potential participants using previous beta data ?	Unk	10 (20.0%)
P |Sa| C |Sc	2	Have you selected the relevant variables for mapping potential customers?	Yes	10 (20.0%)
Last Updated	3	Have you used Descriptive Statistics to map potential participants?	No	10 (20.0%)
06/21/2008	4	Have you performed a basic statistical analysis (mean, average, quartiles, median) for the variables you have selected among potential customers?	No	10 (20.0%)
Monitor	5	Have you identified outliers among potential customers?	Yes	10 (20.0%)
(none)				

Likelihood

Impact Factor (Cf)

Page: 1
Printed: 13 Oct 2008

Figure 7.12 A sample list of questions used to assess risk elements.

probability and high impact are likely to require further analysis, including quantification, and aggressive risk management. The risk rating is accomplished using a matrix and risk scales for each risk. Figure 7.13 represents such a matrix from TRIMS assessing the current risks of an ongoing beta test.

A risk's probability scale naturally falls between 0.0 (no probability) and 1.0 (certainty). Assessing risk probability may be difficult because expert judgment is used, often without benefit of historical data. An ordinal scale, representing relative probability values from very unlikely to almost certain, could be used. Alternatively, specific probabilities could be assigned by using a general scale (e.g., .1 / .3 / .5 / .7 / .9).

The risk's impact scale reflects the severity of its effect on the project's objectives. Impact can be ordinal or cardinal, depending on the culture of the organization conducting the analysis. Ordinal scales are simply rank-ordered values, such as very low, low, moderate, high, and very high. Cardinal scales assign values to these impacts. These values are usually linear (e.g., .1 / .3 / .5 / .7 / .9), but are often nonlinear (e.g., .05 / .1 / .2 / .4 / .8), reflecting the organization's desire to avoid high-impact risks. The intent of both approaches is to assign a relative value to the impact on project objectives if the risk in question occurs. TRIMS provides four dimensions for assessing consequences, or impact: time, quality, cost, and safety. The organization must determine which combinations of probability and impact result in a risk's being classified as high risk (red condition), moderate risk (yellow

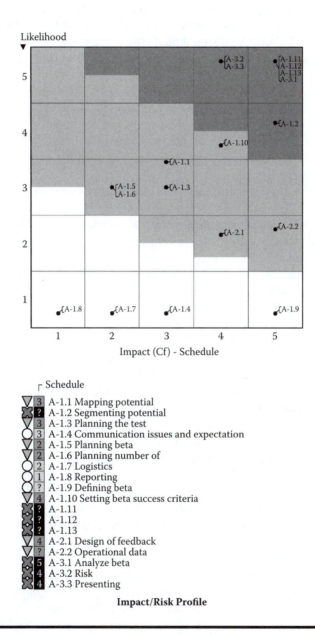

Figure 7.13   Risk matrix of beta testing elements.

condition), and low risk (green condition) for either approach. The risk score helps put the risk into a category that will guide risk response actions.

A quantitative risk analysis process aims to analyze numerically the probability of each risk and its consequence on project objectives, as well as the extent of overall project risk. This process uses techniques such as Monte Carlo simulation and decision analysis to:

- Determine the probability of achieving a specific project objective
- Quantify the risk exposure for the project, and determine the size of cost and schedule contingency reserves that may be needed
- Identify risks requiring the most attention by quantifying their relative contribution to project risk
- Identify realistic and achievable cost, schedule, or scope targets

Quantitative risk analysis generally follows qualitative risk analysis such as the one described above. It requires risk identification. The qualitative and quantitative risk analysis processes can be used separately or together. Considerations of time and budget availability and the need for qualitative or quantitative statements about risk and impacts will determine which method(s) to use. Interviewing techniques are used to quantify the probability and consequences of risks on project objectives. A risk interview with project stakeholders and subject matter experts may be the first step in quantifying risks. The information needed depends on the type of probability distributions that will be used. For instance, information would be gathered on the optimistic (low), pessimistic (high), and the most likely scenarios if triangular distributions are used, or on mean and standard deviation for the normal and log-normal distributions.

Sensitivity analysis helps determine which risks have the most potential impact on the project. It examines the extent to which the uncertainty of each project element affects the objective being examined when all other uncertain elements are held at their baseline values.

Several risk response strategies are available. The strategy that is most likely to be effective should be selected for each risk. Then, specific actions should be developed to implement that strategy. Primary and backup strategies may be selected. Typical risk responses include:

1. *Avoidance:* Risk avoidance is changing the project plan to eliminate the risk or condition, or to protect the project objectives from its impact. Although the project team can never eliminate all risk events, some specific risks can be avoided. Some risk events that arise early in the project can be dealt with by clarifying requirements, obtaining information, improving communication, or acquiring expertise. Reducing scope to avoid high-risk activities, adding resources or time, adopting a familiar approach instead of an innovative one, or avoiding an unfamiliar subcontractor may be examples of avoidance.

2. *Transference:* Risk transfer is seeking to shift the consequence of a risk to a different program element or to a third party together with ownership of the response. Transferring the risk simply gives another party or program element responsibility for its management; it does not eliminate the risk. Transferring liability for risk is most effective in dealing with financial risk exposure. Risk transfer nearly always involves payment of a risk premium to the party taking on the risk. It includes the use of insurance, performance bonds, warranties, and guarantees. Contracts may be used to transfer liability for specified risks to another party. The use of a fixed-price contract may transfer risk to the seller if the project's design is stable. Although a cost-reimbursable contract leaves more of the risk with the customer or sponsor, it may help reduce cost if there are mid-project changes. Similarly, transference to a different program element accomplishes the same thing. For example, if a reliability requirement has been imposed that originally called for a software design solution that involves implementing a heretofore unused, hence risky, set of algorithms, but a hardware solution is available that is perhaps more costly because it involves hardware redundancy, but is known to work, a decision could be made to implement the requirement as a hardware solution to lower the risk. This is another instance of transference.

3. *Control:* Control seeks to reduce the probability and/or consequences of an adverse risk event to an acceptable threshold. Taking early action to reduce the probability of a risk's occurring or its impact on the project is more effective than trying to repair the consequences after it has occurred. Control costs should be appropriate, given the likely probability of the risk and its consequences. Risk control may take the form of implementing a new or alternate, parallel course of action that will reduce the problem—for example, adopting less-complex processes, conducting more seismic or engineering tests, or choosing a more stable vendor. It may involve changing conditions so that the probability of the risk occurring is reduced—for example, adding resources or time to the schedule. It may require prototype development to reduce the risk of scaling up from a bench-scale model.

   Where it is not possible to reduce probability, a control response might address the risk impact by targeting linkages that determine the severity. For example, designing redundancy into a subsystem may reduce the impact that results from a failure of the original component.

3. *Acceptance:* This technique indicates that the project team has decided not to change the project plan to deal with a risk or is unable to identify any other suitable response strategy. Active acceptance may include developing a contingency plan to execute, should a risk occur, for example, budgeting a management reserve to deal with the risk should it develop into a problem. Passive acceptance requires no action, leaving the project team to deal with the risks as they occur. A contingency plan is applied to identified risks that arise during the project. Developing a contingency plan in advance can

greatly reduce the cost of an action should the risk occur. Risk triggers, such as missing intermediate milestones, should be defined and tracked. A fallback plan is developed if the risk has a high impact, or if the selected strategy may not be fully effective. This might include allocation of a contingency amount or having an agreement to accept a resultant schedule slip.

The most usual risk acceptance response is to establish a contingency allowance, or reserve, including amounts of time, money, or resources, to account for known risks. The allowance should be determined by the impacts, computed at an acceptable level of risk exposure, for the risks that have been accepted.

The risk mitigation techniques described above do not fall into absolute, well-defined categories. Some overlap occurs from one category to the next. The categories were presented to give you an idea of what kinds of things can be done to mitigate risks. It is not so important whether the risk mitigation technique was control or avoidance, for example. What is important is that some action was taken to address risk mitigation.

7.3.4 Risk Monitoring and Control

Risk monitoring and control is the process of keeping track of the identified risks, monitoring residual risks and identifying new risks, ensuring the execution of risk mitigation plans, and evaluating their effectiveness in reducing risk. Risk monitoring and control records the risk metrics associated with implementing contingency plans. Risk monitoring and control is an ongoing process for the life of the project. The risks change as the project matures, new risks develop, or anticipated risks disappear. Good risk monitoring and control processes provide information that assists with making effective decisions in advance of the risk's occurrence. Communication to all project stakeholders is needed to periodically assess the acceptability of the level of risk on the project. The purpose of risk monitoring is to determine if:

- Risk responses have been implemented as planned
- Risk response actions are as effective as expected, or if new responses should be developed
- Project assumptions are still valid
- Risk exposure has changed from its prior state, with analysis of trends
- A risk trigger has occurred
- Proper policies and procedures are followed
- Risks have occurred or arisen that were not previously identified

Risk control may involve choosing alternative strategies, implementing a contingency plan, taking corrective action, or replanning the project. The risk response owner should report periodically to the project manager and the risk team leader on the effectiveness of the plan, any unanticipated effects, and any mid-course correction needed to mitigate the risk. The project manager should make periodic reports to higher-level management on the status of the risk management activities on the project.

7.4 Usability Analysis

7.4.1 Introduction to Usability

Operational failures caused by human errors are a main factor of system failure. The Federal Aviation Authority reports that 80% of airplane accidents are attributed to human errors [27]. Operational failures are not always identified correctly. For example, many calls for technical support about TV system malfunctioning turn out to be the result of user errors. The term "Decision Support System for User Interface Design" (DSUID) was introduced in [14] and expanded in [21] to represent procedures employed during system development for preventing user interface design errors and for improving system resistance to such errors. DSUID is about identifying patterns of operational failures and redesigning the user interface so that these failure patterns are eliminated. We can identify three main phases for the application of DSUID:

1. *Design phase:* At this phase, DSUID provides system architects and designers with guidelines and operating procedures representing accumulated knowledge and experience on preventing operational failures. Such knowledge can be packaged in "patterns" or design standard operating procedures (SOPs).
2. *Testing phase:* Operational reliability can be completely and effectively assessed only when testing with real users doing their real tasks, as opposed to testing by testing experts. Therefore, a key testing phase is beta testing, when new versions are installed in real customer environments for evaluating the way they are actually being used.
3. *Tracking phase:* Over time, systems are exposed to new user profiles. The original operational scenario might not be predictive of the evolving demands and, typically, ongoing DSUID tracking systems are required to adapt the system to the changing operational patterns. Statistical process control (SPC) can be employed for monitoring the user experience, by comparing actual results to expected results and acting on the gaps. Tracking and ongoing analysis provide opportunities for continuous improvements in the ways the system responds to unexpected events (see [21]).

This section presents the principles for setting up a DSUID using diagnostic reports based on time analysis of user activity. The diagnostic reports enable product managers to learn about possible sources of dissatisfaction of site visitors. Different types of operational design deficiencies are associated with different patterns of user activity. The methods presented here are based on models for estimating and analyzing the user's mental activities during system operation. Building on this approach [14], a seven-layer DSUID model for data analysis is proposed, and an example of a log analyzer that implements this model is presented. For a general introduction to usability, see [16].

7.4.2 A Case Study: Web Usability Analytics

The goal of usability design diagnostics is to identify, for each site page, all the design deficiencies that hamper the positive navigation experience in each of the evaluation stages. This is a highly important in e-commerce applications. To understand the user experience, we need to know the user activity, compared to the user expectation. Both are not available from the server log file, but can be estimated by appropriate processing. A DSUID system flagging possible usability design deficiencies requires a model of statistical manipulations of server log data. The diagnosis of design deficiency involves certain measurements of the site navigation, statistical calculations, and statistical decision. How can we tell whether visitors encounter any difficulty in exploring a particular page, and if so, what kind of difficulty do they experience, and what are the sources for this experience? We assume that the site visitors are task driven, but we do not know if the visitors' goals are related to a specific website. Also, we have no way to tell if visitors know anything a priori about the site, if they believe that the site is relevant to their goals, or if they have visited it before; it may be that the visitors are simply exploring the site, or that they follow a procedure to accomplish a task. Yet, their behaviors reflect their perceptions of the site contents, and estimates of their effort in subsequent site investigation. Server logs provide time stamps for all hits, including those of page HTML text files, but also of image files and scripts used for the page display. The time stamps of the additional files enable us to estimate three important time intervals:

1. The time the visitors wait until the beginning of the file download is used as a measure of page responsiveness.
2. The download time is used as a measure of page performance.
3. The time from download completion to the visitor's request for the next page, in which the visitor reads the page content, but also does other things, some of them unrelated to the page content.

The DSUID challenge is to decide, based on statistics of these time intervals, whether the visitors feel comfortable during the site navigation, when they feel that they wait too long for the page download, and how they feel about what they see on the screen. How can we conclude that a time interval is acceptable, too short, or too long for page visitors? For example, consider an average page download time of 5 seconds. Site visitors may regard it as too lengthy if they expect the page to load fast, for example, in response to a search request. However, 5 seconds may be quite acceptable if the user's goal is to learn or explore specific information if they expect it to be related to their goal. The diagnostic-oriented time analysis is done by observing the correlation between the page download time and the page exits. If the visitors are indifferent about the download time, then the page exit rate should be invariant with respect to the page download time. However, if the download time matters, then the page exit rate should depend on the page download time. When the page download time is acceptable, most visitors may stay in the site, looking for additional information. However, when the download time is too long, more visitors might abandon the site and go to the competitors. The longer the download time, the higher the exit rate.

To enable statistical analysis, we need to change our perspective on these variables and consider how the download time depends on the exit behavior. We compare the download time of those visits that were successful, with that of those visits that ended in a site exit. If download time matters, we should expect that the average download time of those visitors who abandoned the site will be longer than the average of those who continued with the site navigation. Otherwise, if the site visitors are indifferent about the download time, we should expect that the download times of the two groups would not be significantly different.

To decide about the significance of the usability barrier, we compare the download times of two samples: one of page visits that ended in site exit and the other of all the other page visits. The null hypothesis is that the two samples are of the same population. If the null hypothesis is rejected, we may conclude that the page visitors' behavior depends on the download time: if the download time of the first sample exceeds that of the second sample, and the difference is statistically significant, then we may conclude that the download time of the particular page is significantly too long. A simple two-tailed t-test is sufficient to decide whether the visitors' behavior is sensitive to page download time. The approach described above can be generalized in a comprehensive DSUID model: the Seven Layers Model is described next.

7.4.3 A DSUID Implementation Framework: The Seven Layers Model

In setting up a decision support system for user interface design, we identify seven layers of data collection and data analytics:

- *The lowest layer—user activity:* The records in this layer correspond to user actions screen changes or server-side processing.
- *The second layer—page hit attributes:* This layer consists of download time, processing time, and user response time.
- *The third layer—transition analysis:* The third layer statistics are about transitions and repeated form submission (indicative of visitors' difficulties in form filling)

- *The fourth layer—user problem indicator identification:* Indicators of possible navigational difficulty, including:
  - Time elapsed until the next user action
  - Backward navigation
  - Transitions to main pages, interpreted as escaping current subtask
- *The fifth layer—usage statistics:* The fifth layer consists of usage statistics, such as:
  - Average time to upload website (i.e., the average time from selection of the website until the page is loaded on the user's computer)
  - Average download time
  - Average time between repeated form submission
  - Average time on screen (indicating content related behavior)
  - Average time on a previous screen (indicating ease of link finding)
- *The sixth layer—statistical decision:* For each of the page attributes, DSUID compares the statistics over the exceptional page views to those over all the page views. The null hypothesis is that (for each attribute) the statistics of both samples are the same. A simple two-tailed t-test can be used to reject it, and therefore to conclude that certain page attributes are potentially problematic. A typical error level is set to 5%.
- *The seventh layer—interpretation:* For each of the page attributes, DSUID provides a list of possible reasons for the difference between the statistics over the exceptional navigation patterns and that over all the page hits. However, it is the role of the usability analyst to decide which of the potential sources of visitors' difficulties is applicable to the particular deficiency.

The discussion presented above about the relationships between time intervals and exit rates demonstrates our methodology. Setting up a DSUID involves integration of two types of information:

- Design deficiencies, which are common barriers to seamless navigation, based on the first part of the model described above—visitor's page evaluation
- Detectors of these design deficiencies, common indicators of possible barriers to seamless navigation, based on the second part of the model—visitor's reaction

The way to conclude that the download time of a particular page is too long is by a measure of potentially negative user experience, namely, the site exit rate. Exit rate is a preferred measure for deciding on the effect of download time, but is irrelevant to the analysis of the visitor's response time. The reason for it is that server logs do not record the event of the visitor leaving the site, which means that we cannot measure the time intervals of terminal page views. Therefore, we need to use other indicators of potential negative navigation experience.

A model of a user's mental activities in website navigation lists the most likely visitors' reaction to exceptional situations. Based on this model, we can list the following indicators about the visitor's tolerance to Web page design deficiencies:

- The visitor returning to the previous page may indicate that the current page was perceived as less relevant to the goal, compared to the previous page.
- The visitor linking to a next website page may indicate that the link was perceived as a potential bridge to a goal page.
- The visitor activating a main menu may indicate that the visitor is still looking for the information, after not finding it in the current page.
- The visitor exiting the site may indicate that either the goal has been reached, or the overall site navigation experience became negative.

Accordingly, in addition to site exit, other indicators of potential usability problems are the rates of navigation back to a previous page and the visitor escaping from the page to a main menu. Once we have an estimate of the task-related mental activities, we can adapt the method for deciding about problematic performance, to decide about problematic task-related mental activities. Visitors who feel that the information is irrelevant to their needs are more likely to respond quickly, and are more likely to go backward, or to select a new page from the main menu. Therefore, the average time on a page over those visitors who navigated backward or retried the main menu should be shorter than that of the average of all visitors. On the other hand, visitors who believe that the information is relevant to their needs, but do not understand the page text very easily, are likely to spend more time than the average reading the page content, and the average time on the page should be longer.

The time that visitors spend reading a page depends on various perceptual attributes, including the page relevance to their goals, the ease of reading and comprehending the information on the page, the ease of identifying desired hyperlinks, etc. Assume that for a particular page, the average time on screen before backward navigation is significantly longer than the average time over all page visits. Such a case may indicate a readability problem, due to design flaws, but it can also be due to good page content, which encouraged users who spent a long time reading the page to go back and reexamine the previous page.

7.4.4 DSUID Modeling Techniques

In implementing the seven layers DSUID system, we consider four types of modeling techniques described in [14]:

Model 1—Markov processes: Markov processes model the time dimension in server requests, in which the states represent the Web pages and the transitions between states represent the hypertext activity.

Model 2—Mental activity: To draw cause-and-effect conclusions, we need a model of how certain situations evolve to certain effects. To identify barriers to seamless navigation, we need to have a model of normal mental activities involved in page handling, and possible deviations from the normal activities, due to design deficiencies.

Model 3—Bayesian networks: Bayesian networks map cause-and-effect relationships between key variables.

Model 4—Statistical analysis: Specific measurements of the site navigation, statistical calculations, and rules for statistical decision.

We illustrate our approach with an implementation example of Model 3. A Bayesian network is a graphical model representing cause-and-effect relationships between variables (see [6, 19]). As an example, we show in Figure 7.14 and Figure 7.15 a Bayesian network derived from the analysis of Web log analyzers. The network indicates the impact of variables on others derived from an analysis of conditional probabilities. The variables have been discretized into categories with uniform frequencies. The network can be used to represent posterior probabilities, reflecting the network conditioning structure, of the 1 through 5 categorized variables. We can use the network for predicting posterior probabilities after conditioning on variables affecting others or, in a diagnostic capacity, by conditioning on end

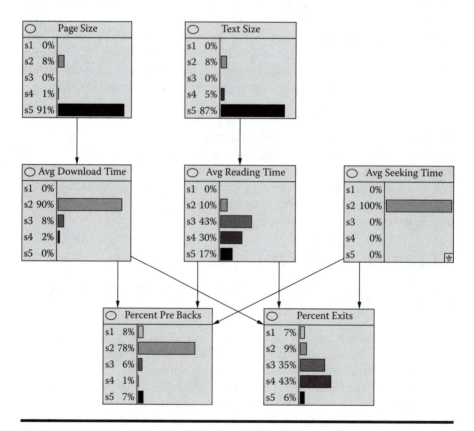

Figure 7.14 A Bayesian network of Web log data, conditioned on low seek time.

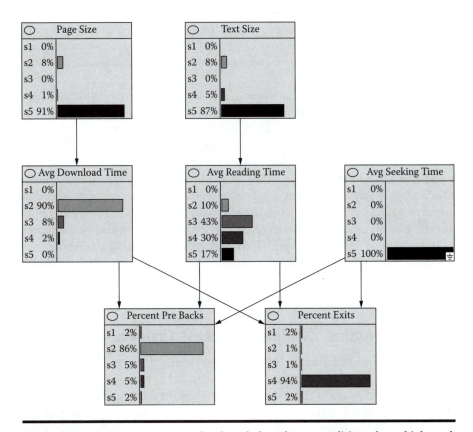

Figure 7.15 **A Bayesian network of Web log data, conditioned on high seek time.**

result variables. In our example, we condition the network to low and high seek time on the Web page, Figure 7.14 and Figure 7.15, respectively.

With low seek time we experience 49% of high and very high exit rates. With high seek time these numbers jump to 96%. Clearly, seek time has affected the behavior of the users. We can even quantify this phenomenon and link, quantitatively, the impact of a second of seek time on exit rate percentages.

The proposed approach is flexible because it can handle problems either by different models (as discussed above) or by combining different statistical approaches. For example, a t-test was described previously to compare download times of successful and unsuccessful visits. The problem leads naturally to a Bayesian approach that combines data and expertise. Along with analysis of the log, we think it is important to elicit the opinion of experts on possible user's reactions when navigating through websites. Independent usability engineers can provide opinions on the usability of the Web pages (from used language to graphical presentation), and describe the most likely reactions to uneven events like high download times.

These different, nonhomogeneous sources of information can be combined by statisticians into an a priori assessment of probability of failure/success when some patterns are followed by the users and under different system characteristics (e.g., download times).

As an example of a Bayesian approach, consider the Markov processes mentioned previously. Transactions, as described in [14], can be thought of as Markov chains with states given by the Web pages in the website. Two absorbing states are present (fulfillment of the task and unsuccessful exit), and our interest is in their probability. The Web log provides the number of transitions from state i to state j, and data can be combined via Bayes' theorem, so that a posterior distribution is obtained. Based on the posterior distribution on the transition probabilities, operational reliability can therefore be analyzed by looking at the predictive probability of ending in the unsuccessful state, given a prespecified pattern, that is, a sequence of transitions between Web pages. With such techniques, one can combine, in a DSUID, expert opinions with data to model transitions between states and describe predicted cause-and-effect relationships.

7.5 Determining Readiness Success Criteria

Readiness is a system attribute that can be determined at various points in a system's development life cycle. Criteria can be specified at these various milestones that can be used to assess the system's progress toward meeting the users' needs, thus lowering the risk that a system will be delivered with which the customer will be dissatisfied. The criteria, referred to as *system readiness levels* (SRLs) in some circles, can be used as a project management tool for capturing evidence of system readiness. Based on this evidence, system maturity can be assessed and communicated in a consistent manner to the users and other interested stakeholders. This provides data on how well the organization's standard development process, as tailored, worked for this project.

7.5.1 Mapping Goals to Strategic Objectives

A methodology for capturing system readiness and a corresponding Microsoft Excel workbook tool to facilitate determining readiness has been developed by the Ministry of Defence (MOD) in the United Kingdom. A description of the methodology can be found at, and the tool can be downloaded at no fee from, http://www.aof.mod.uk/aofcontent/tactical/techman/content/srl_whatarethey.htm. In the methodology developed by the MOD, SRLs are defined at nine steps from initial concept definition to deployment across a set of systems engineering disciplines as a means of assessing system maturity. Each of the nine steps aligns to the key outputs produced by the systems disciplines involved in these phases, such as training, safety and environmental, or reliability and maintainability. These nine steps are identical to the nine phases of a system's development shown in Figure 6.1 in Chapter 6. (The names of the phases used in Chapter 6 differ from the terminology used by the MOD, but the activities performed in these phases are the same.) In our discussion of the methodology, we use the notation used in Figure 6.1. Note also that the MOD methodology was developed for use in military systems, but it can be adapted for use in commercial systems development. The methodology also provides for major milestone reviews. In the methodology illustrated by the MOD, a review is held after the PRD phase, and again after low-level design is completed. Obviously, any adaptation of the methodology can provide for more or less milestone reviews at these or other points in the development process.

The disciplines in the MOD methodology represented in the matrix in the Excel workbook tool include systems engineering, training in how to use and operate the system, reliability and maintainability of the system, safety and environmental considerations, human factors that impact how users interact with the system, information technology, and airworthiness. A placeholder is provided for any project-specific disciplines that may be of importance. Although this set of disciplines applies to the acquisition of an aircraft system, adaptations of these disciplines can clearly be made to apply the methodology to other kinds of systems.

For each of the phases, for each of the disciplines, a number of artifacts and questions to be answered about the status of the effort are listed as being indicators of the successful implementation of that discipline for that phase and the readiness to continue development (or release the product). In any development organization's implementation of this methodology, these artifacts would derive from the standard process in use by that organization, as tailored to meet the unique needs of the project to which it is applied. The process definition and corresponding artifacts are based on satisfying the organization's business needs.

In the first phase, PRD, the focus in all the disciplines is in capturing all the external requirements that affect each discipline. For systems engineering, this would focus on ensuring that all the user requirements had been elicited and captured correctly along with the concept of operations. For training, it would be identifying all the external requirements that impact the scope of the training for the users of the system. The objectives for the phase just completed can be intuited for each of the disciplines from the artifact produced in that phase and the questions to be answered. For each of the disciplines in each phase, multiple outputs and questions to be answered are established that define the criteria that mark completion of that phase of the effort for that discipline. At the end of this phase, a major milestone review would be held, typically with the customer, to ensure that all of the customer's requirements for development of the system, training, reliability, maintainability, etc. have been captured correctly, and to ensure that there is a common understanding of all the requirements between the customer (if different from the user), the user, and the development organization.

Note that the customer does not always have to be an external customer. The same approach can be used in the case of internal customers, for example, the marketing organization when developing a product for commercial sales, or the accounting department when developing an accounting software application.

In the next phase, the one that results in the development of a functional specification, the focus is on developing the top-level requirements for the development of the system based on the user's requirements as defined in the PRD phase. Examples (but not a complete list) of the kinds of outputs of this phase include a system specification, system acceptance criteria, and system reliability and maintainability requirements. System requirements have usually been partitioned between the hardware and the software. For human factors, a number of questions are listed that focus on ascertaining the status of the human factors definition at this point in the development effort.

In the next phase, the focus is on completing the high-level design of the system. Example outputs that indicate satisfactory completion of this phase include the definition of the system architecture and corresponding integration and test strategy. Reliability and maintainability requirements have typically been allocated to the major subsystems of the deliverable product. Software requirements have been allocated to the major subsystems, and a definition of the software applications to be produced would be established.

In similar fashion, the criteria for acceptable completion of each phase are laid out. Successful phase completion in accordance with these criteria indicate a readiness to proceed to the next phase of activity and increasing confidence in the readiness of the delivered product to satisfy user needs. The major milestone reviews permit involvement of the major stakeholders in evaluating the readiness of the system at that point in time.

As noted earlier, the MOD methodology is based on developing military systems; however, it is reasonable to be able to adapt the approach to the development of commercial systems. For example, a manufacturer of medical devices could modify the set of disciplines above to include systems engineering, training in how the medical profession implements the devices, reliability and maintainability of the device, safety and environmental issues (obviously important in a device such as a pacemaker or a stent), human factors related to the device having minimal impact on the patient's daily life, information technology, and regulatory agency compliance. For an accepted standard on project management, see [15].

7.5.2 Using QFD to Establish Readiness Criteria

Quality Function Deployment (QFD) is a process for systematically converting customers' demands into design quality. It deploys quality over design targets and major milestones used throughout the product development process. QFD was conceived in Japan in the late 1960s while Japanese industries were departing from their post-war mode of product development, which had essentially involved imitation and copying, and were moving toward original product development. At that time, there was recognition of the importance of designing quality into new products, but there was a lack of guidance on how to achieve it. Companies were already inspecting for quality, but it was happening at the manufacturing site after new products had been produced. Yoji Akao [1] first presented the QFD concept and method to solve these problems, and QFD has continued to spread since Shigeru Mizuno and Akao published their first book on the topic in 1978 [25]. To ensure smooth product development, Akao constructed a comprehensive QFD system that considers quality, technology, cost, and reliability simultaneously. The objective of each deployment is as follows:

- *Quality deployment* is to systematically deploy customers' demanded quality over the design targets and major milestones used throughout the product development process.
- *Technology deployment* is to extract any bottleneck engineering that hinders quality and to solve it at the earliest possible time.
- *Cost deployment* is to achieve the target cost while keeping a balance with quality.
- *Reliability deployment* is to prevent failures and their effects through early prediction.

In the design stage, the product must effectively prevent a recurrence of the existing product's design problems as well as the new product's potential design problems. To that end, a fault tree analysis (FTA) and failure mode effects analysis (FMEA) can be employed. With respect to parts and materials needed for constructing subsystems, quality deployment is required, and an evaluation of the suppliers of the parts or materials feasibility and capability should be conducted, whether the supplier is internal or external to the company.

Products are made through processes and completed by assembling semi-finished products. Therefore, a semi-finished product and its specifications defined by the subsystem's quality deployment are made using process deployment and design, and by deciding specifications of process conditions. Like product design, process design must effectively prevent a recurrence of the existing process's design problems and the potential design problems of the new process. This can be done by drawing on equipment FTA and process FMEA.

The foregoing quality deployment, which includes technology and reliability, can realize customers' demanded qualities and failure-free qualities, yet it might increase the cost as a result. Using market evaluation information to decide the target cost of the finished product, and corresponding to quality deployment flows to set up cost targets for materials and labor, a balance between QA and cost reduction can be achieved.

The QFD methodology can also be used as a means of specifying readiness at each phase of the development life cycle. To adapt the SRL methodology described in the previous section, QFD can be used to derive the factors to use as the criteria to be applied for each discipline in each phase.

7.6 Managing Readiness Reviews

A development organization can follow a process faithfully and still not produce a system that meets user needs. The proof of the pudding for the process, so to speak, resulting in a system that meets user requirements and is ready for delivery is some sort of a demonstration that the system is ready, usable, and has satisfied all the requirements that the user believes have been specified. Minimally, a review involving key stakeholders that addresses these concerns around the time of delivery is needed. Pragmatically, there should be a number of reviews throughout the development cycle. Thus, readiness reviews are essential as part of the development process in order to determine if the deliverable system will function as intended. Readiness reviews are built into the SRL process described in previous sections. Readiness reviews should be held incrementally during the course of the development life cycle. As indicated previously in this chapter, incremental readiness reviews provide insight along the way as to whether or not the deliverable system will function in the way the user wanted it to function. Leaving the readiness review until the time of delivery will only increase the risk of failure.

The philosophy of incremental reviews was a hallmark of weapon systems development for the U.S. DoD. A number of reviews were held during the development cycle, such as a System Specification Review (SSR), Critical Design Review (CDR), and Test Readiness Review (TRR), as a means of assessing progress toward achieving the delivery of a system that met user needs. Such reviews were highly formalized, with specific criteria for their conduct (see MIL-STD-1521B [24]). With the advent of acquisition reform in the DoD, MIL-STD-1521B was canceled, and the requirement to do such reviews changed. The focus shifted to doing "asynchronous, in-process, interim reviews" [13]. Nonetheless, the requirement for major milestone reviews for major critical defense systems remained part of DoD policy. The difference now was that the reviews to be held and the ground rules for their conduct were something to be negotiated between the acquisition agent and the development contractor. There were arguments pro and con for the cancellation of MIL-STD-1521B. Without getting into value judgments concerning the arguments on either side, suffice it to say that the conduct of such reviews during the lifetime of the standard did not always yield the kind of results and information intended from the reviews. Yet, the need at key points in the development process to step back and get a more global perspective on progress toward achieving the desired system is unquestioned. Banking completely on a readiness demonstration at the end of the development process is very risky.

Clearly, the need for a review process at the end of development cannot be denied. We have discussed beta testing as one form of demonstration of readiness. Usability analyses are also a key ingredient. A formal review, chaired by both the development organization's program manager and a representative of the user organization, should be held after the completion of beta testing and usability analyses. The results of these tests should be reviewed, together with the support and design documentation. User manuals, operator manuals, training materials, and the like are included in the support documentation. Requirements specifications, drawings, source code listings, and the like are included in the design documentation. When there is concurrence between the development organization and the user that the system will function as intended and that the development contractor has performed in compliance with the contract, statement of work, and customer specifications, the system is ready to be used operationally.

Other reviews that should be conducted along the way to reduce risk would, as a minimum, include:

- Customer Requirements Review after the PRD phase to ensure that all the user's requirements have been captured and that there is a common understanding of the requirements between the customer and the development organization
- Functional Requirements Review after the Functional Specification phase to ensure that the requirements elaborated by the development organization from the customer requirements implement the agreed-to customer requirements
- Design Review after the Low Level Design phase to determine if the completed designs implement the requirements
- Test Readiness Review after the completion of the Integration and Functional Testing phases to evaluate the test results to date to determine if the design has been properly implemented and that the system is ready to proceed to System Testing

These recommended reviews are not intended to impose a waterfall development cycle model, and can even be conducted incrementally. The intent is to assess progress, and the reviews can be inserted into the schedule to accommodate the availability of the needed data, whether it is available all at once (which is probably highly unlikely except for the most noncomplex systems), or available over time. The criteria for the determination of the adequacy (readiness) can be extracted from the Excel tool that accompanies the SRL methodology, as described earlier in this chapter.

7.7 Summary

This chapter integrated a number of the concepts and tools presented in previous chapters in the context of the beta test phase of a system. The chapter discussed the design and implementation of beta testing, beta test survey analysis, risk management, usability analysis, and readiness assessments. A key objective of these efforts is to establish the level of readiness of a system or software product prior to official release by the Readiness Review Board. Most organizations require a beta testing phase prior to product release.

Development organizations at CMMI-DEV Maturity Level 5, with well-designed and -executed processes, might still release a system that doesn't quite meet the user's needs. Sometimes this results from an inability of the user to adequately articulate the real requirements for the system at the start of the development process. Clearly, checks and balances must exist during the development process to compensate for any such failures. Beta testing, satisfaction surveys, risk management, usability analysis, and readiness reviews are, in a way, the final assurance that the process has worked properly. These approaches are consistent with the practices found in the CMMI-DEV. The use of formal reviews is the subject of Specific Practice 1.7 in the Project Monitoring and Control Process Area. The use of statistical methods in conducting usability analyses is another application of the high maturity practices in the CMMI-DEV. The analysis of beta testing data is a prime source of information for management to decide on the release of a product. Monitoring the post-introduction phase (after release) is also obviously very important. We discussed monitoring this phase with a focus on usability. The chapter also showed how risk management can be used as a methodology for assessing the readiness status of a system or software product.

References

1. Akao, Y. and Mazur, G., The Leading Edge in QFD: Past, Present and Future, *International Journal of Quality and Reliability Management,* 20(1), 20–35, 2003.
2. Baker, E. and Hantos, P., GQM-R_x — A Prescription to Prevent Metrics Headaches, *Fifth Multi-Conference on Systemics, Cybernetics and Informatics,* Orlando, FL, July 22–25, 2001.
3. Basili, V. and Weiss, D., A Methodology for Collecting Valid Software Engineering Data, *IEEE Transactions on Software Engineering,* November 1984, p. 728–738.
4. Bellcore, In-Process Quality Metrics (IPQM), GR-1315-CORE, Issue 1, September 1995.
5. Bellcore, Reliability and Quality Measurements for Telecommunications Systems (RQMS), GR-929-CORE, Issue 2, December 1996.
6. Ben Gal, I., Bayesian Networks, in *Encyclopedia of Statistics in Quality and Reliability,* Ruggeri, F., Kenett, R.S., and Faltin, F., Editors in Chief, New York: Wiley, 2007.
7. Fenton, N.E. and Shari, N.P., *Software Metrics,* Boston, MA: PWS Publishing Company, 1997.
8. Fisher, R.A., *The Design of Experiments, 1st ed.,* Edinburgh: Oliver & Boyd, 1935.
9. Fisher, R., The Arrangement of Field Experiments, *Journal of the Ministry of Agriculture of Great Britain,* 33, 503–513, 1926.
10. Green P.E. and Rao V.R., Conjoint Measurement for Quantifying Judgmental Data, *Journal of Marketing Research,* 8, 355–363, 1971.
11. Green P.E. and Srinivasan V., Conjoint Analysis in Consumer Research: Issue and Outlook, *Journal of Consumer Research,* 5, 103–123, 1978.
12. Gustafsson, A., Hermann, A., and Huber, F., *Conjoint Measurement: Methods and Applications,* Berlin: Springer, 2001.
13. Hantos, P., System and Software Reviews since Acquisition Reform, presented to the *Los Angeles SPIN,* April 2, 2004.

14. Harel A., Kenett R., and Ruggeri, F., Modeling Web Usability Diagnostics on the Basis of Usage Statistics, in *Statistical Methods in eCommerce Research*, W. Jank and G. Shmueli, Editors, New York: Wiley, 2008.

15. Institute of Electrical and Electronics Engineers, IEEE–1490, A Guide to the Project Management Body of Knowledge, 2002.

16. International Standards Organization, ISO 9241-11, Guidance on Usability, 1998.

17. Kenett, R. and Baker, E., *Software Process Quality: Management and Control*, New York: Marcel Dekker, 1999.

18. Kenett, R.S., On the Planning and Design of Sample Surveys, *Journal of Applied Statistics*, 33(4), 405–415, May 2006.

19. Kenett, R.S., Cause and Effect Diagrams, in *Encyclopedia of Statistics in Quality and Reliability*, Ruggeri, F., Kenett, R.S., and Faltin, F., Editors in Chief, New York: Wiley, 2007.

20. Kenett, R.S., Software Failure Data Analysis, in *Encyclopedia of Statistics in Quality and Reliability*, Ruggeri, F., Kenett, R.S., and Faltin, F., Editors in Chief, New York: Wiley, 2007.

21. Kenett, R.S., Harel, A., and Ruggeri, F., Controlling the Usability of Web Services, *International Journal of Software Engineering and Knowledge Engineering*, World Scientific Publishing Company, Vol. 19, No. 5, pp. 627–651, 2009.

22. Kenett, R.S. and Steinberg, D.M., New Frontiers in Design of Experiments, *Quality Progress*, August 2006, p. 61–65.

23. Kenett, R.S. and Zacks, S., *Modern Industrial Statistics: Design and Control of Quality and Reliability*, San Francisco: Duxbury Press, 1998. (2nd edition, 2003; Chinese edition, 2004.)

24. MIL-STD-1521B, Technical Reviews and Audits for System, Equipments, Munitions, and Computer Programs, Department of Defense, June 4, 1985.

25. Mizuno, S. and Akao, Y., Eds., *Quality Function Deployment: A Company-Wide Quality Approach*, Tokyo: JUSE Press, 1978.

26. Rossi, P.E., Allenby, G.M., and McCulloch, R., *Bayesian Statistics and Marketing*, West Sussex, England: John Wiley & Sons, 2005.

27. Shappell, S. and Wiegmann, D., U.S. Naval Aviation Mishaps 1977–1992: Differences between Single and Dual-Piloted Aircraft, *Aviation, Space, and Environmental Medicine*, 67, 65–69, 1996.

28. TRIMS, Technical Risk Identification and Mitigation, http://www.bmpcoe.org/pmws/download/trims4_v403_setup.exe, 2008.

29. Wall, J.K. and Ferguson, P.A., Pragmatic Software Reliability Prediction, *Proceedings of the Annual Reliability and Maintainability Symposium*, 1977, p. 485–488.

MANAGING AND REPORTING DATA, AND A COMPREHENSIVE CASE STUDY

IV

Chapter 8

Data Reporting and Analysis

Synopsis

Measurement is an essential element of a product's life cycle and the governance of an organization's business. For a product, it starts at the initial planning phase and continues beyond the conclusion of development and manufacture of the deliverables. It covers the entire life cycle of a system—from concept to retirement. At the organizational level, it goes on continuously, with periodic reviews to ensure that the measurement program is efficient and effective, providing information that the organization needs to survive and grow. In previous chapters we discussed the identification of measurement goals and objectives and the associated definition of measures that it is necessary to collect. In this chapter we discuss designing a data collection and reporting system for data collected during development and testing, as well as from the field, following product shipment and installation. It expands on the material in Chapter 4 that focused on setting up and implementing a measurement program. The chapter discusses the operation of data reporting systems, the data analysis, the reporting on findings and consequences, and concerns related to product sustainment.* Field data reporting systems, sometimes called

---

* Sustainment is defined [1] as "The processes used to ensure that a product can be utilized operationally by its end users or customers. Sustainment ensures that maintenance is done such that the product is in an operable condition whether or not the product is in use by customers or end users."

"field tracking systems" or "product performance surveys," are discussed with an example in Section 8.5. In this section we also introduce an advanced reporting model based on a Bayesian statistical approach called Bayesian Estimates of the Current Mean (BECM). We conclude with a summary section.

8.1 Designing a Data Reporting System

Field reporting provides a critical feedback mechanism for determining how well an organization has done the job of implementing effective development and maintenance processes. In the final analysis, the successful use of a product by its customers, and their satisfaction with it, is the proof of the pudding that the effort in implementing and following the process has paid off. If the customers are dissatisfied or are having difficulties in using the product successfully, that is a clear signal that the process has not converted customer and/or user needs into something that is going to satisfy them. At the very least, it may signal an ineffective requirements development process. Accordingly, the need to design a field reporting system that accurately taps user satisfaction and captures defects with the product is essential.

In planning a data reporting system, there are a number of factors to consider. These include defining the data to collect and analyze, establishing the data collection and analysis system itself, methods for collecting the data, data storage, establishing data analysis procedures, and considerations for performance of the data analysis. In this section we discuss these factors in detail.

8.1.1 Defining the Data to Collect

The starting point for the design of any data collection and analysis system is to identify the organization's informational needs. Techniques such as the Goal/Question/Metric (GQM) methodology discussed in Chapter 1 are often used for establishing such informational needs. Data reporting and field tracking systems are set up for:

- Detecting problems
- Quantifying known problems
- Establishing problem causes
- Measuring the impact of system upgrades or design changes
- Evaluating the product or system over time

Many organizations try to copy what other organizations do in designing their data reporting system; however, such efforts can have limited success. What works well for one organization, with respect to the measures they collect, may not work so well for another, even if both are producing similar kinds of systems. The data reporting system of one organization may not satisfy the informational needs of another organization. Information needs are driven, to a large extent, by the nature of the business the organization is in. An IT organization that is developing and operating information systems for internal use will have different informational needs than an organization producing television sets for commercial sale. In turn, the company producing television sets will have different field data informational needs than a company producing commercial airliners, or an airline company operating these planes. An airframe manufacturer, for example, will have informational needs relative to the operation of a new wide-body jet, as well as informational needs driven by requirements of regulatory agencies such as the FAA. Moreover, informational needs may be constant throughout the life cycle, or may change and address needs that are specific to individual phases of a system life cycle. In establishing an overall data reporting program, consideration should be made for the information needs of all relevant stakeholders, including those who need data from the field. In setting up such a program, key personnel from marketing, development, manufacturing, maintenance, and customer service, among others, should participate in order to identify the measures necessary to satisfy their informational needs. The scope of the measurement program should be revisited about once a year to ensure that the spectrum of measures being collected are still relevant and to identify any new measures that might be necessary to address new challenges.

In identifying informational needs, a top-down approach is typically applied. As a first step in such an approach, the goals and objectives at the organizational level for data reporting are stipulated. From these goals, by applying a methodology such as GQM, questions related to how one would know if the goals and objectives are satisfied, or not, are stated. The metrics and measures that answer these questions are then derived (see Chapter 1), and operational definitions are established for each identified measure. This activity involves all the relevant stakeholders and needs to ensure that the measures that have been specified are consistent with the organizational information needs. At the very least, the concurrence of the stakeholders in the final results should be solicited.

8.1.2 Establishing the Data Center

Having established measures and metrics, the mechanics of how the data collection and analysis effort is to be accomplished can then be addressed. Data collection procedures need to be established, including the operation of a data center. Data centers can exist at various levels in an organization. A small organization may have one entity that is responsible for all the data collection and first-level analysis. Larger organizations may have data centers that exist at different levels of organizational hierarchy. Take, for example, the organization depicted in Figure 8.1. Responsibility for managing projects related to similar domains or client bases will usually be vested at the lowest level, which, in this example, is a department. The projects will typically collect measures that are mandated from the corporate level, measures required by the projects' customers, measures related to the implementation

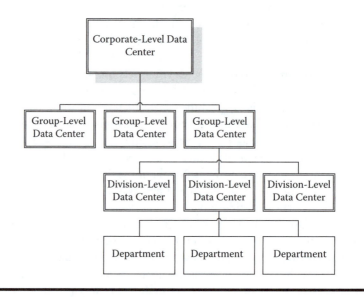

Figure 8.1   Organizational data center structure.

of process improvement, as well as other measures that may be imposed by organizational entities above the department level. In many organizations, a data center will be established at the division level to support the departments within it and the data collection and analysis efforts for the projects they manage. The projects will often report results to directly to division management. Smaller and less mission-critical projects will generally report their results to department management, but their data collection and analysis efforts will still be supported by the data center at the division level.

Generally, the first level of roll-up for reporting measurement results to higher-level management will occur at the division level. Because the divisions report to groups, a subset of the measures collected by the projects and departments will be reported up to group management. Whereas division management may in some cases look at detailed project-level data, group management generally will not (unless there is an overriding need to look at how a specific project is doing). Accordingly, the measures presented to group management are aggregated across the divisions within the group, and reflect how the group is doing relative to corporate-mandated metrics, as well as on measures related to group-unique business objectives. In similar fashion, a subset of the data collected at the group level is aggregated and reported to corporate management. At that level, the interest is in metrics that provide insight into how well the business is performing, and how well the corporation's business objectives are being met.

In implementing data centers in larger organizations such as the one described above, it is important to have compatible data collection and analysis tools at all levels. In addition, forms and templates should also be consistent for data collected at lower organizational levels, rolled up, and reported to higher levels. This helps facilitate data collection and analysis at all levels and reduces errors caused by the need to physically reenter data provided by lower levels in the organization.

8.1.3 Data Collection

Data collection itself should be accomplished in the most unobtrusive manner possible. Procedures that involve a lot of manual input create severe difficulties for the data collection effort, and ultimately the data reporting program. Personnel in the field, such as sales support and technical assistance, are often busy with their daily activities, and data collection procedures with significant manual input will generally be given low priority. In such cases, valuable data may be either lost or compromised. A more effective way to collect the data is to extract the raw data from the work processes themselves, such as collecting online data used to resolve issues, problem reports contact data, etc.

In some cases, data must be entered in a prescribed format, such as paper forms or electronic data repositories. Regulatory agencies that oversee sustainment or development activities, like the Federal Aviation Administration (FAA) or the Food and Drug Administration (FDA), require that. The FAA, for example, requires maintenance personnel to fill out mandated forms to document repairs made or inspections performed. Often, the data collected in these circumstances is specified by the regulatory agency and is intended to fulfill their measurement needs. It will sometimes be necessary to collect additional data on other forms (screen forms or paper) to satisfy other internal measurement needs.

In designing data reporting systems, data collection must be taken into consideration. Online data collection templates are extremely useful. For example, templates for entering data from peer reviews facilitate the peer review data collection process. As issues, action items, and defects are noted during the course of a peer review, they can be entered by the moderator or review chairperson into the template on a laptop or desktop computer that is networked into the organizational data collection system. Connecting the computer with a projector for an online live display enables peer review participants to review the wording of issue statements and action items to ensure that they have been correctly worded. Relevant data concerning preparation time, number of pages or lines of code reviewed, and time spent in the review can also be similarly captured.

8.1.4 Data Storage and Analysis

The means for storing both the raw data and the calculated results must also be specified. Some organizations use tools that are centrally controlled and available for storing the raw data and resultant calculations. Other organizations have a decentralized policy, and each part of the organization uses a different approach. In many cases, the responsibility for the various measurements is distributed to a number of different personnel. Some use spreadsheets that are stored on the hard disks of personal computers, essentially causing the data to be generally unavailable to the organization as a whole. Some of the calculations are performed by hand on printouts. For example, control charts are sometimes annotated in pencil on graphical printouts. While such an analysis of the control chart is extremely useful, hand annotations do not make the information available to the organization as a whole.

Automated tools for collecting and storing raw data and calculated results are strongly recommended. A number of tools for data collection and analysis are available on the market. Popular personal computer statistical analysis tools include Minitab® and JMP®. Web-based tools, such as DataDrill® and VisualCalc™ dashboards, allow for multiple users to enter and retrieve data on a non-interference basis, and are preferable to spreadsheet-based tools like Sigma XL® that are not designed for multiple users entering data concurrently. Figure 8.5 was prepared with JMP,* and Figures 8.6 through 8.10 with Minitab.† Access rules to the data for both entry and retrieval should be specified, and one person and an alternate should be identified as the controller of the data. In large organizations, where numerous measures are being taken, more than one person may be designated as centrally responsible, with the responsibility for specific measures allocated to specific individuals. If the organization chooses to defer acquiring a tool and wishes to use spreadsheets and other statistical tools that work with spreadsheets (e.g., Sigma XL, Crystal Ball®), there is still a necessity for the organization to control the input and use of the data. Relevant stakeholders need to use the data, and these stakeholders need to know that the data they are using is valid. They also need the information on a timely basis. Accordingly, controls need to be established, as do reporting requirements.

As noted earlier, if the organization is large and has instituted data centers at various reporting levels within the organization, it is important that tools, templates, data formats, forms, and procedures be consistent across the various levels of data centers. If data provided at a lower level within the organization must be manually re-input into another spreadsheet or form for data manipulation at a higher level within the organization, there is a nontrivial probability that transcription errors will occur, thus giving erroneous results.

8.1.5 Data Analysis Procedures

The final piece of the planning process is defining and documenting the data analysis procedures. There are two levels of analysis. The first is the analysis of the raw data, which is essentially the manipulation of the raw data into a form where further analysis can be performed by the stakeholders. This latter analysis is identified as second-level analysis. In establishing the data analysis procedures, we are mostly

---

* www.jmp.com.
† www.minitab.com.

concerned with the analysis at the first level. The purpose of establishing procedures at this level is to ensure that the format of the manipulated data, whether in graphical form, bar charts, tabular form, or whatever, is provided in a consistent format where meaningful analyses by others can be performed. In some cases, the first-level analysis may be more complex, and may actually cover concerns of the second level. To illustrate, in analyzing control charts, several rules can be applied for identifying special causes of variation. Some parts of the organization might apply only four signaling rules, whereas other parts might apply as many as eight rules (for information on rules for detecting special causes or sporadic events, see [12]). In order to ensure that special causes of variation are identified consistently from case to case, procedures must be in place to establish what signals are of concern and how to recognize them. Other personnel may be responsible for determining corrective action, such as fixes to the process or rebaselining the process, but there need to be procedures in place for that as well. For example, the process for performing root cause analysis should be codified to ensure that it is performed to the same level of detail in every case. Wide variations in the level of detail that organizations go to in performing root cause analysis affect data quality and organizational knowledge. Statistical hypothesis testing may be necessary after corrective actions have been applied in order to ensure that the corrective action applied has resulted in a significant change to the process. All of this implies that the personnel performing these functions should be trained in statistical analysis and the fundamentals of statistical process control (SPC).

An experience at one client company illustrates the importance of codifying how the data analyses are to be performed. A focused assessment of the data reporting system revealed that a significant special cause of variation was not identified. The assessment findings were that the project manager responsible for analyzing the data did not (1) recognize that the first-level analyst had missed the signal and (2) recognize that the control chart that was being used was incorporating data from what appeared to be two separate and distinct processes. The impact was a significant delay in identifying the problem and at considerable expense to the organization. Had there been more detailed procedures, and more rigorous application of the SPC conventions, it is likely that the correct interpretations would have been made, and separate control charts established for the two distinct processes. This would have prevented the expensive delays in the company's deliveries.

8.2 Operating a Data Reporting System

8.2.1 Staffing

The operation of a data reporting system involves dedicating personnel to this role and making sure they are properly trained to perform it. Their training should cover the operation of automated data collection systems and the analytic tools used, statistical analysis to be performed on the data, use of spreadsheet features, and understanding how to evaluate the inputs and results for validity and effectiveness. Depending on the size of the organization, this function can range from a part-time assignment in the smallest organizations to a number of full-time staff in very large organizations. In some companies, especially where Web-based tools are used to collect and analyze data, one or more persons are assigned the task of collecting the data and performing the computerized analyses of the data (a data coordinator). In this case, access permission profiles are established at different levels for the various stakeholders who interact with the system. Those who enter data may be granted permission to only input data with the specific data entities identified for which they have permission. Others may be granted permission to see only certain reports. Typically, the data coordinator will have the full range of permissions, and will establish the permission levels for all the other personnel interacting with the system.

8.2.2 Data Entry

The data may be entered directly into a database by the individuals responsible for their assigned data entities, or may be entered in spreadsheets or other forms and submitted to the data coordinator for input or for performing calculations and graphing. Input data should be checked against specified validity ranges. Data entered manually should be scanned by the data coordinator for validity. For data entered into a tool, validity ranges should be preset in the system, if possible. In any case, the data coordinator should perform quality checks on the data to ensure data validity. When anomalous results are observed, the data coordinator should discuss the related data with relevant stakeholders to determine what to do. This may involve review of all the input data and calculations to determine why the results came out the way they did. In some cases, the results may be valid; in others, the data may have to be reentered. In some cases, such as data from clinical trials, manual data is always entered twice and a cross-check is performed to ensure data quality. When it is determined that the results are acceptable, the data coordinator will prepare the necessary reports and send them to the appropriate stakeholders.

The data coordinator needs to manage the data entry process. It is up to the data coordinator to remind the people responsible for the inputs to get them in on time. When the inputs are not being provided at the required times, the data coordinator needs to escalate the information to the proper levels of management to ensure that the participants are performing their roles. The data coordinator is also responsible for maintaining a schedule and ensuring that all reports are provided at the specified time. Follow-up reports to the various stakeholders are required at prespecified intervals.

8.2.3 Data Backups

Two types of data backups should be a part of every data center's strategy. They serve two different functions: (1) on-site restoration to a previous checkpoint or recovery from a system failure, and (2) recovery from a disaster such as a fire or an earthquake that destroys the data center. Data backups are another responsibility of the data coordinator. On-site backups may occur on a daily basis, hourly basis, or even transaction by transaction. Backups may simply be a tape copy of the critical data taken once per day, and then stored in a safe location. Another strategy is for backups to be made to disk on-site and automatically copied to an off-site disk, or uploaded directly to an off-site disk with the data transmitted over a network to the off-site location. Data can also be mirrored to one or more other computer systems, depending on the defined backup strategy. The key consideration for any backup is the frequency at which the data is likely to change and the criticality of it. Data that is updated once per week will not need backing up any more frequently than that. On the other hand, data that is being entered daily should be backed up once per day, at a minimum. Credit card data centers that are continuously receiving large numbers of transactions need to backup or mirror the input data on a continuous, seamless basis.

Also of importance is the architecture of the data collection system. If all the elements of the data collection and analysis system are located on one machine, together with other data that is being collected by the organization such as personnel records, online document storage, financial data, purchase orders, order status, etc., the backup strategy and frequency will be dictated by the organizational data center policies. In such a case, the backup strategy needs to consider the needs of all the affected parties and ensure that the organization's management structure is responsive to all the user requirements. Backups are necessary for a number of reasons: data re-creation, disaster recovery, "what-if" analyses, etc. In the case of data re-creation, it is sometimes necessary to go back and re-create data from some point in the past. For example, an error may have been noted for some data point in the past, and all subsequent data points that depended on that value are wrong. The backup strategy must consider how far back the organization wants to be able to have the capability to retrieve old data; consequently, how long a period old data will be archived and stored is a major consideration. There may be governmental and regulatory agency requirements that set mandatory requirements on data archiving. Department of Defense (DoD) contracts, for example, will often specify a requirement for data retention for a minimum period of time extending beyond delivery of the final item on the contract. Accordingly, data storage facilities must be adequate to accommodate documents and backup tapes or disks for the required period of time.

Disaster recovery can be of two types: restoring from a system crash or reestablishing the databases in the event of a natural disaster such as destruction to the physical data center from fire or earthquake. In the event of a system crash, the backed-up data must of course be complete and readily available so that the system can be restored in a hurry. To recover from a crash, the backup data should either be on-site or, if at an off-site location, it should be capable of being downloaded from that site rapidly. In the event of a disaster, a complete physical rebuild of the data center will likely be necessary. Accordingly, the backed-up data should be stored at an off-site location far enough away so as to be unaffected by the natural disaster. The Los Angeles 6.7 magnitude earthquake in Northridge in 1994 caused severe damage in hit-or-miss fashion (fires and collapsed buildings) over a wide area up to approximately 20 miles from the epicenter. An area of approximately 75 miles was affected. A point 75 miles from the epicenter of an earthquake of that magnitude is unlikely to experience severe damage, and it would seem that an off-site location approximately 50 to 75 miles from the data center is reasonable. On the other hand, a 7.9 magnitude earthquake such as was experienced in Sichuan Province in China in 2008 would cause such severe damage over such a wide area that 50 miles wouldn't be far enough for an off-site location. Fortunately, earthquakes of that magnitude are very rare.

The bottom line on storing backup copies is that off-site storage is mandatory, and on-site storage may also be necessary, depending on how the data is being backed up and restored. The site for off-site storage should be selected on the basis of what kinds of natural disasters (hurricanes, earthquakes, etc.) are likely to occur, what extent of area is likely to be affected, and the vulnerability of the off-site location to the effects of the natural disasters.

8.2.4 Power Failure Considerations

To prevent loss of data in the event of a power failure, computers used for the data center operations should be powered by an uninterruptible power supply (UPS). The UPS should have surge protection and the capability to gracefully degrade; that is, in the event the power fails, the UPS should have battery backup and software that begins to shut down all applications running at the time of the power failure if the power is off beyond a specified time threshold. Without these features, valuable data can be lost or corrupted. Power shortages should be tracked in order to identify trends or significant deterioration in the quality of the power. Such tracking should be included in the data reporting system. For an advanced analysis of computer crashes due to power failures at Bell Laboratories, see [1]. While many of the precautions that we discuss here may seem self-evident, it is amazing the number of organizations that fail to take even some of the most basic steps. As an example, following the Northridge Earthquake, a large number of businesses were incapacitated for quite some time afterward for failure to provide readily available off-site backup storage.

8.3 Analyzing Data

The data center is also typically involved in analyzing several different categories of data. It supports the collection and analysis of data for the organization as a whole, for various product lines or service categories, and for individual projects.

Organizational data is used as benchmarks by the various projects for estimation, monitoring and control, and for establishing and maintaining process performance baselines. An example of organizational-level data is productivity data for specific types of tasks, normalized by an appropriate parameter such as size, or a process performance baseline established for defects per unit-level test normalized for unit size. Product line or service category data is similar in character but established for the product line or service category, rather than for the organization as a whole. Project-level data typically includes comparisons for each individual project of actual performance versus planned or budgeted performance. The higher levels of aggregation will generally compare actual results versus product line or organizational-level benchmarks or baselines.

As already noted, data analysis occurs at two levels: (1) basic preparation of the data for more detailed analysis and (2) detailed analysis for making engineering and management decisions. The sophistication of the analyses will depend on the maturity of the organization, using the CMMI as an example framework. Organizations functioning at Maturity Level 2 or 3 will typically produce data showing trends that indicate the status of the various processes being implemented by the organization. An organization functioning at Maturity Level 2 will be typically tracking:

- Requirements volatility (number of new, modified, and deleted requirements) by month for each project
- Planned versus actual hours spent in planning each project
- Actual staffing versus plan
- Actual time spent in performing individual tasks versus plan
- Action item status by month (number open versus number closed)
- Number of QA audits performed versus number planned by month
- Defect density by project
- Number of problem reports open and closed by month by defect severity
- Time to resolve problem reports as a function of problem severity
- Number of baseline configuration audits performed per month versus the plan

This list is not a recommendation concerning what should be measured or even a complete list of typical measures. It is only intended to illustrate the functioning of a data center.

The data center would be collecting data in support of specific projects, but would also be aggregating data in order to establish organizational benchmarks. For example, if the organization is collecting actual time spent in performing individual tasks versus the planned budget, the data will be collected from each project, formatted in tabular or graphical form to show trends over time, and then normalized by an appropriate size parameter to provide an organizational-level benchmark. For example, an organization developing Web-based applications might normalize the data by the number of screens created for each project. The project data would

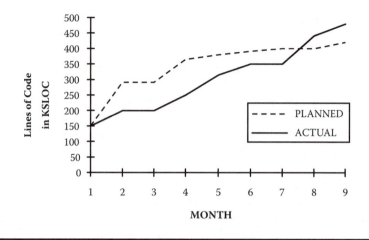

Figure 8.2   Software size. Kenett, R.S. and Baker, E.R., *Software Process Quality: Management and Control*, New York: Marcel Dekker, Inc., 1999.

be used by the project manager for his/her management of the project's activities, and the organizational-level data would be used for other purposes, such as facilitating estimation for upcoming projects.

The data center will also format data for use by the project, as illustrated by the example given in Figure 8.2. In this example, the data center presented the variation of the size of the software in terms of estimated or actual lines of code per month versus the planned size, by month. It is up to the project to make any necessary interpretations of the data in terms of whether or not the actual size at any point in time is a problem. In Chapter 4 we discussed various interpretations that the project could make of the data, which is a level of detailed analysis that usually occurs outside the operation of the data center.

Note that the planned number of lines of code in Figure 8.2 is not constant. One would think that the planned size of the software would be a constant; however, contractual changes to the scope of the effort would cause changes to the planned size, either up or down. In addition, the organization may have been keeping historical records that show how the planned size of the software typically grows over time, as the organization learns more about the scope of the effort as they got deeper into the details of the development in the later stages of the development cycle. Accordingly, the growth factor might be integrated into the planned profile, rather than to establish a higher fixed limit at the beginning of the project, because a large cushion, so to speak, during the early stages of the development effort might give the developers a false sense of security when in fact the size was already growing too big.

Similar kinds of graphics can be generated by the data center in support of product lines or the organization as a whole. At the organizational level, the organization may be tracking service response time as a measure of service quality once a product is on the market (see Figure 8.3). While this measure alone will not

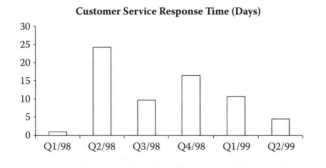

Figure 8.3   Customer service response time.

pinpoint specific problems, tracking the amount of time it takes to resolve service issues can indicate if there are problems with the functional entity performing service or problems with the product, in general. Here again, the data center presents the basic data, and it is typically up to a different part of the organization to make the interpretations regarding the data.

8.4 Reporting on Findings and Consequences

The data center supports its internal customers by reporting the basic metrics from which the findings and consequences are derived. Basic metrics are reported to various levels of management. The basic data is typically reported to a project manager, product line manager, or an individual designated within the organization to be responsible for various types of organizational data. This person, or his/her designee, will generally do a more detailed analysis of the data that is then reported at specified intervals up the organizational hierarchy. These reports contain the findings, consequences, and recommendations resulting from such an analysis. Sometimes, when the analysis is very straightforward, the output from the data center will be sufficient to address findings and consequences. For example, in Figure 8.4 we see a graphical portrayal of project personnel hours expended per month versus the planned expenditure. The project identified to the data center the threshold levels that were superimposed on the graph. Note that in Month 5, the project started overrunning, and it continued through the last reporting period in Month 8. The consequences of that, and recommendations concerning the resolution of the overrun, are generally contained in a PowerPoint presentation or textual report made by the cognizant manager to the next level of management. The graphic, however, would be provided by the data center for inclusion in the report.

In this example we looked at a situation where a project manager had to make a decision regarding the allocation and utilization of personnel resources to ensure that the project didn't finish in a cost overrun. A number of decisions could have been made; as an example, reduce the scope of the project. In this case, a number

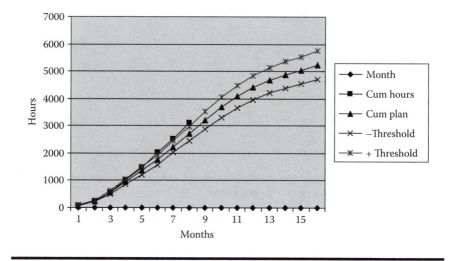

Figure 8.4   Tracking of aggregated person-hours.

of stakeholders are involved in the decision-making process: the customer (internal or external), the project manager, and his/her chain of command. The customer would have been involved from the perspective of agreeing or disagreeing with a reduction in the scope of the project, whereas management would have been involved with ratifying or vetoing the project manager's decision. In any event, in this example, the findings, consequences, and recommendations are confined to the project and its stakeholders. For other types of data reported, the consequences may be more far-reaching.

Organizations that are implementing Statistical Process Control (SPC) or that are operating at the higher maturity levels on the CMMI will be utilizing control charts. The CMMI high maturity organizations will also be utilizing process performance models. The data center will be producing the control charts, and qualified users will generally be interacting with the available statistical tools to perform process performance modeling calculations (e.g., Crystal Ball or @Risk to perform Monte Carlo analyses). In these kinds of circumstances, the analysis of the data may affect a development project and the organization as a whole, or a product line and the organization as a whole. As an example, consider an organization producing chemical solutions and controlling the chemical processes with SPC. After the data center has created a control chart showing the level of contaminants by production run of a chemical solution for a specific product, a control chart will indicate possible special causes of variation. The control chart is sent to the product manager for further analysis and action. He/she would then produce a report for upper-level management that shows the results of that analysis and the corrective action implemented. Typical issues addressed by such a report are: Did a special cause of variation exist? What was determined to be the underlying cause? Were the contaminants that were identified harmful, or were they not critical with respect to the use of the finished product? What corrective action was identified and applied? If the contaminants were not harmful, was it necessary to recalculate the process performance baseline for that chemical process for that product?

Clearly, there are implications for the organization as a whole, and the report to upper management might have to address that. For example, if the contaminant was in a chemical that was supplied by a vendor, and that chemical was used to make other products as well, all product lines using that input chemical would have to take some action concerning this. On the other hand, if the contaminant was considered not harmful, the process performance baseline (i.e., allowable contaminants for that type of chemical process) might be recalculated for the organization as a whole, if deemed to be a low risk option.

The reporting structure and process should take into consideration the fact that what happens locally within an organization can have consequences at a larger level within that organization as addressed by the Causal Analysis and Resolution (CAR) Process Area of CMMI. One of the practices in CAR has a link to the Organizational Innovation and Deployment Process Area for considering the applicability of a corrective action implemented within a project to the organization as a whole as a process change or improvement.

8.5 Field Data Reporting

We already noted that a measurement program is a "womb-to-tomb" effort. It begins with the first concept for a product or a service, and it continues until the product is retired or until the service is no longer performed. It is supplemented by the measures needed to facilitate the continued effectiveness and growth of the business as a whole. As such, the concerns of a variety of stakeholders, such as field support and maintenance services, sales and marketing, accounting, etc., must be considered in the initial development of a measurement program, in addition to the concerns of the product or service development effort. Some measures that will be taken will be organizational in nature, in that the measures provide the organization with information concerning such diverse considerations as sales, quality of service of all the company's products, maintainability, and effectiveness of field support services, such as help desks. Other measures may be unique to a product itself, such as the number of problem reports as a function of model or version. The product-unique measures will typically supplement the organizational measures. In some respects, these measures may be considered project-level measures in that separate and distinct projects may be organized around the initial release and subsequent updates to that specific product.

When focusing on field data in particular, the informational needs for field data and reporting are both organizational and product-unique. Two categories of informational needs with respect to field data become apparent: information for internal use and, in some cases, information required by regulatory agencies. Information required by regulatory agencies may be the easy part of specifying the informational needs in that the regulatory agency will specify what data is required. The harder part is specifying what is needed within the organization.

Organizational information needs address a number of areas: operational issues such as quality of their fielded products, maintainability, customer satisfaction, effectiveness of their customer support operations (help desk, repair service, etc.), and business-related issues, such as the proportion of the market penetrated, potential for future growth in any one market area, and new market segments to explore. Some of the business-related informational needs can be met from field data, while others cannot. For example, point-of-sale data can provide some insight into market penetration. Feature requests can give insight into the continued viability of that market segment, as well as possibly new products to develop.

As pointed out in previous chapters, the CMMI [4] is a collection of best practices. Several Process Areas discuss the establishment and use of an organizational process database. Field data is just another element of this database, and should be subject to the same kinds of controls that the organizational process database should be subject to.

In establishing the requirements for the collection and analysis of field data, we go through the same process for defining the measurement goals and objectives; the measures themselves; the data collection, storage, and analysis procedures; and reporting requirements that we have described in this and other chapters. Only the purpose for the measures, and what the source of the raw data might be, may be different from other measures. The needs for this data should not be treated differently from any other data, nor should they be treated as an afterthought.

As a final example in this chapter, we consider a company operating a data center tracking released software versions. At the end of the year, the data center produced a basic data report as presented in Table 8.1. The report includes six variables for forty-two software versions developed in nine different units:

1. Unit Number: The software development team that developed the software version.
2. PM: The total number of person months used to develop and test the software version.
3. Escaping Defects: The number of Severity 1 through 3 defects detected during the first 3 months after release of software version divided by PM—the lower the better.
4. Defect Distribution: The proportion of defects detected during the various internal testing phases versus defects detected by customers during the first 3 months after release of software version—the higher the value of this parameter, the better it is.
5. Effort Variance: The effort variance, in PM, from the budget baseline of the software version during the release development period.

Table 8.1   Data Center Basic Data Report

Unit	PM	Escaping Defects	Defect Distribution	Effort Variance	Scope Instability
1.0	727	0.31	0.71	−0.03	0.01
1.0	250	1.63	0.37	−0.09	0.66
1.0	773	0.12	0.92	0.04	0.14
1.0	49	0.05	0.89	−0.07	0.14
1.0	52	0.23	0.74	−0.09	0.03
1.0	923	0.08	0.92	−0.04	0.00
1.0	21	1.27	0.52	0.11	0.00
2.0	7	0.43	0.76	0.00	0.00
2.0	60	0.20	0.61	0.05	0.05
2.0	105	0.40	0.74	0.02	0.15
3.0	747	0.60	0.53	0.31	0.80
3.0	25	1.87	0.57	0.09	0.21
4.0	230	1.27	0.16	0.36	0.00
4.0	553	3.31	0.14	0.92	0.00
4.0	57	0.93	0.66	0.06	0.20
4.0	29	0.75	0.58	−0.10	0.03
4.0	60	0.63	0.66	−0.18	0.73
4.0	16	0.06	0.92	−0.03	0.00
4.0	36	0.90	0.41	0.00	0.00
4.0	37	0.38	0.39	0.09	0.01
4.0	86	0.69	0.73	0.08	0.00
5.0	157	0.31	0.68	0.02	0.57
5.0	182	0.99	0.57	0.12	2.25
5.0	35	0.46	0.56	0.09	3.46
5.0	214	0.30	0.86	0.11	0.21

Continued

Table 8.1   Data Center Basic Data Report (*Continued*)

Unit	PM	Escaping Defects	Defect Distribution	Effort Variance	Scope Instability
5.0	24	0.16	0.90	−0.08	0.00
6.0	53	0.78	0.75	0.07	0.54
7.0	29	0.41	0.70	−0.13	0.07
7.0	533	0.12	0.83	0.06	0.00
7.0	73	0.19	0.89	−0.02	0.13
7.0	87	0.51	0.69	0.02	0.07
7.0	118	0.42	0.70	−0.07	0.34
7.0	120	1.26	0.42	−0.26	0.00
8.0	74	1.73	0.45	0.04	0.14
8.0	42	0.50	0.76	−0.18	0.02
8.0	53	1.10	0.61	−0.22	0.04
8.0	10	1.33	0.35	0.10	0.16
8.0	96	0.51	0.67	−0.08	0.28
8.0	137	0.52	0.71	0.02	0.63
9.0	230	1.27	0.16	0.36	0.00
9.0	249	3.31	0.14	0.92	0.00
9.0	40	0.12	0.83	−0.06	0.00

6. Scope Instability: Represents changes to requirements scoped to the release and stability of the software development baseline, that is, number of changed, deleted, and added requirements after the scope sign-off date, divided by the number of requirements in the release scope at the release sign-off date.

Table 8.2 presents basic statistics of the variables tracked in the field report. We observe that for the 42 software versions being tracked, the invested Person-Months ranged from 176.2 to 923. Escaping Defects ranged from 5 defects per 100 PM to 33.1 defects per 100 PM. Defect Distribution ranged from 13.9% to 92% and Scope Instability from 0 to 3.46. The Effort Variance ranged from −26% to +92%. The reports also presented the lower quartiles, Q1, the Median and the upper quartiles, Q3, which indicate to us the values covering the central 50% of the values of these variables. Table 8.3 presents the report, by unit.

Table 8.2 Basic Statistics of Field Report Variables

Variable	Mean	Std. Dev.	Min.	Q1	Median	Q3	Max.
PM	176.2	235.9	7.0	36.8	73.5	218.0	923.0
Escaping Defects	0.772	0.751	0.050	0.282	0.510	1.140	3.315
Defect Distribution	0.6227	0.2207	0.1394	0.5025	0.6750	0.7600	0.9200
Effort Variance	0.0555	0.2352	−0.2600	−0.0725	0.0200	0.0900	0.9201
Scope Instability	0.2874	0.6373	0.0000	0.0000	0.0600	0.2275	3.4600

Table 8.3   Basic Statistics of Field Report Variables, by Unit

Variable	Unit	Mean	Standard Deviation	Minimum	Maximum
PM	1.0	399.0	394.0	21.0	923.0
	2.0	57.3	49.1	7.0	105.0
	3.0	386.0	511.0	25.0	747.0
	4.0	122.7	173.7	16.0	553.0
	5.0	122.4	87.3	24.0	214.0
	6.0	53.0	*	53.0	53.0
	7.0	160.0	185.8	29.0	533.0
	8.0	68.7	44.4	10.0	137.0
	9.0	173.0	115.6	40.0	249.0
Escaping Defects	1.0	0.527	0.645	0.050	1.630
	2.0	0.343	0.125	0.200	0.430
	3.0	1.235	0.898	0.600	1.870
	4.0	0.991	0.935	0.060	3.310
	5.0	0.444	0.323	0.160	0.990
	6.0	0.780	*	0.780	0.780
	7.0	0.485	0.408	0.120	1.260
	8.0	0.948	0.521	0.500	1.730
	9.0	1.570	1.616	0.124	3.315
Defect Distribution	1.0	0.7243	0.2128	0.3700	0.9200
	2.0	0.7033	0.0814	0.6100	0.7600
	3.0	0.5500	0.0283	0.5300	0.5700
	4.0	0.5167	0.2622	0.1400	0.9200
	5.0	0.7140	0.1593	0.5600	0.9000
	6.0	0.7500	*	0.7500	0.7500

Table 8.3   Basic Statistics of Field Report Variables, by Unit (*Continued*)

Variable	Unit	Mean	Standard Deviation	Minimum	Maximum
	7.0	0.7050	0.1621	0.4200	0.8900
	8.0	0.5917	0.1596	0.3500	0.7600
	9.0	0.3740	0.393	0.1390	0.8280
Effort Variance	1.0	−0.0243	0.0744	−0.0900	0.1100
	2.0	0.0233	0.0252	0.0000	0.0500
	3.0	0.200	0.156	0.090	0.310
	4.0	0.133	0.331	−0.180	0.920
	5.0	0.0520	0.0835	−0.0800	0.1200
	6.0	0.0700	*	0.0700	0.0700
	7.0	−0.0667	0.1159	−0.2600	0.0600
	8.0	−0.0533	0.1282	−0.2200	0.1000
	9.0	0.4070	0.4900	−0.0560	0.9200
Scope Instability	1.0	0.1400	0.2376	0.0000	0.6600
	2.0	0.0667	0.0764	0.0000	0.1500
	3.0	0.5050	0.4170	0.2100	0.8000
	4.0	0.1078	0.2422	0.0000	0.7300
	5.0	1.2980	1.4980	0.0000	3.4600
	6.0	0.5400	*	0.5400	0.5400
	7.0	0.1017	0.1267	0.0000	0.3400
	8.0	0.2117	0.2252	0.0200	0.6300
	9.0	0.0000	0.0000	0.0000	0.0000

From Table 8.3 we notice that the largest software version (923PM) was developed by Unit 1, and Unit 9 experienced the highest escaping defects (3.315). Both Tables 8.2 and 8.3 were generated with Minitab version 15.1.

Histograms of the data summarized in Table 8.2 are presented in Figure 8.5, generated with JMP version 8.0. The darkened areas in the histograms correspond to the group with lowest escaping defects. We notice that it corresponds to the versions with highest defect distribution. Figure 8.6, Figure 8.7, and Figure 8.8 present

Figure 8.5   Histograms of variables in field report.

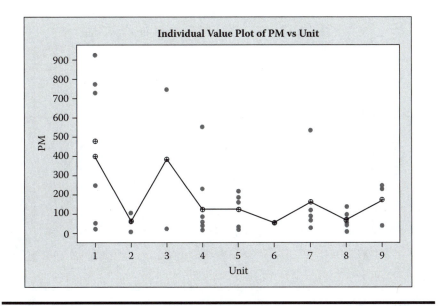

Figure 8.6   Individual value plot of PM, by unit.

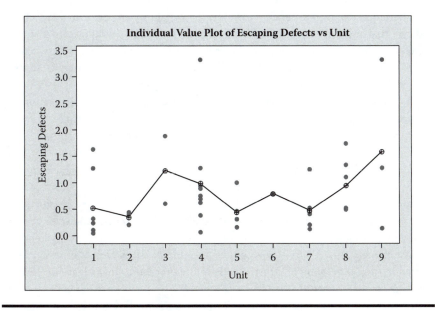

Figure 8.7   Individual value plot of escaping defects, by unit.

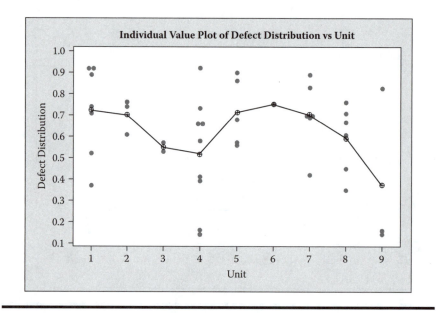

Figure 8.8   Individual value plot of defect distribution, by unit.

plots of individual values of PM, Escaping Defects, and Defect Distribution, by unit, for the forty-two software versions presented in the report. The average value for each unit is indicated by connected circles.

From these figures (generated with Minitab version 15.1) we can identify units with high and low performance. However, the variability in performance seems to indicate that there are no units with unusual performance. In fact, formal statistical testing shows that there are no statistically significant differences between the units. The main implication of this finding is that improvements must be considered at the organizational level, and not focused at specific units.

As a follow-up, the data center decided to investigate the overall relationship between Escaping Defects and Defect Distribution. A simple scatterplot shows, as expected, that the higher the Defect Distribution, the lower the Escaping Defects (see Figure 8.9).

To quantify this relationship, a linear regression was fitted to the data. Figure 8.9 shows the fit and 95% Prediction Intervals (PI), indicating the expected level of Escaping Defects for individual Defect Distributions. We observe a reasonable fit with R-Square of 64.3%. On the left of the plot we have several software versions with very low Defect Distribution, implying that most defects there were reported by customers (about 90%!). What could cause this? Perhaps Defect Distribution is related to total development effort. We therefore proceed to investigate the data by focusing on the impact of PM.

If we quantify software versions as "Average Size" for PM < 180, "Intermediate Size" for 180 < PM < 500, and "Very Large" for PM > 500, we can redraw Figure 8.9

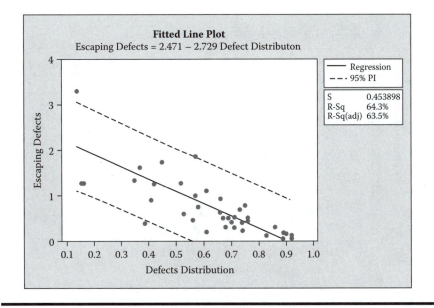

Figure 8.9 Scatterplot of escaping defects by defect distribution, with regression line.

with coded data. Figure 8.10 presents the resulting scatterplot with added "jittering" of the points in order to distinguish software versions having identical values. From this figure we notice that the software versions on the left are all related to Intermediate Size or Very Large versions. We do, however, see Very Large versions also on the bottom right of the plot so that PM alone does not predict low Defect Distribution (and high Escaping Defects). To further investigate this point, we need to break down the PM figures into various components such as design, coding, and testing. This data was not available, however, and a change in the reporting system was suggested to allow for such tracking.

As discussed in previous chapters, measurement programs track variables over time. A basic tool for such tracking is the control chart first presented in Chapter 4. Another tool, with faster detection of changes is the CUSUM chart presented in Chapter 6. We proceed to describe yet another procedure, the Bayesian Estimate of the Current Mean (BECM), for tracking a key indicator such as escaping defects or effort variance. The procedure is more informative than the control chart or CUSUM chart and provides effective diagnostics. BECM is described in detail in [12, 14]; here we focus only on a general introduction with a case study.

Consider a Key Indicator such as Effort Variance, which is reported at fixed time intervals such as weekly or monthly. Let us label these reporting times $i = 1, 2,$ .... A general term describing these observations is X_i. The reported observations, X_i, are determined by the value of a base average level μ_0 and an unbiased error e_i. In order to provide a concrete example, we refer to these observations as *effort*

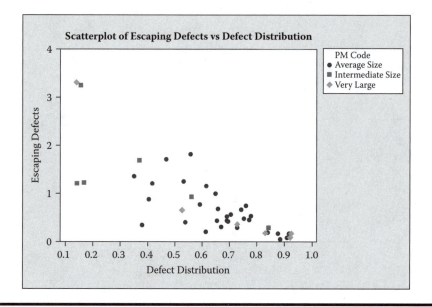

Figure 8.10 Scatterplot of escaping defects by defect distribution with coded data, by PM.

variance. Effort variances are due to main causes that can be modeled statistically. We want to track changes over time that reflect upward or downward shifts in order to trigger appropriate management actions. Possible deviations in effort variance are modeled by having μ_0 randomly determined according to a normal distribution with known mean μ_T (a specified target value) and a known variance σ^2. At some unknown point in time, a disturbance might change the process average level to a new value, $\mu_0 + Z$, where Z models a change from the current level. The probability of shifts in effort variance is considered a constant, p. Accordingly, we apply a model of at most one change point at an unknown location. The corresponding statistical model, for at most one change point, within n consecutive observations is as follows:

$$X_i = \mu_i + e_i, \quad i = 1, 2, \ldots, n$$

where μ_i is the observation's mean and e_i the measurement error; the at-most-one-change point model is:

$$\mu_i = \mu_0, \quad i = 1, \ldots, j - 1$$

and

$$\mu_i = \mu_0 + Z, \quad i = j, \ldots, n$$

where j indicates that the shift occurred in the time interval between the $(j - 1)$st and j-th observations; $j = 1$ indicates a shift before X_1; and $j = n + 1$ indicates no shifts among the first n observations.

The Bayesian framework is based on the following distributional assumptions:

$$\mu_0 \sim N(\mu_T, \sigma^2), \quad Z \sim N(d, t^2), \quad e_i \sim N(0, 1), \quad i = 1, 2, \ldots$$

The parameters μ_T, σ^2, and t^2 are assumed to be known based on previous or benchmark data, and $N(\cdot, \cdot)$ designates a random variable having a normal distribution. For n observations, let J_n denote the random variable corresponding to the index of the time interval at which a shift occurred. The BECM model provides an estimate of the current mean μ_n and its distribution percentiles. As an example, take fifteen effort variance monthly observations, from July to January.

Table 8.4 lists the posterior distributions of J_i for this time frame. The maximal probabilities are indicated in bold and point to an estimate of where the shift has occurred (i.e., October). The probability of shift in the next interval represents the probability of a future shift.

Up to October, a shift is indicated as potentially occurring in the future. After the November observation, the maximal probability of a shift is consistently pointing at the interval between October and November. Table 8.5 shows the percentiles of the distribution of the effort variances. As expected, the procedure shows a decrease in the mean in November, that is, a significant improvement.

To present the results graphically, we can use a boxplot display. Figure 8.11 presents a boxplot of the "effort variance" distribution in December, generated from the BECM model using data up to December. The statistics of the distribution are presented below the graph, indicating a mean (average) of 0.26 (i.e., +26%), a standard deviation of 0.5811, a lower quartile (Q1) of –0.0892, and an upper quartile (Q3) of 0.6237. The boxplot consists of a box with boundaries at Q1 and Q2, a partition at the position of the median, whiskers up to 3 standard deviations of the median, outliers beyond these limits indicated as stars, and a crossed circle indicating the position of the mean. Figure 8.12 presents boxplots from July to January. The shift in November is clearly visible, indicating a significant change from an average effort variance of –200%, indicating possible over-budgeting to a level around zero, which represents a more realistic plan and better forecasting in project efforts.

A variation of BECM was developed and implemented by Bruce Hoadley at AT&T in the form of the Quality Measurement Plan (QMP). For more on BECM, QMP, and other tracking methods, see Chapter 12 in [12].

8.6 Summary

In previous chapters we discussed how measurement programs are defined and established, and what the data tells us about the status of processes in use and the

Table 8.4 Probability of a Shift in Effort Variance at Any Given Month

i	July	August	September	October	November	December	January
Xi	-2.081	-2.837	-0.731	-1.683	1.289	0.5650	0.2560
Jan	0.0027	0.0028	0.0023	0.0025	0.	0.	0.
Feb	0.0057	0.0066	0.0037	0.0037	0.	0.	0.
Mar	0.0022	0.0023	0.0016	0.0016	0.	0.	0.
Apr	0.0072	0.0085	0.0029	0.0026	0.	0.	0.
May	0.0050	0.0054	0.0019	0.0017	0.	0.	0.
June	0.0047	0.0043	0.0017	0.0016	0.	0.	0.
July	0.0194	0.0039	0.0051	0.0055	0.0004	0.0001	0.
Aug	**0.9606**	0.0079	0.0078	0.0074	0.0011	0.0003	0.0002
Sept	—	**0.9583**	0.1583	0.0643	0.0964	0.0794	0.0927
Oct	—	—	**0.8149**	0.0139	0.0155	0.0149	0.0184
Nov	—	—	—	**0.8952**	**0.8833**	**0.9052**	**0.8886**
Dec	—	—	—	—	0.0032	0.0001	0.0001
Jan	—	—	—	—	—	0.	0.
Feb	—	—	—	—	—	—	0.

Table 8.5 Percentiles of the Distribution of the Mean Effort Variance at Any Given Month

i	July	August	September	October	November	December	January
X_i	−2.081	−2.837	−0.731	−1.683	1.289	0.5650	0.2560
$\mu_{.01}$	−2.937	−2.979	−2.746	−2.697	−1.789	−1.044	−0.911
$\mu_{.25}$	−2.328	−2.401	−2.193	−2.181	−0.0587	−0.101	−0.094
$\mu_{.50}$	−2.085	−2.172	−1.947	−1.966	−0.113	+0.292	0.243
$\mu_{.75}$	−1.839	−1.941	−1.627	−1.735	+0.368	0.690	0.584
$\mu_{.99}$	−0.831	−1.233	0.872	−0.263	1.581	1.668	1.425
μ_i	−2.069	−2.161	−1.731	−1.918	−0.108	0.296	0.246

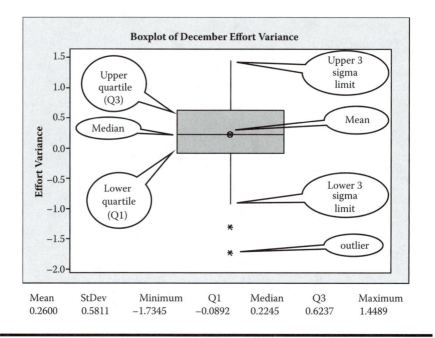

Mean	StDev	Minimum	Q1	Median	Q3	Maximum
0.2600	0.5811	−1.7345	−0.0892	0.2245	0.6237	1.4489

Figure 8.11   Boxplot of effort variances in December.

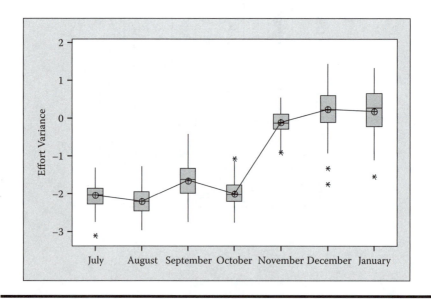

Figure 8.12   Boxplots of effort variances in July–January.

products that they produce. In this chapter we discussed the mechanics of how that data is produced, analyzed, and reported. We discussed the planning aspects, which involve:

- Specifying the measures
- Defining the data collection and analysis procedures
- Training of the data center personnel
- The need for tools, templates, and other forms for ensuring consistency in how the data is collected
- Repositories for the data

In discussing the operation of the data center, we pointed out the:

- Need for dedicated personnel
- Need for access control
- Role of the data coordinator
- Need for an intelligent backup and restore strategy
- Need for backup power

We discussed the analysis of the data and the kinds of analyses that are performed. This included the responsibility for the various levels of analysis. We also discussed reporting in terms of who is responsible for what aspects of reporting, and to whom the reports go. We concluded the chapter with a discussion of field data reporting, including a comprehensive example. For more general aspects of *data reporting systems*, we refer the reader to [3, 4, 11]. For specific examples from telecommunication systems, see [1, 2]. For examples of data reporting and analysis systems in risk management, see [7–9]. Unique characteristics of tracking systems with a focus on usability of Web applications and Service-Oriented Architectures are presented in [5, 11]. Aspects of tracking defects for assessing the effectiveness of testing and inspections are discussed in [10]. Other tracking techniques applied to system and software applications, such as the Shiryayev-Roberts control chart and statistical comparisons of Pareto charts, are described in [6, 13].

This chapter also presented the basic elements of field reporting systems, supplemented by an example of primary, basic analysis and second-level, more sophisticated statistical analysis. In Chapter 10 we provide a case study where process and organizational improvements have been carried out in the context of an organization implementing Agile Scrum development.

References

1. Amster, S., Brush, G., and Saperstein, B., Planning and Conducting Field-Tracking Studies, *The Bell System Technical Journal,* 61(9) Part 1, 2333–2364, November 1982.

2. Bellcore, Reliability and Quality Measurements for Telecommunications Systems (RQMS), GR-929-CORE, Issue 2, December 1996.
3. Chrissis, M., Konrad, M., and Shrum, S., *CMMI: Guidelines for Integration and Product Improvement, 2nd edition,* Reading, MA: Addison-Wesley, 2007.
4. CMMI Product Team. CMMI for Development, Version 1.2 (CMU/SEI-2006-TR-008), Pittsburgh, PA: Software Engineering Institute, Carnegie Mellon University, 2006.
5. Harel A., Kenett R., and Ruggeri, F., Modeling Web Usability Diagnostics on the Basis of Usage Statistics, in *Statistical Methods in eCommerce Research*, W. Jank and G. Shmueli, Editors, New York: Wiley, 2008.
6. Kenett, R.S. and Pollak, M., Data-Analytic Aspects of the Shiryayev-Roberts Control Chart: Surveillance of a Nonhomogeneous Poisson Process, *Journal of Applied Statistics*, 1(23), 125–137, 1996.
7. Kenett, R.S. and Raphaeli, O., Multivariate Methods in Enterprise System Implementation, Risk Management and Change Management, *International Journal of Risk Assessment and Management*, 9(3), 258–276, 2008.
8. Kenett, R.S. and Salini, S., Relative Linkage Disequilibrium Applications to Aircraft Accidents and Operational Risks, *Transactions on Machine Learning and Data Mining* 1(2), 83–96, 2008.
9. Kenett, R.S. and Salini, S., Relative Linkage Disequilibrium: A New Measure for Association Rules, in P. Perner, Editor, *Advances in Data Mining: Medial Applications, E-Commerce, Marketing, and Theoretical Aspects*, ICDM 2008, Leipzig, Germany. Lecture Notes in Computer Science, Springer Verlag, Vol. 5077, 2008.
10. Kenett, R.S., Assessing Software Development and Inspection Errors, *Quality Progress*, p. 109–112, October 1994, with correction in issue of February 1995.
11. Kenett, R.S., Harel, A., and Ruggeri, F., Controlling the Usability of Web Services, *International Journal of Software Engineering and Knowledge Engineering*, World Scientific Publishing Company, Vol. 19, No. 5, pp. 627–651, 2009.
12. Kenett, R.S. and Zacks, S., *Modern Industrial Statistics: Design and Control of Quality and Reliability*, San Francisco: Duxbury Press, 1998. (2nd edition, 2003; Chinese edition, 2004.)
13. Schulmeyer, G., *Handbook of Software Quality Assurance, 4th edition*, Boston, MA: Artech House, 2008.
14. Zacks, S. and Kenett, R.S., Process Tracking of Time Series with Change Points, in *Recent Advances in Statistics and Probability (Proceedings of the 4th IMSIBAC)*, J.P. Vilaplana and M.L. Puri, Editors, Utrecht: VSP International Science Publishers, 1994, p. 155–171.

Chapter 9

Systems Development Management Dashboards and Decision Making

Synopsis

In Chapter 8 we discussed establishing a measurement program and the scope of the measurements included in such a program. As part of that discussion, we talked about reporting the results of the measurement and analysis effort, and the necessity for disseminating that information to the relevant stakeholders. In this chapter we discuss an approach for disseminating that information in a handy, easy-to-view format: the Systems Development Management Dashboard. We discuss the approach and how a tool for it can be obtained.* As discussed in Chapter 1, metrics are used to provide answers to questions that help us make decisions. This chapter discusses project-level and organizational-level measures, and provides

---

* The authors wish to thank American Systems for granting permission to incorporate their copyrighted material concerning the Control Panel and Risk Radar in this chapter. Any use of this material without the written permission of American Systems is forbidden. The authors of this book have no financial interest in the sale or use of these tools.

an introduction to decision-making models. We also review the impact of "Black Swans"—that is, rare events with significant consequences—that modern managers need to know how to identify and address. In that context we present the Taleb Quadrants [9], and a basic review of mathematical and statistical decision theory.

9.1 Introduction to Systems Development Management Dashboard

A Systems Development Management Dashboard (SDMD) is a tool for visualizing and monitoring the status of a project. (A project might be a development project for a product, a system, or software, or it could be a process improvement project.) It visually provides information vital for predicting its future course. Visualization of information is a critical feature of management systems. For more on this topic, see [5]. The SDMD metrics and charts, when accessible to all members of the team, allow the entire project team to determine the status of a project quickly and identify areas for improvement. An SDMD can help project managers keep their projects on course when data for the control panel is updated regularly and gauges are maintained within acceptable ranges. Organizations that implement a process improvement model such as the CMMI must operate such a dashboard. In general we distinguish five major categories of project data:

1. Progress reports
2. Requirements and configuration
3. Staffing levels
4. Risk analysis
5. Quality information

These categories cover the primary areas that every project manager should track on large-scale systems development projects. For each category, we describe a set of generic metrics that are routinely used by organizations to manage projects. These can, of course, be supplemented with additional metrics, as needed, to provide better insight into the status of the project. The CMMI Level 2 Process Areas (PAs) cover these five categories so that organizations at Level 2 or beyond should be able to operate an SDMD without difficulty. In the next section we present an example of an SDMD.

9.2 A Systems Development Management Dashboard Case Study

In this section we describe a case study implementation of a Systems Development Management Dashboard. The SDMD Control Panel we refer to was developed originally by the Software Program Managers Network (SPMN)* as a comprehensive shareware Excel implementation (see [2]). This Control Panel is now offered as a free download and is a product of American Systems [2]. It is a tool for visualizing and monitoring the condition of a project, and predicting its future course. The SDMD Control Panel helps project managers keep their projects on course when data is updated regularly and gauges are maintained within acceptable ranges.

In Chapter 3 we presented the structure of the CMMI model [1] and pointed out that each PA is comprised of specific and generic goals. Simply put, *Specific Goals* are comprised of practices that implement the objectives of the process area, while the *Generic Goals* are comprised of practices that create an infrastructure to ensure that the specific practices of the PA will be performed in a consistent manner every time the specific practices in the PA are performed. These are referred to as generic, in that they are common to every PA in the model. In particular, every PA has a Generic Practice (GP) 2.8, which is "Monitor and Control the Process." Implementation of this practice involves collecting quantitative data to evaluate the status and effectiveness of the performance of the PA. The data shown in this case study is an example of metrics that can be used to implement GP 2.8. Note that in the example we provide, there is often more than one metric described for a given PA. In some cases, a single metric may suffice to evaluate status and effectiveness; in other cases, however, more than one may be needed.

The Control Panel† software allows the user to add or delete gauges as the user sees fit. Consequently, if there are other indicators that a project leader would like to include, they can be added.

The SDMD Control Panel, as depicted here, has five major categories and two subcategories of project data, namely:

- Progress
  - Productivity
  - Completion
- Change
- Staff
- Risk
- Quality

These categories were chosen to cover the primary areas that every project manager should track on medium- and large-scale development projects. The project data reported by the gauges are derived from basic data contained in an Excel

---

* The Control Panel is now available from American Systems. A demo version can be downloaded from the American Systems website [2].
† The description of data categories are excerpted from copyrighted material provided by American Systems.

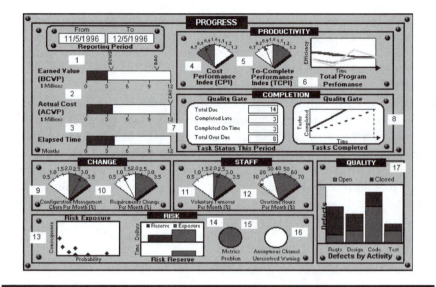

Figure 9.1 The SDMD Project Control Panel (The graphic of the Control Panel is copyrighted by American Systems Corporation. The descriptions on the following pages of the seventeen gauges contained in the Control Panel shown are excerpted from American Systems Corporation's copyrighted materials. Use of these descriptions without specific authorization from American Systems is prohibited.)

workbook, which forms the basic part of the tool.* The data is for a hypothetical project, but a number of the gauges described here relate to measures that we have discussed in previous chapters as those that are recommended for managing projects. The tool includes a spreadsheet (worksheet) of project data. That worksheet also provides a definition of project boundary conditions, for example, total schedule, budget, etc. Table 9.1 illustrates that portion of the spreadsheet. (The control panel, which depicts how all the information is tied together, is illustrated in Figure 9.1). We describe below how the information in the worksheet is used as input to the gauges.

The gauges are numbered in the figure in accordance with the descriptors below. The data in this case is for a mythical project, and is intended to illustrate the kinds of information that can be displayed in a Control Panel.

9.2.1 Progress

The basic data for Gauges 1, 2, 4, 5, and 6 is derived from the data contained in the portion of the spreadsheet shown in Table 9.2

---

* The description of the tool is excerpted from copyrighted material provided by American Systems.

Table 9.1 Project Boundary Conditions

Project Date Data				Project Baseline Data	
Type of Reporting Period (M = Monthly or W = Weekly)	Project Start Date (month/day/year)	Total Length of Project (in reporting periods)	Current Reporting Period	Budget at Completion (BAC)	Original Number of Project Requirements
Monthly	8/6/2006	12	4	10,000,000	85

Table 9.2 Cost/Schedule Performance Data

Reporting Period Data		Earned Value Measures			
Reporting Period No.	Period End Date	Cumulative Planned Value (BCWS)	Cumulative Earned Value (BCWP)	Actual Cost (ACWP)	Estimate at Completion (EAC)
1	9/5/2006	500,000	500,000	650,000	10,000,000
2	10/5/2006	1,000,000	970,000	1,550,000	11,000,000
3	11/5/2006	2,000,000	2,000,000	2,500,000	11,500,000
4	12/5/2006	3,500,000	3,000,000	3,700,000	12,000,000
5	1/5/2007	4,000,000			
6	2/5/2007	5,250,000			
7	3/5/2007	6,150,000			
8	4/5/2007	7,000,000			
9	5/5/2007	7,750,000			
10	6/5/2007	8,500,000			
11	7/5/2007	9,250,000			
12	8/5/2007	10,000,000			

1. In Chapters 1 and 4 we discussed *earned value*. Earned value is a means of determining the degree to which the project is on schedule and meeting budgetary constraints. It also provides a rough estimate of cost at completion. The Earned Value or Budgeted Cost of Work Performed (BCWP) gauge (Gauge 1) shows the cumulative earned value delivered to date. The cumulative Earned Value indicator shows the amount of work that has been completed on the project. This metric is based on the notion that, at the beginning of a project, every task is allocated a budget, which then becomes its planned value. The tasks are generally defined as elements of a work breakdown structure (WBS), and costs and schedule are estimated based on the definition of the work packages (WBS elements) contained within the WBS. As work is completed on a task (work package), its budget (or planned value) is "earned" as a quantitative measure of progress. The maximum value on the gauge is the total original budget for the project known as Budget at Completion (BAC). Note that the BAC is constant for the life of the project (if no additional funds are authorized for the project) and represents the total value of work to be performed. The other indicator on the gauge shows the

cumulative planned value or Budgeted Cost of Work Scheduled (BCWS), which is the total value of work that was originally scheduled for completion by the end of this reporting period.

The BCWP, BCWS, and BAC indicators can be compared with one another to make critical observations about progress on the project. By comparing the BCWP indicator with the BCWS indicator, you can determine if the project is ahead of or behind schedule. This metric is referred to as the Schedule Performance Index (SPI). It is calculated by dividing BCWP by BCWS. This is a good measure of schedule deviation because it takes into account the amount of work that was planned to be completed. A value of 1.0 indicates that the project is on schedule; a value less than 1.0 indicates that the project is behind schedule.

In the example shown in Figure 9.1, SPI indicates that the project is behind schedule. Establishing a planned value and a completion criterion for each task before work begins is critical for using the Earned Value metric successfully to measure progress. BCWP is the sum of the planned values for all completed tasks. The best completion criteria for a software task will require that no planned value credit can be taken until all work is completed and tested. These completion criteria are known as quality gates.

2. The Actual Cost or Actual Cost of Work Performed (ACWP) gauge (Gauge 2) shows the cumulative actual cost incurred on the project to date. Estimate at Completion (EAC) is the maximum value on this gauge, which represents the current best estimate for total cost of the project. Note that EAC might have a different value from BAC in the above Earned Value gauge because better total cost estimates can be made as the project progresses. At times, EAC may actually exceed BAC, and EAC may even change for different reporting periods.

   By dividing ACWP into the BCWP in the above Earned Value gauge, you can estimate how your project is performing against its budget. This metric is known as the Cost Performance Index (CPI, Gauge 4); it shows how well the project is turning actual costs (ACWP) into progress (BCWP). Although the scales for this gauge and the Earned Value gauge are the same, cumulative actual cost can be compared with BAC to determine project status toward overrunning the original budget, and with EAC to determine project status toward overrunning the current estimated total cost. In the example shown in Figure 9.1, it would indicate that the project is overrunning its budget.

3. The Elapsed Time gauge (Gauge 3) shows the end date for the current reporting period. It indicates the number of months that have elapsed since the project started.

4. The Cost Performance Index (CPI) gauge (Gauge 4) shows how efficiently the project team has turned costs into progress to date. It is calculated by dividing the cumulative Earned Value by the cumulative Actual Cost (BCWP/ACWP). It is a historic measure of average productivity over the life of the project. CPI represents how much work was performed for each dollar spent,

or "bang for the buck." When CPI has a value of 1.0, the project team is delivering a dollar of planned work for each dollar of cost. When CPI is less than 1.0, there is the potential for a productivity problem. For example, a CPI of 0.82 means that you got 82 cents worth of planned work for each dollar you paid in cost. A CPI of less than 1.0 may indicate that the project team didn't perform as well as expected, or that the original budget was too aggressive for the amount of work to be performed.

5. The To-Complete Performance Index (TCPI) gauge (Gauge 5) shows the future projection of the average productivity needed to complete the project within an estimated budget. It is calculated by dividing the work remaining by the current estimate of remaining cost [(BAC – BCWP)/(EAC – ACWP)].

   The TCPI gauge must be used in conjunction with the CPI gauge. TCPI should be compared to CPI to determine how realistic the most recent estimated total cost (EAC) is for the project. Note that CPI measures the average historic productivity to date. If TCPI is greater than CPI, then the project team is anticipating an efficiency improvement to make it more productive. The estimated total cost of the project (EAC) can therefore be calibrated by comparing TCPI to CPI. Always question claims of future productivity improvement that result in a 20% or greater increase in TCPI over CPI in order to ensure that they are based on sound reasoning. This is especially true of "silver bullets" like new tools, languages, or methodologies, which may actually decrease productivity due to training and start-up costs. The line on this gauge should be about 20% above the current value of the CPI gauge to show the relationship and warning level between the two gauges.

6. The Abba* Chart, also known as a Total Program Performance Efficiency chart, (Gauge 6) is composed of four different performance indicators showing trends in historic and projected efficiency to date. The indicators are:
   - TCPI (Gauge 5): The curve in Gauge 6 is TCPI.
   - Completion Efficiency (CE): The curve is a ratio calculated by dividing BAC by EAC to estimate the productivity required to complete the project within a projected total cost (EAC). Note that this curve lies essentially on top of the TCPI curve, effectively making it look like there is only one curve.
   - CPI (Gauge 4): The dark gray curve in Gauge 6 is CPI.
   - Monthly CPI: The light gray curve; it is a ratio calculated by dividing the monthly Earned Value by the monthly actual cost (as opposed to cumulative values for the CPI calculation).

   The data for the graphs is also derived from the hypothetical cost/schedule data contained in the Excel workbook.

7. Quality Gate Task Status This Month (Gauge 7) shows the completion status of tasks during the current reporting period. A quality gate is a predefined completion criterion for a task. The criterion must be an objective yes/no

---

* Named for Wayne Abba of the U.S. Department of Defense.

indicator that shows a task has been completed (see discussion on Gauge 1 above). The task completion data is derived from the earned value data to the extent that the tasks are the work packages in the WBS used to calculate earned value. The indicators are:

- – *Total Due* is the total number of tasks scheduled for completion during this reporting period plus any overdue tasks from previous periods. This indicates the total quantity of work required for the project to keep pace with the schedule.
- – *Completed Late* is the number of tasks completed late during this reporting period. This number includes those tasks scheduled for this period that were completed late, as well as any overdue tasks from previous periods that were completed in this period. The Completed Late indicates how well the project is completing work, even if it is late according to the original schedule.
- – *Completed On Time* is the number of tasks originally scheduled for completion during this reporting period that were completed by their original scheduled due date. This number indicates how well the project is keeping up with scheduled work.
- – *Total Overdue* is the total number of tasks for all previous reporting periods that are overdue by the end of the current reporting period. This is an indicator of the quantity of work needed to get the project back on schedule.

The total number of tasks completed in this reporting period (Gauge 8) is the sum of Completed On Time and Completed Late. Total Overdue then is equal to Total Due minus Completed On Time and Completed Late.

The data for Gauges 7 and 8 are taken from the portion of the worksheet illustrated in Table 9.3.

8. The Quality Gate Tasks Completed graph (Gauge 8) shows the cumulative number of tasks completed by the end of each reporting period to date plotted with the cumulative number of tasks scheduled for completion. If the number of tasks completed falls below the number planned, then the horizontal distance on the time axis gives an idea of the current schedule slip to date.

As noted previously in our discussion of the CMMI models, the data in these gauges are metrics that can be used to implement GP 2.8. Gauges 1 through 8 are examples of metrics that can be utilized to implement GP 2.8 for the Project Monitoring and Control (PMC) process area.

9.2.2 Change

9. CM (Configuration Management) Churn per Month (Gauge 9) is calculated by taking the number of items under configuration control (baselined items) that have been modified and rechecked into the configuration management system over the last reporting period, and dividing that number by the total number of baselined items in the system at the end of the period. It is expressed

Table 9.3 Quality Gate Task Data

Reporting Period Data			Quality Gate Tasks			
Reporting Period No.	Period End Date	No. Scheduled This Period	Total Due This Period (no. scheduled + no. overdue from last period)	No. Completed On Time This Period	No. Completed Late This Period	Cumulative No. Overdue at End of Period
1	9/5/2006	5	5	3	1	1
2	10/5/2006	7	8	2	1	5
3	11/5/2006	7	12	2	2	8
4	12/5/2006	6	14	3	3	8
5	1/5/2007	10	18			18
6	2/5/2007	12	30			30
7	3/5/2007	15	45			45
8	4/5/2007	13	58			58
9	5/5/2007	11	69			69
10	6/5/2007	16	85			85
11	7/5/2007	12	97			97
12	8/5/2007	5	102			102

as a percentage. A modified baselined item is one that was previously in the system, but was reviewed sometime later and modified or replaced.

The worksheet data from which the CM Churn per Month for Figure 9.1 is calculated is shown in Table 9.4. It is also the source of the data for Gauge 10. This gauge serves as an indicator of the architectural soundness of the system. If the rate of "churn" begins to approach the 2% per month level, this shows a lot of rework is going on, which could point to deeper problems in the project. A high churn rate may mean that the original design was not robust enough. It could also be a symptom of changing requirements (see Gauge 10), which could indicate that the project is drifting toward disaster.

10. Requirements Change Per Month (Gauge 10) is calculated by dividing the number of new, changed, or deleted requirements specified in this reporting period by the total number of requirements at the end of this period. It is expressed as a percentage. Typical projects experience a requirements change of 1% per month. Some requirements growth is to be expected, particularly on large projects. However, a high rate of requirements change can indicate that the customer is not sure of what is wanted, or the original requirements definition was poor. A high rate often predicts disaster for the project.

Table 9.4   Configuration Management Data

Reporting Period Data		Configuration Items	
Reporting Period No.	Period End Date	Total No. in CM System at End of Period	No. Modified and Rechecked into CM This Period
1	9/5/2006	50	1
2	10/5/2006	52	0
3	11/5/2006	60	3
4	12/5/2006	65	2
5	1/5/2007		
6	2/5/2007		
7	3/5/2007		
8	4/5/2007		
9	5/5/2007		
10	6/5/2007		
11	7/5/2007		
12	8/5/2007		

As noted previously, these gauges can be related to the implementation of GP 2.8 of the CMMI Process Areas. In this case, Gauges 9 and 10 are useful for the Configuration Management and the Requirements Management Process Areas. High CM churn is never good. A baseline should never be established if the configuration item is unstable. High CM churn would indicate that perhaps certain items were baselined too early, or that the development went down the wrong path, necessitating change. The later in the development cycle this occurs, the more expensive making the changes will be. A similar statement can be made about requirements changes. Clearly, requirements changes after software has been coded or prototype hardware components have been built will be very expensive.

9.2.3 Staff

11. Voluntary Turnover Per Month (Gauge 11) is calculated by dividing the number of staff leaving during this reporting period by the number of staff at the beginning of this period. It is expressed as a percentage. The target range is less than 2% per month. A person can leave the project in a number of ways, such as by quitting the organization, requesting reassignment to another project, or being reassigned to another project at management's discretion. Turnover is an important measure for risk assessment. Every project lasting six months or longer should expect and prepare for some staff turnover. Each project member who leaves the team causes a productivity drop and schedule disruption. Bringing on new team members, regardless of their skills and experience, does not necessarily solve the problem; they require time to become familiar with the project and processes. In addition, a productive team member will usually have to devote time to orient the new hire, thus taking away additional resources from the project. Appropriate allowances should be included in the productivity resource estimates to allow for staff turnover. The basic data for Gauge 11 of Figure 9.1 is derived from this extract from the worksheet, and is illustrated in Table 9.5.

   Recall that in Chapter 4, we had shown staffing profiles of planned staffing versus actual staffing as one means of tracking progress. We also showed how deviations from the planned profile could be an indication of impending problems, but had indicated that the plot of actual versus planned staffing needs supplemental information in order to provide more clarity into what is actually happening. The information provided by the Progress gauges is richer in information and enhances the capability to determine if problems exist, and the nature of them.

12. Overtime Per Month (Gauge 12) is calculated by dividing the overtime hours by the base working hours for all project staff in this reporting period. It is expressed as a percentage. The target range is less than 10%. When the overtime rate approaches 20%, the ability of staff to respond effectively to crises suffers significantly. It may also indicate a problem with estimation. If an

Table 9.5   Staffing Data

Reporting Period Data		Voluntary Turnover	
Reporting Period No.	Period End Date	No. of Staff at Beginning of Period	No. Staff Leaving Voluntarily This Period
1	9/5/2006	75	5
2	10/5/2006	70	2
3	11/5/2006	80	1
4	12/5/2006	85	2
5	1/5/2007		
6	2/5/2007		
7	3/5/2007		
8	4/5/2007		
9	5/5/2007		
10	6/5/2007		
11	7/5/2007		
12	8/5/2007		

organization does not have a good database from which to estimate productivity, a project will most likely be underestimated in terms of schedule and effort, requiring unplanned overtime. If the productivity data is good, then perhaps the factors used to estimate the effort and schedule for the project were not as solid as originally thought. For example, on a software project, if the basis for estimating cost and schedule is based on function points, and excessive unplanned overtime occurs, the count of function points in the beginning of the project was undoubtedly way off, or requirements changed significantly, and the requirements change was not properly factored in to the schedule and effort estimates. As a means of mitigating risk, depending on the type of risk and the method of mitigation selected, some funds may be set aside as a contingency to account for unplanned overtime. On some forms of government contracts, however, setting aside contingency funds may not be permissible in bidding on the project.*

The data for this gauge is also derived from the worksheet, and that portion of the data in the worksheet is illustrated in Table 9.6.

---

* Discussion of this issue is outside the scope of this book.

Table 9.6   Overtime Hours

Reporting Period Data		Overtime Hours	
Reporting Period No.	Period End Date	Base No. Staff Hours This Period	No. Overtime Staff Hours This Period
1	9/5/2006	13,000	1,000
2	10/5/2006	13,500	2,000
3	11/5/2006	13,500	3,375
4	12/5/2006	13,200	500
5	1/5/2007		
6	2/5/2007		
7	3/5/2007		
8	4/5/2007		
9	5/5/2007		
10	6/5/2007		
11	7/5/2007		
12	8/5/2007		

The data for Gauges 11 and 12 are also metrics that can be utilized to implement GP 2.8 for the PMC process area.

9.2.4 Risk

13. The Risk Exposure chart (Gauge 13) shows each risk plotted by its cost consequence and probability. Each data point in this chart is associated with a specific risk, and has an identifier associated with it. The probability is expressed in terms of expected occurrence over the life of the project. The regions on the graph show where risks fall into areas of low, moderate, or high risk exposure. Clearly, high probability/high consequence risks indicate high risk exposure, while low probability/low consequence risks indicate low risk exposure.

American Systems has developed a Web-based program called Risk Radar Enterprise (RRE) [3].* RRE is a risk management database that helps project managers identify, prioritize, and communicate project risks in a flexible

---

* The description of Risk Radar Enterprise is excerpted with permission from material copyrighted by American Systems.

and easy-to-use form. RRE provides standard database functions to add and delete risks, together with specialized functions for prioritizing and retiring project risks. Each risk can have a user-defined risk mitigation plan. A set of standard short- and long-form reports and viewgraphs can be easily generated to share project risk information with all members of the development team. The number of risks in each probability/impact category by time frame can be displayed graphically, allowing the user to visualize risk priorities and easily uncover increasing levels of detail on specific risks. RRE also provides flexibility in prioritizing risks through automatic sorting and risk-specific movement functions for priority ranking. RRE features include:

- A *set-up screen* allows the project leader to set project specific information, such as the title of the project, in one place.
- A *custom risk matrix settings screen* allows the project manager to establish what is tantamount to a risk management strategy. The project leader can set up a 3×3, 4×4, or 5×5 matrix of probability or likelihood against impact or consequence, with customizable values for high medium or low risks.
- A *risk records screen* allows the project leader to add new risks, modify existing risks, delete risks, and retire risks.
- A *risk state screen* allows the project leader to present all the information for all the risks on a single screen without scrolling. Details of any individual risk category can be called up by clicking on any one square that shows the number of risks in any individual risk category. The details of any individual risk can then be called up by clicking on the action icon.
- A *screen that permits the project manager to select among 23 different report formats with 18 different filters that he/she may want to generate.* It also has the capability to generate ad hoc reports.
- *RRE data can be exported in .cvs format.* Because Control Panel is an Excel application, it can accept data in this format.

14. Risk Reserve (Gauge 14) shows the total cost risk exposure and schedule risk exposure compared to the current cost and time risk reserves for the project. Risk exposure for a risk is calculated by multiplying the probability by its potential negative consequence. Although the consequences (and therefore the risk exposure) for different risks are not necessarily independent, a first approximation to the total cost risk exposure of the project can be made by adding together the individual cost risk exposures for all risks. The same approximation to total schedule risk exposure in time can be made by adding together the individual schedule risk exposures for all risks. The data for this gauge is also derived from the worksheet, and that portion of the data in the worksheet is illustrated in Table 9.7.

Risk reserves for cost and schedule should be established at the beginning of the project to deal with unforeseen problems, although as noted previously, on some forms of government contracts, setting up such a reserve in advance

Table 9.7   Risk Reserve Data

Reporting Period Data		Risk Reserve	
Reporting Period No.	Period End Date	Total Budget Risk Reserve ($)	Total Schedule Risk Reserve (days)
1	9/5/2006	30,000	20
2	10/5/2006	28,000	19
3	11/5/2006	28,000	19
4	12/5/2006	28,000	18
5	1/5/2007		
6	2/5/2007		
7	3/5/2007		
8	4/5/2007		
9	5/5/2007		
10	6/5/2007		
11	7/5/2007		
12	8/5/2007		

may not be permitted. The cost and time risk reserve for a project will change over time as some of these reserves are used to mitigate the effects of risks that actually occur and affect the project.

15. The Metrics Problem indicator (Gauge 15) shows that management has received either a warning or bad news about some of the metrics on the project.

16. The Anonymous Channel Unresolved Warning indicator (Gauge 16) shows that project management has received either a warning or bad news about the actual status of the project. An open-project culture in which reporting bad news is encouraged is conducive to a healthy project. Warnings from anonymous or known project personnel should be welcomed and tracked.

9.2.5 Quality

17. Defects by Activity displays the number of detected defects open (i.e., yet to be fixed) and the number of defects closed in each phase of the project (Gauge 17). Defects are problems that, if not removed, could cause a program to fail or to produce incorrect results. Defects are generally prioritized by severity

level, with those labeled 1 being the most serious. In Chapter 4, Section 4.4.2, we discussed beneficial measurement examples. There we discussed problem report tracking and presented several different examples of reports that are useful. Gauge 17 is yet another form of report useful for tracking defects. The quality indicators on this chart help you answer the question: What is the quality of the product right now? In the example illustrated by the control panel in Figure 9.1, the histogram depicts requirements, design, code, and test defects that have been open and closed. The raw data for the histogram is also contained in the worksheet and is illustrated in Table 9.8.

9.3 Additional Metrics at the PMO Level

The SDMD control panel presented in Section 9.2 focuses at the project level. Improvement projects, such as those following the DMAIC Six Sigma process (see Chapter 2), have, inherently, measures for success. Each project starts by identifying the quantified way its progress and achievements will be determined. For example, a Six Sigma improvement project focused on reducing "escaping defects," such as problems in software versions reported by customers, will track escaping defects to determine progress (see Chapter 8). An aggregation of improvement projects will have an impact at the organizational level. To determine this impact, more "macrolevel" measures are needed. In this section we describe such metrics that are typically tracked at the Project Management Office and Research and Development (R&D) level.

The measures we discuss here include:

M1: Profit from New Products and Systems
M2: Cost of Late Launch
M3: R&D Spend
M4: New Product/System Quality
M5: Uniqueness of Products
M6: Idea Generation Rate
M7: New Product Break-even Time
M8: New Product Introduction Rate
M9: Product Standardization
M10: Cost of Quality

Each measure's motivation, the questions it can help provide answers to, and the data it requires are discussed next. Such measures, in addition to a specific project's SDMD, provide a global view of the product portfolio and R&D processes.

In discussing these measures, we define the following processes as being critical for a product's life cycle:

Table 9.8 Defect Data by Activity and Category

Reporting Period Data		Defects by Activity								
Reporting Period No.	Period End Date	Requirement Defects Open	Requirement Defects Closed	Design Defects Open	Design Defects Closed	Coding Defects Open	Coding Defects Closed	Test Defects Open	Test Defects Closed
1	9/5/2006	20	10	12	4	2	2	1	0
2	10/5/2006	22	5	12	6	8	3	5	6
3	11/5/2006	10	15	15	4	12	3	4	2
4	12/5/2006	5	20	8	12	25	10	8	5
5	1/5/2007								
6	2/5/2007								
7	3/5/2007								
8	4/5/2007								
9	5/5/2007								
10	6/5/2007								
11	7/5/2007								
12	8/5/2007								

- Generate demand
- Develop products and systems
- R&D
- Introduction
- Fulfill demand
- Plan and manage operations

9.3.1 M1: Profit from New Products and Systems

Profit from new products and systems assesses the contribution that they make to the profitability of the organization. This is an important measure as it assesses the success of new products and systems in terms of market response (customer orders), as well as considering the efficiency and effectiveness with which the systems are developed and subsequently brought to market. This not only assesses whether there is a market for the products that are developed, but also the costs incurred in developing the products and systems. It assesses how well the "generate demand" process generates demand for products that have been developed and the efficiency with which products and systems are produced and distributed. Monitoring the Profit from New Products and Systems gives a company an idea of the contribution of new products to its overall profitability. It is most relevant for companies that frequently introduce new products; the more frequently new products are introduced, the more important it is that the whole process, from idea generation to end-product delivery, is profitable. The data from this measure provides an important input for decisions on new product or system development, as it allows a full understanding of where revenues can most effectively be generated and where costs are incurred during the process, thus providing an informed view as to the potential profitability of future options.

M1 is computed using the following data:

- Profit generated from new products/systems in a given period
- Percent (%) of profit that is generated from new products/systems in a given time period

where:

Profit is the excess of sales turnover over expenditure. In the case of profit from new products and systems, it relates the revenue generated from the sale of new products or systems in comparison to the cost of developing, producing, marketing, and distributing the product or system. Defining "new" products and systems should consider to what extent modifications result in a "new" product or system. The given period over which this measure should be used and analyzed will vary depending on the nature of the market in which the new products compete. When defining the time period, consideration should be given to the product life cycle because new products launched in 1 year may not have generated adequate sales within that year to reflect their impact on new revenue generation. So, measuring new product or system sales per year may not take into account the time lag involved in new sales revenue generation. Analysis of the measure by product or product group provides an understanding of which are the most profitable products or product groups, and hence which are the markets on which the organization should focus. This is an important feedback mechanism for reviewing the assumptions made when deciding which products to introduce and which markets to aim for.

Data for this measure will be available from the organization's financial accounting information systems, which record revenues and costs incurred. It is important that the accounting systems include an appropriate basis for the allocation of costs to ensure that they represent as accurately as possible the costs incurred in developing, producing, and selling each new product or system.

9.3.2 M2: Cost of Late Launch

The Cost of Late Launch measures the additional costs incurred as a result of the failure to get products to market when planned. This includes measurement of the extra cost incurred due to late launch of a product, resulting in development projects going over budget. Additional costs have a direct implication on the break-even point and target cost compliance of projects and overall profitability. In addition to actual costs incurred (i.e., exceeding budgets for labor costs, overhead, etc.), there are also additional costs that in many cases have greater implication for profitability. These costs include lost revenue or customer penalties as a result of failure to deliver new products when expected, as well as the need for increased promotion to compensate for the late launch. Boeing, for example, faced the potential of severe cost penalties in its 777 aircraft program if it failed to meet delivery dates to various airlines. In turn, penalty clauses were inserted into supplier contracts for delays in delivery from them. In many organizations the most significant cost is in terms of lost sales. This includes the sales that are lost during the period when the new product is not available, lost opportunities for getting the product or system to market first, and lost reputation of the product and organization. In the pharmaceutical industry, for example, a drug patent has a finite length, and a late launch means that sales and profits are effectively lost forever.

M2 is computed from the following data:

■ Total costs incurred as a result of the late launch of new products/systems in a given period
■ Number of man-hours per month consumed after planned date of launch

- Total value of customer penalties as a result of late launch in a given period
- Total value of customer lost sales as a result of late launch in a given period

where:

Customer penalties are penalties imposed by customers as a result of the failure to deliver the new product or system when promised.

New products/systems are those that have been recently introduced. The definition of "recently introduced" is completely dependent on the nature of the market and product or system being introduced. In highly dynamic markets, products and systems will be considered new for a shorter period of time than in more stable, established markets. Defining "new" products and systems should also consider to what extent modifications result in a "new" product or system.

Lost sales are expected sales that are not realized because of late launch of the product. Use of the measure should include analysis of each stage of the Develop Products and Systems process to identify which parts of the processes are causing the late launch. This will provide important feedback to the process so that the improvement of planning can be targeted on the most appropriate areas.

Analysis of costs should consider the differences in costs incurred between different products, product groups, and customers. This will identify where the greatest penalties for late introduction will be incurred, which will be important information when deciding on prioritization and resourcing of product introduction projects. Data for this measure is readily available from the list of tasks to be completed within the Gantt chart of projects. After the date of intended launch, all outstanding activities are translated into "man-hours" and costed. The sales department should make an assessment of the expected sales that are lost as a result of failure to get the project to market and should record penalties that customers impose.

9.3.3 M3: R&D Spend

Research and Development (R&D) Spend reflects the amount of resources directed toward the systematic search for innovations and indicates the commitment of the organization to developing new products and processes in order to maintain competitiveness. The amount spent has a direct impact on the innovative outputs (products, processes, or patents) of the business unit. It can be regarded as one of the key inputs to the innovation process. The level of spend on R&D will vary considerably between industries, depending on the rate at which new products and processes need to be introduced in order to maintain competitiveness and the cost of developing these products and processes. This measure is an input to the R&D process, as the level of resources available will have a significant effect on its output.

As a result, it will assess the Develop Products and Systems process and the strategic importance of developing new processes, products, and systems.

M3 is measured from the following data:

- R&D expenditure as percentage of sales turnover
- R&D expenditure as a percentage of operating costs

  where:
  R&D expenditure is expenditure that relates to research and development activities. R&D will relate to the development both of new products and processes.
  Turnover is income from the sale of goods and systems.

One should be aware that R&D is only one of the key elements within the innovation process. Other non-R&D inputs include new product marketing, patent-related work, financial and organizational change, design engineering, industrial engineering, and manufacturing start-up.

It is important to note that the amount of R&D Spend does not always translate into innovative products or processes. Specifically, there are eight factors that are critical for new product success:

1. The product has a high performance-to-cost ratio.
2. Development, manufacturing, and marketing functions are well coordinated and integrated.
3. The product provides a high contribution to profit margin for the firm.
4. The new product benefits significantly from the existing technological and marketing strengths of the business units.
5. Significant resources are committed to selling and promoting the product.
6. The R&D process is well planned and coordinated.
7. A high level of management support for the product exists from the product conception stage through to its launch in the market.
8. The product is an early market entrant.

Where possible, the amount an organization spends on R&D (including ratios) should be compared to that of competitors. As spend does not guarantee results, comparison of the level of spend should include consideration of the value that is created with that expenditure. The amount of R&D expenses is typically reported in the consolidated statement of income. Some conglomerates specify a fixed R&D budget (1% to 5% of sales, for instance) based on the profit line for each business unit.

9.3.4 M4: New Product/System Quality

New Product/System Quality measures the extent to which the new product or system launched is fit for its intended purpose. This includes the conformance of new products and systems to design specifications and also broader concepts of quality that relate to the fitness of the product or system to the purpose for which it was designed. It is commonly acknowledged that in addition to the conformance to design specifications, there are other dimensions of product quality (performance; reliability; usability, maintainability, etc.). While it may be possible to deliver a product that 100% conforms to specification, it might still be possible to improve the product or system quality with regard to one of the other dimensions of quality performance.

It should also be kept in mind that not all the dimensions of quality will be critical to each product or system, depending on customer or market requirements and expectations. This is a very fertile area for measures, and one must be careful to identify the correct mix.

The quality of new products/system has the following implications:

■ Success of new product/system launch
■ Ability to capture first mover advantage
■ Reputation of firm

The measure of New Product/System Quality measures how well the Develop Products & Systems process identifies customer's new product requirements and then converts them into designs and products that conform to them.

M4 is measured using:

■ Percent (%) of new products/systems that are 100% "fit for purpose"
■ Percent (%) of new products delivered to the customer that are defect free
■ Percent (%) deviations of technical parameters from design specifications
■ Percent (%) deviations of marketability parameters from launch program

where:

Fit for purpose relates to the degree to which the product or system meets the customer's requirements.

Defects are departures from the intended quality level or state, or a product or system's non-fulfillment of an intended requirement or reasonable expectation for use.

Technical parameters cascade into a few key variables that relate to the performance of the product/system in the field.

Marketability parameters relate to key variables that serve as guideposts to successful product launch (gathered from previous new product/system experience, market research, focus groups, consulting advice). For example, the quality of a vehicle includes both the technical parameters that define the class of vehicle (size, degree of luxury, weight) and its marketability (ride, handling, aesthetics), as perceived by the customer.

All defects or complaints from customers should be analyzed to identify their cause so that they can be eliminated to prevent recurrence. Analysis of trends over time will identify specific problem areas. There should be quality control processes in place to collect data relating to defects. The organization should also proactively seek feedback from customers regarding the new product or system's fitness for purpose. For more on quality control, see [4].

9.3.5 M5: Uniqueness of Products/Systems

This measure assesses the extent to which a new product or system launched differentiates both from competitor's products and the organization's existing portfolio of products and systems. Higher uniqueness is in line with product differentiation strategy. The objective of developing unique products and systems is to generate sales in new markets or to increase market share in existing markets. Adding a unique product is a way of stimulating mature markets. Therefore, the measure assesses how well the Develop Products and Systems process generates innovative and unique ideas, and converts them into innovative and unique products. Introduction of unique products will also affect the level of customer satisfaction, especially where there is strong competition in the market based on product or system features. When developing unique products, it is important to ensure that new and innovative features are required by the customer. It will be counter-productive to incur the cost of including additional features if the customer doesn't require or value them. As a result, this measure should be used in conjunction with a measure of new product and system sales to assess whether uniqueness creates value.

M5 is measured by:

■ Percent (%) of new products and systems introduced in a given period that are considered unique
■ Percent (%) of product or system features and attributes that are considered unique

where:

Unique refers to products or systems, or features and attributes of products or systems that cannot be found elsewhere in the market.

New products/systems are those that have recently been introduced (see also Measure M2). The definition of "recently introduced" is completely dependent on the nature of the market and product or system being introduced. In highly dynamic markets, products and systems will be considered new for a shorter period of time than in more stable, established markets. Defining "new" products and systems should also consider to what extent modifications result in a "new" product or system.

Uniqueness can be captured as a basket of measures that is spread across product function, features, and aesthetics. This should ideally be benchmarked against a product in the same category. Measurement of uniqueness can only be made by subjective assessment of features and attributes. This can be done as assessment of the effectiveness of the Develop Products and Systems process but should have considerable input from the marketing function, who have knowledge of the market, and from consumers. As measurement will be subjective, care must be taken when using the data. The integrity of data collection will be critical. This measure is related to the measure of patents, which will also indicate the uniqueness of products.

9.3.6 M6: Idea Generation Rate

The front end of any innovation process lies in ideas. For a company that supports innovation, the ability to introduce new products, systems, and processes depends on the pool of ideas available. This measure accounts for this ability to generate new ideas. New ideas might relate to new products and systems, improvement of current products and systems, or improvement of processes. Although employees tend to be the main source of new ideas, they can also come from the other stakeholders. As a result, this is also a measure of stakeholder contribution. As such, it is strongly linked to the measures of Feedback and Suggestions from each of the stakeholders.

The measure should be used to encourage the generation of ideas where there is considered to be insufficient input into product development and improvement of the performance of the business. In such circumstances the measure could be used in conjunction with incentive schemes, which might include recognition or rewards. It is important that organizations proactively encourage ideas and suggestions. To be most effective, this measure should be part of an evolving measurement of idea generation and implementation. Measurement of the number of ideas or suggestions should encourage increased volume of ideas and suggestions. Once the volume of ideas and suggestions has increased, it is desirable that the sophistication of the measure is increased to encourage implementation of ideas and suggestions and their positive impact.

The data for computing M6 is:

■ Number of ideas/suggestions for improvement received monthly

where:
Ideas/suggestions are contributions made with the objective of improving the operations of the organization. New ideas might relate to new products and systems, improvement of current products, or systems or improvement of processes.

For companies with suggestion schemes and a structured new ideas assessment system, data for this measure can be readily obtained from such a system, typically monitored by the human resources department or processes responsible for managing relations with stakeholders. This measure is to be distinguished from Idea Conversion Rate, which measures the effectiveness of new ideas that can be implemented. Idea Generation Rate only measures the number of new ideas generated monthly; it does not say anything about the ease of implementation. For this, the new ideas assessment system would act as a filtering gate. All new ideas generated must be weighed within the context of a company's resources and capabilities. It is not always easy to collect data regarding the generation of new ideas. It is often difficult to define what constitutes a discrete idea. For this measure to be effective, there is need for a mechanism (i.e., new ideas assessment system) to identify and record the generation. Such a system is important to ensure that ideas are monitored and their effect is maximized.

Use of this measure should include analysis of the source of the ideas being generated. This will include employees from the R&D process, employees from other processes or functions of the organization, and other stakeholders in the organization. The results of this measure must be used with caution if there are incentive schemes in place to encourage the generation of ideas. These may bias the results to give a false impression of idea generation. For example, certain employees might have one big idea but only submit it as a number of smaller sub-ideas in order to take advantage of rewards that are made available. Similarly, there should be consideration of the usefulness of the idea before offering rewards.

9.3.7 M7: New Product/System Break-Even Time

Break-Even Time (BET) measures the time it takes for sales turnover from a new product or system to equal the cost of bringing that new product or system to market, as measured from the beginning of the project. This is an important measure as it assesses the speed with which a new product or system makes a positive contribution to the profitability of the organization. The measure not only considers the costs of bringing the product to market, including promotion, but also considers the success of the product in the marketplace in terms of the turnover it generates. As a result, it is an important measure of the development process as it assesses the efficiency and the effectiveness of that process. The measure underlines an important principle of not just developing new products or systems, but also producing them with competitive costs, compelling quality, and bringing them to market ever more quickly. The measure is appealing because it accounts for the entire new product development process—the assessment of customer needs, the effectiveness of R&D, the speed of ramp-up to volume production, the efficiency of distribution efforts, the adequacy of training programs, and related issues. The measure is useful in serving as a "reality check" against the portfolio of development projects.

M7 is computed from:

■ Time elapsed when cumulative sales turnover from the new product or system equals the cost of bringing it to market

where:

Turnover is the total income from the sale of goods and systems.

New products/systems are those that have been recently introduced (see also Measure M2). The definition of "recently introduced" is completely dependent on the nature of the market and product or system being introduced. In highly dynamic markets, products and systems will be considered new for a shorter period of time than in more stable, mature markets. Defining "new" products and systems should also consider to what extent modifications result in a "new" product or system.

Cost of bringing product to market includes all the related costs from concept to launch (R&D, product design costs, prototyping costs, production costs, distribution costs, training costs, advertising costs, etc.).

Analysis should include comparison by product or product group, which will indicate which are the most profitable and on which products or product groups effort should be placed to improve the effectiveness and efficiency of product introduction and marketing. The actual break-even time will vary considerably between industries and products and systems within industries. Key factors affecting the break-even time include the complexity of the product, the cost of development involved, and the profit margin included in the final selling price. The actual break-even point will be affected by the "Target Cost versus Actual Cost" measure. The development process should include a process to monitor the progress of new products and systems that are developed. This should enable evaluation of the output of the process, including the success of the products that are introduced into the market. The process will require data from accounting information systems regarding the costs incurred and sales revenue generated. This monitoring process should provide important and valuable feedback regarding the performance of the process.

9.3.8 M8: New Product/System Introduction Rate

The objective of the measure of New Product/System Introduction Rate is to capture information on the rate at which a company introduces new products or systems. While New Product/System Introduction Time measures the length of time taken to proceed from initial concept of product or system through to actual launch, the New Product/System Rate measures the rate of change (refresh rate) of the existing portfolio of products and/or systems. In new and dynamic markets, the rate at which new products or systems can be introduced will be a key determinant of competitiveness. In markets where new products and time-to-market are order-winning criteria, ensuring that the introduction rate is better than competitors is essential.

M8 is computed using:

■ The total number of new products or systems launched over a given period
■ New product/system introduction rate versus competition
■ New product/system introduction rate versus industry standard

where:

New products/systems are those that have been recently introduced (see also Measure M2). The definition of "recently introduced" is completely dependent on the nature of the market and product or system being introduced. In highly dynamic markets, products and systems will be considered new for a shorter period of time than in more stable, established markets. Defining "new" products and systems should also consider to what extent modifications result in a "new" product or system.

Competition consists of providers that produce products or systems that are in direct competition with those of the organization. This measure is concerned with the rate at which competitors introduce new products or systems, as well as the rate at which the organization introduces them. In markets in which time-to-market is an order-wining criterion, it is essential that performance is better than that of the competition.

Industry standard is the introduction time that is demonstrated throughout the industry in which the organization competes. This should be considered the minimum acceptable introduction rate in order to compete in the market. The given period should be defined based on the characteristics of the market in which the products and systems compete. In a dynamic market where new products and systems are frequently introduced, the measure should be reviewed more frequently to pick up performance trends quickly. The R&D process should maintain a monitoring system to collect data regarding the rate at which new products and systems are developed and introduced. Analysis of new product or system introduction rates should compare performance for different products and product groups to identify which are the most competitive and for which performance improvement is important. There should also be analysis of each stage of the introduction process to identify where improvement can be made. Each stage of the process could be benchmarked with other organizations to see how they can be improved. Analysis must also include a comparison of introduction rate versus that of competitors and standard rates within the industry.

9.3.9 M9: Product Standardization

The measure of Product Standardization provides an indication of the variety of products or product variants that are offered by an organization. It assesses the amount of product that is standard, as opposed to the number of product variants or products that are customer specific. This is an important measure of the attractiveness of the products offered to the market and also has a significant impact on the operations of the organization. Provision of products or systems that are standard can reduce the variety of features that are offered by the organization, reducing product quality and potential customer satisfaction. As with the measurement of product complexity, increasing the variation of products and systems offered (reducing standardization) might increase product quality and customer satisfaction, but it can also reduce responsiveness and increase the cost of delivering products and systems. Increasing the variation of products and systems can be achieved by producing standard core products or systems that can be customized later in the production or system delivery process. This will contain the costs and difficulties of increased product complexity while also allowing for increased variation in the products offered to the customer. This will be achieved by designing the product or system, and the production or delivery processes, in a way that enables the production and modification of standard products or systems. This might include taking a modular approach to production.

Although such a measure would typically be applied to products, it is also possible to define standardization in a way that allows measurement of systems. This measure should be used where there are variations in the requirements of specific products and where customization of products or systems is necessary. When standardizing products, there is a need to ensure that the customer is still getting the product that is required and that their range of choices is not reduced.

To increase responsiveness to customer requirements, the amount of standardization should be as great as possible; and where possible, customization of standard products should be undertaken as late as possible. This enables inventory of standard products to be built, with customization taking as little time and resources as possible.

M9 is computed from:

- Percent (%) of products that are standard
- Percent (%) of products that are customer specific
- Number of product variants offered
- Percent (%) of production lead time/production process/parts that are standard
- Average lead time/cost/manufacturing process for the skeleton product

where:

Standard products are products that are the same; that is, there are no variations to products for different customers or markets.

Customer-specific products are those that are tailored to customer requirements.

Product variants are variations of standard products that suit specific customer needs. Product variants might include limited variations from which customers can choose, or the ability of customers to have standard products or systems customized to their requirements.

Production lead time/production process/components that are standard measures the amount of the product that is standard, indicating the degree of standardization. Where possible, customization should be undertaken as late as possible to increase flexibility, allowing stock of standard products to be built, with standardization undertaken late in the process to respond to specific requirements.

Skeleton product is the standard product that can be varied or customized to specific requirements.

Analysis should compare the variation between different products offered by the organization, as well as a comparison with competitors or others in the industry. Comparison of the amount of product that is standard and the point in production that products are customized should be used to redesign products and production processes to increase responsiveness.

Data should be collected by identifying the key constraint(s) or variable(s) in the production process, monitoring their performance through the monitoring and control of operations.

The unit of measure will vary depending on the priorities of the organization and the marketplace, as well as the organization's critical resources or variables.

9.3.10 M10: Cost of Quality

Cost of Quality aims to financially quantify the activities in the prevention and rectification of defects. It measures the total cost (cost of conformance and cost of nonconformance) of ensuring that products of the correct quality are delivered. It is critical to measure the cost of quality for the following reasons:

■ To display critical quality-related activities to management in meaningful terms
■ To illustrate the impact of quality-related activities on key business performance criteria (profitability, operational costs, inventory levels, customer satisfaction, etc.)
■ To enable benchmarking against other divisions or companies
■ To establish bases for budgetary control over the entire quality operation
■ To provide cost information to motivate employees across all levels of company

The measure of Cost of Quality provides an indication of how well products and processes are designed to produce high-quality products, and does so in the common unit of cost to allow comparability of data. It also measures how well all activities are executed (Fulfill Demand) to ensure that products and systems are delivered to specification, as most of the failure costs will come from operations. In addition, the costs of prevention are managed in the Plan and Manage Operations process.

To compute M10, you need to compute:

■ Total cost of quality conformance
■ Total cost of nonconformance to quality requirements

where:
Cost of conformance (or cost of control) is the cost of providing new products and systems to specified standards. For example, prevention costs (preventing defects and nonconformities—quality planning, process control, quality training, etc.) and appraisal costs (maintaining quality levels through formal evaluation—inspection, testing, quality audits, etc.).

Cost of nonconformance (or cost of failure to control) is the cost of time wasted, materials, and capacity in the process of receiving, producing, dispatching, and reworking unsatisfactory products and systems. The cost of nonconformance includes internal failure costs (e.g., scrap and rework costs) and external failure costs (which include warranty claims, field repairs, returns, product recall, lost customers, etc.).

Quality costs are those costs associated with the definition, creation, and control of quality as well as the evaluation and feedback of conformance with quality, reliability, and safety requirements, and those costs associated with the consequences of failure to meet the requirements both within the factory and in the hands of the customers.

Performing process modeling to identify subprocesses in the production of a new product or system is an effective first step in building a big-picture view of the cost of quality. Having identified opportunity areas for reduction in cost of conformance and nonconformance, measures can be introduced to monitor the Cost of Quality. Operations reviews should also be established to look into root causes of nonconformance. Organizations should have a quality control or quality assurance function that is responsible for ensuring that products and systems that are delivered are of appropriate quality. This process should include recording and analyzing data regarding the costs of quality. The function should also receive feedback from other departments or processes, such as the sales department regarding missed sales.

9.4 Risk Management and Optimizing Decisions with Data

9.4.1 Uncertainty Models and Risk Management

In Chapter 7 we provided an introduction to risk management in the context of evaluating outcomes from a beta test. In this section we take a more general view of risk management and expand it in the context of statistical decision theory. This formalizes the intuitive decision-making process of project leaders and managers who rely on information in SDMD control panels to identify and mitigate risks. We begin with some basic elements of risk management and optimal decisions.

Buying insurance is a typical passive form of risk management; loss prevention and technological innovations are active means of managing risks. Loss prevention is a means of altering the probabilities and the states of undesirable, damaging states. For example, properly maintaining one's own car is a form of loss prevention seeking to alter the chances of having an accident due to equipment failure such as a blowout at high speed on a freeway from a bald tire. Similarly, driving carefully, locking one's own home effectively, installing fire alarms, etc. are all forms of loss prevention. Of course, insurance and loss prevention are, in fact, two means to the similar end of risk protection. Insurance means that the risk is being accepted and cash provisions are being made to handle the risk if it materializes into a problem. Car insurance rates tend, for example, to be linked to a person's past driving record. Certain drivers might be classified as "high-risk drivers," and required to pay higher insurance fees. Inequities in insurance rates will occur, however, because of an imperfect knowledge of the probabilities of damages and because of the imperfect distribution of information between the insured and insurers. Thus, situations may occur where persons might be "over-insured" and have no motivation to engage in loss prevention. Passive risk management is widely practiced also in business and industry. Systems and software development are no different, especially when proactive prevention initiatives are difficult to justify on the basis of return on investment.

In the passive mode of risk management, we try to measure the risk of an event such as not meeting a deadline for shipping a version because of high-severity failures. If the cost of the event occurring is c and the probability of the event occurring is p, then the risk of the event is its expected loss: $R = cp$.

Lets us call the event A and the loss function L. To model the situation, we consider a binary random variable $X = 0; 1$, which is 1 if event A happens and 0 if it does not happen. Think of L as attached to X, with $L(0) = 0$ and $L(1) = 1$: $R = E\{L(X)\} = cp$, where $E\{\cdot\}$ stands for mathematical expectation, or average loss.

Indeed, we typically think of risk as expected loss. Even if A does not happen, we may have loss and then we may have $L(0) = c0; L(1) = c1$, and the method is the same:

$$R = E(L(X)) = c0(1 - p) + c1p$$

Now let us think actively. We observe some data X and we make a decision $d(X)$. That is to say, we do not just sit back and accept our loss; rather, we take some positive action. The loss L (or reward) then depends on the value of the action, and we can attach the loss to $d(X)$. The risk then becomes:

$$R = E(L(d(X))$$

A somewhat more precise version says that a decision maps the data X into a set of possible actions called the action space A: $d(X)$ which is in A. We can also talk about d as being a decision rule, which means that it says what we do if we see X (a rule just means a mathematical function). So we have a paradigm that says: *Find the rule (i.e., make the decision) that has the smallest risk.*

However, we need to know the probability distribution generating the data. In the above example we need to know p.

One approach is to describe uncertainty about the value of p by a probability distribution function. To describe this model of randomness, we introduce yet another parameter, θ. This generates a family of distribution functions, and now the risk becomes $R(\theta) = E(L(d(X)), \theta)$.

So, if we know what θ is, we can compute the risk. What we can do is conduct a thought experiment: If θ is the true value representing the uncertainty about p, what is the best decision rule? Is there a rule that has smallest risk for all values of θ? Is there a decision rule whose worst risk, over all values of θ, is better than (no worse than) other rules about worst risk? Such a rule is called a minimax rule.

With minimax, management considers risks in various decision scenarios and picks the one that minimizes the maximum possible risk. Another approach is to consider Bayesian statistics. The choice of loss function used to make the decision rule depends on what we want to achieve. There are three basic possibilities, depending on our objectives.

Option 1: A loss function that helps us learn about θ. This is the bedrock of statistics. We change notation and say, for example, $d(X) = \theta_{est}$; an estimator for θ. The loss penalizes us according to how distant, in some sense, θ_{est} is from the true θ.

Option 2: A loss that has to do not so much with θ itself but more to do with some special characteristic of the distribution of X. For example, if X is the response time of a Web system in which we plan to submit a request, then if we think $X = 2$ *seconds* is a good maximum response time, we may decide to ask for it. We have a loss if we go to all the trouble of filling in the forms to order the system and it is too slow to respond.

Option 3: In this category we let the decision making extend to the data collection. We can let the distribution of X, or, if you like, the parameter θ, depend on the data collection procedure. At its simplest form, this may refer to the sample size n. In general it depends on the whole *experimental design,* including where, what, and how we measure. It is remarkable how a proper measuring method (such as an effective beta test plan) can improve our decision making.

9.4.2 Bayes Methods in Risk Management

Bayes methods are used for getting around the issue of not knowing θ (for more details on Bayesian methods, see [4]). With Bayesian statistics, our intuitive or data-based knowledge can be represented by a *prior distribution* for θ and then we have a mathematical expression for the Bayesian risk:

$$RB = E(L(d(X); \theta)) = E\theta E_X(L(d(X); \theta \mid \theta) = E_X E\theta_{|X}(L(d(X); \theta \mid X)$$

The internal expectation in the previous expression is the *posteriori risk*; and if we can choose the decision rule to minimize this *all the time*, then we minimize the overall Bayesian risk. This simple statement holds the key to Bayesian risk theory. It can be shown (under some conditions) that using a Bayesian rule is equivalent to behaving rationally. In other words, any decision rule whose risk is not dominated by some other rule for all θ is equivalent to a Bayes rule for some prior distribution.

A point that is both very problematical and very interesting is that we may be forced to be "Bayesian" because of not having enough data to estimate parameters. We should make a distinction between physically unknown parameters and more open physical parameters for which we simply don't have enough data, but for which we can say really physically exist. This is not a black-and-white world; rather, all shades of gray exist with philosophical and psychological implications. Bayes methods address the practical issue of eliciting probabilities when no data is available to obtain a good estimate. Indeed, there is a formal mathematical statistics method, called *decision analysis*, that concentrates on eliciting utilities by presenting alternative *gambles* to subjects.

The statistical methods outlined above are thought to fail when there is perception of a very large risk. This applies when the loss L is extremely large but the probability p is very small so that the risk $R = Lp$ is virtually undetermined. This is the case in enormous catastrophes such as global warming, nuclear war, pandemics, environmental disasters, or threats to safety-critical systems.

We refer to these events as "Black Swans" (see Section 9.4.4).

In the past 10 years, there has been growing interest in risk management, driven partly by regulations related to corporate governance of the private and public sectors. Much of this risk management is based, however, on informal scoring using subjective judgment. In Chapter 7, Section 7.3, we applied risk management to beta testing; here we take a more general approach and refer to any decision or event related to project and process management. As already outlined in Section 7.3, modern risk management systems consist of several activities as described below:

1. *System mapping:* An organizational chart and process map is a good start. The application of EKD presented in Chapter 2 is another.
2. *Risk identification:* What are the main risk events (items)? Classify these by area and sub-area. Risk identification does not stop at the first stage; it should be a continuous process (see below). It is best to reduce dimensionality manually by grouping similar items.

3. *Risk measurement, risk scoring:* This is difficult. Ideally we want the loss (impact) and the probability, but both are difficult to assess. Note that a score is the value of a *metric*.
4. *Risk prioritization:* Prioritization is done on the basis of the risk scores. A single score may be overly simplistic; consequently, there is consideration for being *multi-objective*. Technically this process allows us to identify the main *risk drivers*.
5. *Decision:* Any risk management system is basically a *decision support system*. It should not make the decisions automatically (there are interesting exceptions, such as automatic trading in finance). One needs to make decisions and record the decisions taken (i.e., risk mitigation approaches).
6. *Action:* This is not the same as decision! At the very least, the action will take place after the decision. It is critical to record the actual action.
7. *Risk control:* This is better defined now that we have the above activities 1 through 5. Risk control is the ongoing process of (a) monitoring the effect of the actions, including (i) rescoring the risks, (ii) introducing new risks (more prioritization), (iii) removing old risks from the list, and (iv) new or repeat decisions and actions. This is the process of feedback and iteration. It is a continuous or at least regular process.

Additional general risk management components include:

1. *Risk champions:* Facilitators who have ownership of the process or individual areas of risk.
2. *Stakeholders:* These can be employees, customers, experts, partners ("extended enterprise"), shareholders, patients, students, the general public, special sub-groups of the public (e.g., old, young, drivers), the government, agencies, etc. It is very important to draw all the stakeholders into the risk management process.
3. *Experts (again):* These are key personnel (e.g., engineers). However, it is a mistake to always rely only on the experts. Public risk perceptions are often profound, although somewhat erratic and sensitive to media sensation.
4. *Senior commitment:* This is very similar to what is said about quality. One needs champions at the board level, not just as a back-office risk function. The ideal is to have risk as a regular item at board meetings.
5. *Risk communication:* To all stakeholders.
6. *Can we control?:* There may not be the legal right or it may be physically impossible to take some of the actions: one cannot prevent a hurricane.

9.4.3 *Financial Engineering and Risk Management*

Mathematical economics or financial engineering provides complementary aspects to risk management. It includes discussions on complete/incomplete markets and rational expectation models, and includes *equilibrium theory, optimization theory, game theory,* and several other disciplines. The concept of a complete market has various technical definitions, but in general it means that every player in the marketplace has the same information, and that new information reaches every player at the same time and is instantaneously incorporated into prices. Roughly, no player can get an advantage from an information point of view (as in insider trading). If the market is not in equilibrium, then there is opportunity to make an *arbitrage*, a something-for-nothing profit just by taking a buy or sell action.

Rational expectations assume that an efficient market should be in equilibrium. The proof is that if it is not efficient, some player would have an advantage and drive the market in a particular direction. Under rational expectation, one behaves according to the best prediction of the future market equilibrium point. Note that this is quite close to the Bayes rule, which says that the conditional expectation θ_{est} = $E(\theta \mid X)$ is the best rule for estimation of θ under quadratic loss:

$$L = (\theta_{est} - \theta)^2$$

Rational expectation says that the conditional expectation under all available (past) information should be the basis for (rational) decisions. These economic principles lie behind our desire to minimize some risk metric. Very roughly, if we like a particular risk metric and we fail to minimize it, there is a better rule than the one we are using. Even worse, someone else may be able to take advantage of us. Note the interplay between information and action that we saw in the Bayes methods. We have the same conundrum here: Do we expend effort on getting better information or on taking better action? It is no surprise that some modern economic theories are about operating with *partial information*. A good way to think of the area of mathematical economics is as an extension of the decision theory we discussed above, with specific decisions taken at certain times. These are: buy, sell, swap, and price (the act of setting a price). For more on these topics, see [13].

The freedom to make a decision is called an *option*. Note the connection with risk management, in which we may or may not have this freedom (e.g., we do not have the freedom to stop a hurricane but we may have some prevention or mitigation options for the effect of the hurricane). The key point is that the option will have a time element: a freedom (e.g., to buy or sell at or before a certain date, and perhaps at a certain price). A *real option* is where the option is investment in a "real" object like a factory or technology. When time elapses, certain things happen: the value of assets may change, or we may have to pay interest. The things that happen may, in fact, usually be random (stochastic) and we typically may have no control. Or, rather, we have control in the choice of what to buy or sell and the time at which to do it, but once bought we are in the grip of the market until we sell. A whole body of work in Economics and Portfolio Theory is connected to mean-risk or mean-variance theory. This is when we try to combine means and, for example, variances. The idea has arisen both in economics and in engineering. Think of a portfolio or projects in various stages of development. Ideally, this portfolio would have a high return with low uncertainty (see [13]).

When dealing with operational issues, aspects of operational risks are evaluated using similar techniques, including Value at Risk and related scoring methods. These techniques are beyond the scope of this introductory section. For more on operational risks, see [10, 11].

The next subsection is at the interface between financial engineering, economics, statistics, and operational risk management. We present it here because of its importance to modern risk management methodology and because it has many implications for systems and software development. It relates to the work of Nassim Taleb [9], a hedge fund manager who dared to question the system, before the dramatic events of September 15, 2008, when Lehman Brothers announced that it was filing for Chapter 11 bankruptcy protection and the occurrence of its ripple effects (Iceland declared bankruptcy as a country on October 8, 2008, and CitiBank laid off 51,000 people).

9.4.4 Black Swans and the Taleb Quadrants

In 2007, Taleb published a book that predicted the 2008 economic meltdown [12]. He described the meltdown and similar major impact events, such as the 9/11 terrorist attack, the Katrina hurricane, and the rise of Google, as "Black Swans." A Black Swan is a highly improbable event with three principal characteristics: (1) It is unpredictable; (2) it carries a massive impact; and (3) after the fact, we concoct an explanation that makes it appear less random, and more predictable, than it was. The Western world did not know about the black swan before 1697 when it was first spotted in Australia. Up until then, all swans where white. Why do we not acknowledge the phenomenon of Black Swans until after they occur? Part of the answer, according to Taleb, is that humans are hardwired to learn specifics when they should be focused on generalities. We concentrate on things we already know, and time and time again fail to take into consideration what we don't know. We are, therefore, unable to truly estimate opportunities, too vulnerable to the impulse to simplify, narrate, and categorize, and not open enough to rewarding those who can imagine the "impossible." Taleb has studied how we fool ourselves into thinking that we know more than we actually do. We restrict our thinking to the irrelevant and inconsequential, while large events continue to surprise us and shape our world.

Taleb proposed a mapping of randomness and decision making into a quadrant with two classes of randomness and two types of decisions (see Table 9.9). Decisions referred to as "simple" or "binary" lead to data-driven answers such as "very true" or "very false," or that "a product is fit for use or defective." In these cases, statements of the type "true" or "false" can be stated with confidence intervals and P-values. A second type of decision is more complex, emphasizing both its likelihood of occurrence and its consequences. The other dimensions of the Taleb

Table 9.9 The Taleb Quadrants for Mapping Uncertainty and Decisions

Complex Decisions	Simple Decisions	Domain/Application
Thin-tailed "Gaussian-Poisson" Distributions	I. Classical Statistics	II. Complex Statistics
Heavy-tailed or unknown "Fractal" Distributions	III. Complex Statistics	IV. Extreme fragility (Extreme vulnerability to Black Swans)

quadrant characterize randomness. A first layer is based on "forecastable events," implied by uncertainty described with finite variance (and thus from thin-tail probability distributions). A second dimension relates to "unforecastable events," defined by probability distributions with fat tails. In the first layer, exceptions occur without significant consequences because they are predictable. The traditional random walk, converging to Gaussian–Poisson processes, provides such an example. In the second domain, large consequential events are experienced but are also more difficult to predict. "Fractals" and infinite variance models provide such examples (see [9]). These are conditions with large vulnerability to Black Swans. The reader is referred to Taleb [12] for more insight on Black Swans.

9.5 Summary

This chapter is both specific and general. In Sections 9.1 and 9.2 we described in detail the System/Software Development Management Dashboard and its implementation in the American Systems Control Panel. Section 9.3 provided additional macro-level metrics that are important to managers who deal with several projects, and a research and development organization. In Section 9.4 we introduced the reader to basic elements of mathematical economics and financial engineering. This methodology is necessary for quantifying (and justifying) decisions regarding projects and process, at the economic level. In that section we also introduced the concept of the Black Swan and its implications to risk management and decision making. Additional areas that relate to the use of data and metrics include mapping of cause-and-effect relationships [6], the use of incidents and "near misses" to manage risks of accidents or failures [8], and transforming data into information and knowledge [7]. In the next chapter we conclude the book with a case study where process improvement, metrics, and management initiatives are combined to enhance the competitive position of an organization on its way to Level 5 in CMMI.

References

1. CMMI Product Team. CMMI for Development, Version 1.2 (CMU/SEI— 2006—TR-008), Pittsburgh, PA: Software Engineering Institute, Carnegie Mellon University, 2006.
2. American Systems, http://www.americansystems.com/Systems/ ProfessionalTechnical-ITSystems/RiskManagement/DownloadProjectControlPanelDemo.htm (accessed 17 June 2009).
3. American Systems, Risk Radar *Enterprise*, version 1.0.6.5.3.
4. Kenett, R.S. and Zacks, S., *Modern Industrial Statistics: Design and Control of Quality and Reliability*, San Francisco: Duxbury Press, 1998. (Spanish edition, 2000; 2nd paperback edition, 2002; Chinese edition, 2004.)
5. Kenett, R.S. and Thyregod, P., *Aspects of Statistical Consulting Not Taught by Academia*, *Statistica Neerlandica*, 60(3), 396–412, August 2006.
6. Kenett, R.S., Cause and Effect Diagrams, in *Encyclopedia of Statistics in Quality and Reliability*, Ruggeri, F., Kenett, R.S., and Faltin, F., Editors in Chief, New York: Wiley, 2007.
7. Kenett, R.S., From Data to Information to Knowledge, *Six Sigma Forum Magazine*, November 2008, p. 32–33.
8. Kenett, R.S. and Salini, S., Relative Linkage Disequilibrium Applications to Aircraft Accidents and Operational Risks, *Transactions on Machine Learning and Data Mining*, 1(2), 83–96, 2008.
9. Kenett, R.S. and Tapiero, C., *Quality, Risk and the Taleb Quadrants, Quality & Productivity Research Conference*, IBM, Yorktown Heights, NY, June 2009. http://papers.ssrn.com/sol3/papers.cfm?abstract_id=1433490.
10. Kenett, R.S. and Raanan, Y., *Operational Risk Management: A Practical Approach to Intelligent Data Analysis*, New York: Wiley, to appear in 2010.
11. Panjer, H., *Operational Risks: Modeling Analytics,* New York: Wiley, 2006.
12. Taleb, N., *The Black Swan: The Impact of the Highly Improbable,* New York: Random House, 2007.
13. Tapiero, C., *Risk and Financial Management: Mathematical and Computational Methods,* New York: Wiley, 2004.

Chapter 10

A Case Study:
The Journey of Systems,
Inc. to CMMI Level 5

Synopsis

Throughout this book we have discussed techniques and strategies that help organizations improve their processes, plan for process improvement, and measure their ability to improve their processes and attain their business goals and objectives. These techniques are not disparate, uncorrelated methodologies, but are, in fact, an integrated system of techniques that, when organized together, allow organizations to achieve all these objectives in a rational, coordinated manner. In this final chapter we demonstrate how these techniques tie together. We do this by presenting a case study that utilizes all these techniques in an organized, coordinated, and structured manner. The case study is composed of examples from clients from different industries and different for whom we have consulted. We have adapted the examples to protect anonymity and not reveal proprietary information. The company we visit is called "Systems, Inc." This chapter presents that company's journey to CMMI Level 5.

10.1 Background of Systems, Inc.

Media, telecommunications, and software companies are now competing on each other's traditional turf, with cable companies offering phone service, telephone companies providing satellite TV, and software developers tying it all together. In this competitive environment, service providers need to cut operating costs dramatically, while learning to be brisk innovators of new products and services [10]. Moreover, such service providers are also required to continuously enhance their products and services. Maximum operational efficiency is achieved by enhanced process alignment, innovation through agility and time-to-market, and customer loyalty—throughout the intentional customer experience.

Information systems have a major influence on the agility of a company. They can inhibit or enhance it. Agility means the ability to adapt by quickly changing products, systems, and business processes. The agility of a company's information systems cannot be achieved by accelerating product/system development using traditional development processes and architecture. Agility must be built into the core development processes, roles, and architecture, taking into account the complexity of global development organizations that need to support media and telecommunications service providers.

Systems, Inc. is a typical global supplier of integrated software, hardware, and services solutions to media and telecommunications service providers. The company offers customer relationship management, order management, service fulfillment, mediation, and content revenue management products. These enabling systems support various lines of business, including video-on-demand, cable, and satellite TV, as well as a range of communications services, such as voice, video, data, Internet Protocol, broadband, content, electronic, and mobile commerce.

Systems, Inc. also supports companies that offer bundled or convergent service packages. In addition, the company's information technology services comprise system implementation, hardware and software integration, training, maintenance, and version upgrades. Its customers include media and communications providers, and network operators and service providers. The company was founded in 1990 in the United States. Today, the company's workforce consists of more than 5,000 professionals located in ten countries, with development and support centers in China, India, Ireland, Singapore, and the United States.

As a result of having a CMMI SCAMPI Class A appraisal performed in September 2008 (see Chapter 3), Systems, Inc. management decided to launch focused improvements in its software development capabilities. Specifically, the company decided to focus on the application of agile software and system development through Scrum (see Chapter 2). This was the recommended approach for correcting some of the weaknesses found during the appraisal.

Implementing Scrum, or any other agile approach, in multisite organizations with highly demanding customers is a significant challenge. Governance in such a context, where knowledge management is critical and dissemination of best practices across teams and cultures is required, poses problems of capturing process patterns, storing this knowledge, and ensuring the knowledge reuse that needs to occur. We review the main elements of Scrum in Section 10.3.1.1 and proceed to show how combining Scrum and EKD (the latter introduced in Chapter 2, Section

2.4) has provided Systems, Inc. with a comprehensive framework for managing and improving its software and system development.

10.2 The Systems, Inc. SCAMPI Appraisal

10.2.1 Some Background on the Need for an Appraisal

Systems, Inc. develops and supplies sophisticated business management systems that operate on different platforms. Its products are distributed worldwide through local distributors with new versions coming out almost every year. In the summer of 2006, Systems, Inc. faced severe setbacks. Its new product release had been delayed by almost a year, and competitive pressures were building up. Traditional customers began looking at alternative products with better and more reliable performance. At the end of 2007, Systems, Inc.'s management decided to take a proactive role and perform an internal appraisal, using the CMMI as a benchmark and a trained appraisal team. This was considered a necessary first step in the deployment of the improvement plan of software products and processes. As mentioned, the appraisal was carried out in 2008.

10.2.2 Appraisal Interviews at Systems, Inc.

We join the appraisal team in typical appraisal discussion groups and interviews (see Chapter 3, Section 3.1.5). By this point in the appraisal process, a review of the artifacts comprising the process architecture and the resultant work products has been completed. Interviews with project managers and practitioners are now being conducted to corroborate the initial observations regarding the implementation of the process. The purpose of such interviews and discussions is to provide the appraisal team with sufficient understanding about the practices in use by the systems development division at Systems, Inc. The observations gathered by the team about the process definitions and resultant work products, the input provided by Systems, Inc. personnel, and the experience of the appraisal team are used to compose the appraisal findings, which will then allow Systems, Inc.'s management to launch specific action plans.

10.2.2.1 Middle Management Discussion Group

The managers present in the meeting began by describing some of the tasks for which they are currently responsible. As the discussion began, one problem began to surface almost immediately: the monthly error reports produced by the Quality Assurance Group. The process is as follows: Each released Systems, Inc. system accumulates and collects error conditions that occur while in use. Once a month, the errors that were recorded during the month are downloaded by Systems, Inc. distributors and e-mailed to Systems, Inc. headquarters for processing and reporting. The process of downloading the errors was not explained, but it is assumed that it is a manual process. The files from all the sites are processed when received. Quality Assurance is responsible for overseeing this effort. This process includes the following steps:

- *Data loading:* In this step, error data from each system is loaded into a special-purpose Web application developed with .Net version 2.0.
- *Report generation:* Once the data is loaded, error reports are generated for each Systems, Inc. system version. In addition, a summary report is generated, by distributor, and for all newly installed systems. The report formats are predefined and contain both tables and graphs.
- *Interim report distribution:* During the discussion, it was pointed out that certain distributors require that the generated reports be faxed to them as soon as possible, even prior to completing the statistical analysis.
- *Final report distribution*: When the reports are completed, they are disseminated according to a fixed distribution list that includes management and technical leaders.

There are plans to automate this process, whereby error data will automatically be collected via direct modem interfaces, requiring minimal human intervention. It is believed that when the new process is implemented, the time required to produce the reports will significantly decrease.

Another area of concern is the lack of ability to isolate and filter out errors that were recorded during scheduled maintenance or training sessions. Currently, this type of data must be identified and removed manually, which is a time-consuming process and susceptible to additional human errors.

There were several areas of deficiency related to this process that emerged during the appraisal. These emerged as a consequence of the documentation reviews and the issues brought up during the discussion groups. The following are the deficiencies that were observed:

- There were no additional processes that analyze the errors, their source, or their cause.
- There was no process for tracing the errors reported and establishing their ultimate disposition and resolution. No procedures for this were found during the documentation review.
- There was no interface between the error reporting process and the software maintenance and development teams. Information was thrown over the fence.
- There was no policy for error data retention and/or disposition, and no evidence of the existence of such a policy was found during the documentation review.

These were issues that were good candidates for being addressed as part of the process improvement effort because they currently seemed to be causing a great deal of concern in the organization.

10.2.2.2 Requirements and Customer Interface Functional Area Representative (FAR) Discussion Group

One of the customer representatives indicated that the ABC Company, the largest client in his territory, has thirty-five Systems, Inc. systems, with another six on order. ABC collects statistical data on the performance of these systems. According to the specifications for the Systems, Inc. system, the *mean time between failures* (MTBF) is 5,000 hours. However, in reality, the Systems, Inc. systems operate with an average of approximately 3,000 hours MTBF. At one point in time, the MTBF was even lower. The customer representative explained how he completes problem reports after each service call, including his personal approach to investigating the problems and determining their appropriate disposition. The documentation provided with the Systems, Inc. system was not always comprehensive enough, so that he made many decisions on his own. This customer representative also described how the ABC Company was becoming more demanding with questions such as: What will be done by Systems, Inc. to prevent the latest problem from recurring?

Further discussion revealed that this customer representative's experience was not atypical. It was becoming evident to the appraisal team that there was no clear understanding of the problems experienced by customers like ABC, and that there was no in-depth system error data recorded in the trouble report forms. To accomplish correction of the software, a manual pre-analysis of the problem reports was required in order to isolate and classify errors resulting directly from software-related problems.

10.2.2.3 Software Design and Coding FAR Discussion Group

Based on the review of the design and coding work products, it was decided to formulate some open-ended questions to guide the discussion with the Software Design and Coding FAR Discussion Group. The discussion with the Software Design and Coding FARs was to explore issues related to the software design and coding process, in particular, schedules, definition of tests, and baselines for software versions. The results of these discussions are summarized as follows.

■ *Schedules:* The development of schedules was very difficult for the new system version. The organization primarily used the waterfall model as their life-cycle model, and required the production of large quantities of paper documentation as a means of capturing the results of the development effort. Because of the rather inflexible development cycle model the organization was using, routine "housekeeping" and documentation tasks, such as recording actual time spent on tasks, were often put aside until a later point in time. Often, developers recorded chunks of time spent on performing tasks (e.g., 20 or 40 hours) as the easy way out of having to keep detailed records, whether or not they actually spent that amount of time on those tasks. The development team had to estimate development activities in an environment where little existed in the way of accurate historical data that could provide insight into the amount of effort required to develop systems of similar complexity. In addition, the FARs indicated that currently, there are no procedures or tools for collecting detailed time records by specific activities. The availability of such records could be used for developing more accurate time estimates in the future, and for reporting and comparing actual time to current plans and schedules.

■ *Definition of tests and baselines for software versions:* The FARs stated that the most critical issue in releasing a new version of the system is the verification of Systems, Inc.'s system performance. This is essential to the users because they must have confidence that compatibility of the Systems, Inc. system is strictly enforced between versions. A related area of concern is the integration of software upgrades into the system. Currently, there are no formal procedures to ensure that all the components that should be included in a particular version are indeed included in it. They said that from what they knew, the test team is looking for ways to develop testing procedures that will cover as many conditions as possible to ensure that most (and hopefully all) bugs are detected prior to releasing a version.

A problem related to establishing a baselined reference point for performing testing is the fact that requirements keep changing. Even though there is configuration control, no one seems to be assessing the impact of the changes. Marketing, who is the customer of the development group, continuously submits changes to the requirements, even as late as during the final integration test cycle. Because the configuration management system operates rather slowly, the changes aren't always communicated to the developers and the testers.

These observations were confirmed by the discussion that occurred during the Test and Quality Assurance FAR Group.

The above descriptions are only a sample of the discussions in FAR groups, and do not represent a history of all the discussion groups. The document review and the totality of the discussion groups yielded several findings that later resulted in process improvement actions. We discuss in detail two of these improvements.

After the FAR and project manager discussions, the appraisal team members consolidated their observations and discussed (1) what they heard and (2) what the consequences were of what they heard. The team then prepared the following preliminary findings:

■ Systems, Inc. is delivering systems with a lower level of reliability than the customers want.
■ Customers' problem reports are being actively collected and analyzed. This provides Systems, Inc. with an opportunity to fix and follow up on customer

complaints. The generated reports, however, are not used effectively, and several issues with the process need to be addressed.

- Inadequate schedule and work effort estimation procedures and project plans make the planning and control of software development tasks difficult and chaotic.
- Lack of formal configuration management makes the generation of new versions a high-risk operation with no management visibility of what is actually being shipped to customers.
- The data accumulated by Systems, Inc. on hardware and software products and processes is not effectively used by management.

These findings by the appraisal team later resulted in a number of recommended process improvement projects, one of which was to implement a more flexible, responsive development process. The methodology that was eventually selected for this was Scrum, an *agile* methodology.

10.2.3 *Benchmarking of Systems, Inc.*

How does Systems, Inc. stack up against other development organizations? This subsection provides statistics derived from the SEI's database on appraisals conducted by authorized lead appraisers. To position Systems, Inc. in the context of the capability of other software companies, we refer to the Capability Maturity Model Integration for Development (CMMI-DEV), developed by the Software Engineering Institute (see Chapter 3). As pointed out in Section 3.1.2, the staged representation of the model characterizes the capability of development organizations, using five increasing levels of maturity comprised of twenty-two process areas [8]. The higher an organization is positioned on the maturity scale, the greater the probability of project completion on time, within budget, and with properly working products—or, in other words, the lower the development risks. The lowest maturity level, the initial level, is characterized by ad hoc, heroic efforts of developers working in an environment characterized by few or no project plans, schedule controls, configuration management, and quality assurance, and mostly verbal undocumented communication of requirements. Going up the maturity scale involves establishing proper project management systems, institutionalizing organizationwide practices, and establishing sound software engineering disciplines. Although Systems, Inc. would appear to have implemented elements of Process Areas from Maturity Level 2 to Level 5, the company exhibits the characteristics of a Level 1 organization.

To review, Table 10.1 lists the characteristics of the twenty-two PAs of the five Maturity Levels with respect to their primary focus. Note that an organization going from Level 1 to Level 2 needs to focus primarily on establishing project management and support capabilities, whereas an organization going from Level 2 to Level 3 is putting in place PAs whose primary focus relates not only to project management concerns, but also to establishing and institutionalizing standardized practices for

Table 10.1 Version 1.2 CMMI-DEV Process Areas by Maturity Level

Maturity Level	Name	Focus	Process Areas	Category
1	Initial		None	None
2	Managed	Basic Project Management	Requirements Management	Engineering
			Project Planning	Project Management
			Project Monitoring and Control	Project Management
			Supplier Agreement Management	Project Management
			Measurement and Analysis	Support
			Process and Product Quality Assurance	Support
			Configuration Management	Support
3	Defined	Process Standardization	Requirements Development	Engineering
			Technical Solution	Engineering
			Product Integration	Engineering
			Verification	Engineering
			Validation	Engineering
			Organizational Process Focus	Process Management
			Organizational Process Definition	Process Management

Table 10.1 Version 1.2 CMMI-DEV Process Areas by Maturity Level
(*Continued*)

Maturity Level	Name	Focus	Process Areas	Category
			Organizational Training	Process Management
			Integrated Project Management, Specific Goals 1 and 2	Project Management
			Integrated Project Management, Specific Goals 3 and 4 (apply to IPPD only, in addition to Specific Goals 1 and 2)	Project Management
			Risk Management	Project Management
			Decision Analysis and Resolution	Support
4	Quantitatively Managed	Quantitative Management	Organizational Process Performance	Process Management
			Quantitative Project Management	Project Management
5	Optimizing	Continuous Process Improvement	Organizational Innovation and Deployment	Process Management
			Causal Analysis and Resolution	Support

the organization as a whole in all aspects of development. There is focus on the practices that affect the system and software engineering, support, and process management aspects of the development enterprise. Level 2 organizations, in establishing a project management discipline, are free to specify project-unique system and software development processes. But each project must specify a specific development process, which must be documented in their project's program management plan. Level 3 organizations, on the other hand, have established organizationwide standard practices (which can be product line or application domain-specific). These standard practices cover project management, process management, engineering, and support. (The CMMI-DEV, however, does encourage tailoring the standard process for the unique characteristics of an individual project.) Establishing standardized practices occurs at the organizational level, as does establishing an organizational entity that has the responsibility for the definition and maintenance of the standard process. Five PAs focus on product development, reflecting the engineering aspects of the organization's activities. Two PAs focus on more advanced project management activities, three PAs focus on process management, and the remaining PA at Level 3 focuses on the support aspects of development. Level 4 focuses on establishing and implementing quantitative goals and objectives for the organization and the projects for products and processes (including the use of quantitative and statistical methods). At Level 5, the organization and the projects use their quantitative knowledge of process performance to improve overall quality.

The SEI appraisal database semi-annually reports the results of appraisals over a moving 5-year period, ending in the month and year indicated in the report. The September 2008 report includes findings from appraisals performed in 3,009 organizations in the United States and overseas (see [7]). Of these organizations, 13% of them (like Systems, Inc.) were in the process of moving from the Initial Level (Level 1) to the Repeatable Level (Level 2). An additional 32.2% were rated at Maturity Level 2, and 44.4% were recognized as being at Level 3. While these figures might suggest that a large number of organizations are moving toward high maturity, this is a misleading impression because the organizations reporting are not separate companies. In many instances, these organizations are lower-level entities within a corporate structure, many of which have adopted the CMMI-DEV. For example, one large multinational corporation has performed SCAMPI appraisals in approximately a dozen separate divisions, and in some cases, some of these divisions have had several appraisals performed. Further, these results are from appraisals that were conducted to completion. In many instances, if it begins to look like the organization won't make the target maturity level, it will abort the appraisal. Others won't even start one until they feel reasonably certain they will attain the desired maturity level. And then there are a vast number of organizations that have not or will not implement the CMMI-DEV. So, although these figures would make it appear that the development capability of organizations is greatly improving, the data is skewed in the direction of organizations that have achieved success in process improvement.

In the SCAMPI appraisal method, to rate an organization at Maturity Level 2, all the PAs at Level 2 must have been satisfactorily implemented. That means that all the practices of the Level 2 PAs must have been characterized as Largely Implemented or Fully Implemented. In determining that the organization is functioning at Level 1, the appraisal team determined that these criteria have not been met. Significant weaknesses were observed in almost all Level 2 PAs.

10.3 Strategic Planning for Process Improvement at Systems, Inc.

In Chapter 1 we discussed the basic elements of a general strategic plan, and in Chapter 5 we introduced a focused methodology for performing strategic planning for process improvement. Systems, Inc. implemented that methodology and came up with a number of proposed process improvement projects that were prioritized by top management. Considering the nature of the business and its need to remain competitive, two improvement projects received the highest scores and were considered the highest priority projects to implement. One was a project to implement a more flexible, shorter-duration development methodology, and the other was to implement a measurement program. The organization decided to include in the measurement program some elements of a Maturity Level 4 process as a means of evaluating the effectiveness of the changed process. Because there is considerable risk in changing to a significantly different development process, it was felt that collecting statistical data on the current development process *and* its replacement would help in determining if the risk associated with a new development process was materializing into a significant problem. Consequently, to establish a baseline for the current development process, its introduction was delayed until such time as a stable, realistic baseline of the old development process could be established.

Time-wise, the organization established the measurement program before it implemented the new development process; however, in the following subsections, we first discuss the development process improvement project.

It should be noted that there were significant problems in other areas, such as configuration management, project management, and information available from defect reports. Process improvement projects were also initiated to address some of the other critical appraisal-observed deficiencies; however, the two process improvement projects upon which we focus in the remainder of this chapter best illustrate how all the principles discussed in this book tie together.

10.3.1 The Systems, Inc. Software Development Challenge

Many software projects at Systems, Inc. did not achieve the company's goal with respect to time-to-market, budget, quality, and/or customer expectations. The main

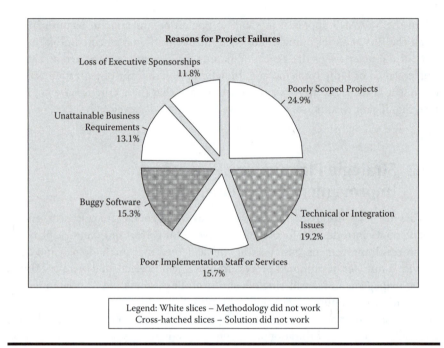

Legend: White slices – Methodology did not work
Cross-hatched slices – Solution did not work

Figure 10.1 Corporate executives' reasons for project failures. (Adapted from Forrester Business Technographics, June 2003, North American Benchmark Study.)

root-cause for this was found not to be in the technology arena, but in the development process methodology (see Figure 10.1). During the appraisal, it was determined that the development process in use was too rigid and inflexible, and Scrum was selected as the process improvement project to be implemented to improve the responsiveness and timeliness of the development process.

One possible response to this phenomenon is to make development methodologies more disciplined and predictive—that is, more planning, greater attention to analysis of requirements, formal sign-off, detailed and documented design before coding, strict change control, etc. This approach is known as the "waterfall process" [4]. Under this approach the organization gets better control, but project success rates do not necessarily improve. Many organizations have found that, by this approach:

■ Methodologies simply prove bureaucratic and slow to deliver products on time.
■ It is difficult for the business to completely conceptualize all requirements in one pass.
■ It is even more difficult to capture all the requirements in a document in a completely detailed and unambiguous form. The customer often does not understand UML representations and the language in the specifications. That leads

to free interpretation of the requirements by the various players downstream in the development cycle (e.g., system analysts, designers, developers, testers).

■ Businesses constantly change—requirements and business priorities constantly change; the longer the project, the greater the risk. If change is successfully suppressed, the business gets a system it can't use.

■ It's difficult to completely design a system in a single pass, and will not be of great value if the requirements change anyway.

■ Developers do not know how to estimate the impact of complex requirements.

Realizing that the waterfall process is inappropriate for system development involving a high degree of innovation led to the development of a new development paradigm, the "agile process" [11]. This has worked very successfully in software development projects and is applicable to the development of any kind of system, within limitations. Boehm [1] defines five attributes of projects that are determinants of the extent to which agile methods would apply. These include:

1. Project size, in terms of number of project personnel
2. Mission criticality, in terms of the effect of a system failure
3. Personnel skills, in terms of a three-level capability scale
4. Project dynamism, in terms of percent of requirements change per month
5. Organizational culture, in terms of seeking to thrive on order versus chaos

To illustrate the limitations, Boehm described a project that began to experience difficulty when the project size reached 50 personnel, 500,000 source lines of code, and nontrivial mission criticality requirements. At that point, the project had to develop a mix of plan-driven and agile methods to manage the project to a successful conclusion.

One of the main ideas behind agile methodologies is that this evolutionary process seeks to move the development effort forward by a large number of small steps, on schedule-based cycles (thirty to ninety days), each of which delivers some benefits—that is, a stable and workable solution. Agile and Scrum processes have been proven effective. A common rule of thumb is that 20% of the features of a system deliver 80% of the benefit; so, more frequent releases can bring significant gains. A particularly important observation about agile processes is that they seek to shorten or eliminate feedback loops. Bringing empowered users in contact with the development team, for example, can remove much longer cycles of trying to describe what is needed, developing it, and trying out the product.

Similar observations can be made about other agile practices, such as establishing a joint cross-discipline team that is accountable as a team to deliver specific working features. The communication is mostly face-to-face, with minimal unnecessary documentation (versus the waterfall model where the main means of communication between the disciplines is through documentation that can lead to the need for interpretation).

Also, other agile practices, such as automatic unit testing, pair programming (versus peer reviews), daily stand-up meetings for planning, and automated build and regression testing, have derived from the notion that getting better feedback sooner improves both quality and control sooner.

Finally, priority-driven planning is an essential feature of most agile methods. Managing requirements individually, rather than as a composite document, allows the high-benefit, low-cost features to be rolled out to users sooner. Also, fixing timescales for deliveries, rather than fixing the required features and allowing timescales to slip, is a far more effective means of ensuring continuous enhancement of business performance.

In systems development, an evolutionary approach requires investment—particularly in test-driven design, continuous integration, and comprehensive automated build facilities from a configuration management system. Such investment is nearly always worthwhile because it can eliminate a separate integration and test phase and allow early release of useful functionality. The traditional *quality gates* are not applicable to the new agile process because now activities are not divided into phases, but rather they are focused around the implementation of individual features or stories. To replace quality gates, the new agile framework proposes a set of repeating checkpoints. Some checkpoints repeat only once in iteration, for example, the checkpoint that marks the end of the iteration planning, or the iteration review at the end. Other checkpoints are associated with individual requirements or features, such as the checkpoint that marks completion of a feature. While these checkpoints occur much more frequently than quality gates, in an agile process they must be much lighter weight in terms of ceremony.

For agile methods to scale up, the most important thing is to establish a suitable framework that is sufficiently generic to apply across the organization while providing detailed guidance. This framework in itself has the power to improve agility by helping developers structure their deliveries into shorter cycles, while also steering them toward a more iterative and collaborative approach [3]. Agile methods scale up, but only within the limitations defined by Boehm [1]. We now proceed to describe the basic elements of the agile process, as it was implemented at Systems, Inc.

10.3.1.1 Scrum at Systems, Inc.

Systems, Inc. decided to implement Scrum for managing agile projects. Scrum was first described in 1986 by Takeuchi and Nonaka [9]. The name is taken from the sport of Rugby, where everyone on the team acts together to move the ball down the field. The analogy to development is that the team works together to successfully develop quality software. In particular, "Scrum" refers to the entire team banding together for a single aim: getting the ball!

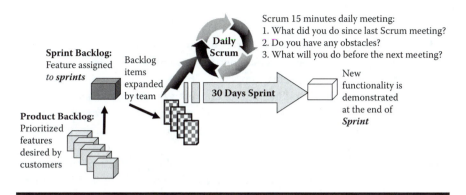

Figure 10.2 The Scrum process.

10.3.1.2 The Scrum Process

Figure 10.2 presents a conceptual picture of the Scrum process. Key Scrum practices include:

- Focus on schedule-based cadence; sprints are iterations of fixed 30-day duration.
- Work within a sprint is fixed. Once the scope of a sprint is committed, no additional functionality can be added except by the development team.
- All work to be done is characterized as product backlog, which includes requirements to be delivered, defect workload, as well as infrastructure and design activities.
- The product backlog is the basis for the sprint backlog as defined by the sprint team and the product owner. The team decides what it can develop.
- A Scrum Master mentors and manages the self-organizing and self-accountable teams that are responsible for delivery of successful outcomes at each sprint.
- A daily stand-up meeting is a primary communication method.
- A heavy focus on time-boxing. Sprints, stand-up meetings, release review meetings, and the like are all completed in prescribed times.
- Scrum also allows requirements, architecture, and design to emerge over the course of the project.

The Scrum life-cycle phases are planning, staging, development, and release. There are typically a small number of development sprints to reach a release. Later sprints focus more on system-level quality and performance as well as documentation and other activities necessary to support a deployed product. Typical Scrum guidance calls for fixed 30-day sprints, with approximately three sprints per release, thus supporting incremental market releases on a 90-day time frame (see [5, 6]).

10.3.1.3 Roles in Scrum

- *Scrum Master:* The main focus of the Scrum Master is to:
  - Safeguard the process, removing the barriers between development and the customer, enable close cooperation between roles and functions across the organization, and facilitating creativity and empowerment.
  - Improve the productivity of the development team in any way possible.
  - Improve the engineering practices and tools so each increment of functionality is potentially shippable.
- *Scrum Team:* This is a cross-functional group of people with all the different skills that are needed to turn requirements into something that is an increment of potentially shippable functionality. The Scrum Team is responsible for the project, that is, is committed to deliver. It has the right to do everything within the boundaries of the project guidelines to reach the iteration goal. The team consists typically of five to ten people, and is a cross-functional team (QA, programmers, designers, etc.). Members should be full-time, as much as possible. Teams are ideally self-organizing, no titles, but (rarely) a possibility. Membership can change only between sprints.
- *Product Owner:* The Product Owner is responsible for the ROI of the project. This is the person who is investing or representing everyone's interest in the project, serves as the ultimate arbiter on any requirements issues, and accepts or rejects work results. The product owner defines the features of the product, decides on release date and content, prioritizes features according to market value, and can change features and priority every sprint. This requires high-bandwidth communication and transparency into the team's progress.
- *Management:* Management is in charge of final decision making, along with the charters, standards, and conventions to be followed by the project. Management also participates in setting the goals and requirements.

10.3.1.4 Meetings/Activities in Scrum

- *Pre-game—High-level design and architecture:* In the architecture phase, the high-level design of the system, including the architecture, is planned based on the current items in the product backlog. In the case of an enhancement to an existing system, the changes needed for implementing the backlog items are identified, along with the problems they may cause. A design review meeting is held to go over the proposals for the solution and decisions are made. In addition, preliminary plans for the contents of releases are prepared.
- *Pre-Game—Release planning:* The product backlog lists all the functionality that the product or solution could have in it. If all of this functionality is built before a release to production, it may waste the opportunity that Scrum provides for seeing an early return on investment and useful production feedback. For these reasons, it is common practice to divide the product

backlog into "releases," or collections of useful system capability that make sense when put into production.

■ *Sprint planning:* Usually every sprint has a goal or main theme that expresses the product owner's motivation for that sprint, embodied as specific measurable exit criteria. Each sprint must include some business functionality. The sprint planning session consists of two segments (usually around 4 hours each). In Segment 1, the product owner selects the ideal backlog for the coming sprint and communicates its meaning and importance to the team. In Segment 2, the team decides what it can commit to delivering in the sprint. The product owner answers questions but does not direct the team's choices. The team decides how to turn the selected requirements into an increment of potentially shippable product functionality. The team self-organizes around how it will meet the sprint goal. The team devises its own tasks and figures out who will do them. The outcome is the sprint backlog.

Clearly, as part of planning the sprint, effort estimation must be performed in order to effectively size the sprint to fit in with the objective of a 30-day duration for each sprint. Various techniques and tools exist for facilitating the estimation process. One such technique is planning poker. "Stories" are created to characterize the functionality desired for the sprint. A deck of cards numbered in a Fibonacci sequence is given to each of the sprint team members. Each team member selects a card to characterize the degree of difficulty in developing a solution for the story. The team members with the highest and lowest cards are called upon to explain the rationale for the number they selected, and the rest of the team members are given the opportunity to change the number they selected, if they wish. Following this reassessment, the features are prioritized for the sprint. More details about this technique can be found at http://www.planningpoker.com/, and a no-cost tool to support planning poker is also available at this site.

■ *Spike:* A spike is an experiment that allows developers to learn just enough about something unknown in a user story (e.g., a new technology) to be able to estimate that user story. A spike must be time-boxed. This defines the maximum time that will be spent learning and fixes the estimate for the spike.

■ *Daily Scrum:* A short status meeting that is time-boxed to 15 minutes and is held daily by each team. During the meeting, the team members synchronize their work and progress, and report any impediments to the Scrum Master.

■ *Sprint review:* The sprint review provides an inspection of project progress at the end of every sprint. The team presents the product increment that it has been able to build. Management, customers, users, and the product owner assess the product increment. Possible evaluation consequences are:

  - Restoring unfinished functionality to the product backlog and prioritizing it
  - Removing functionality from the product backlog that the team unexpectedly completed

- Working with the Scrum Master to reformulate the team
- Reprioritizing the product backlog to take advantage of opportunities that the demonstrated functionality presents
- Asking for a release sprint to implement the demonstrated functionality, alone or with increments from previous sprints
- Choosing not to proceed further with the project and not authorizing another sprint
- Requesting that the project progress be sped up by authorizing additional teams to work on the product backlog

■ *Sprint retrospective:* This is a meeting facilitated by the Scrum Master at which the team discusses the just-concluded sprint and determines what went well and what could be changed that might make the next sprint more productive. While the sprint review looks at "what"" the team is building, the retrospective looks at "how" they are building.

■ *Post-Game Release sprint (integration, system packaging):* When the product owner and stakeholders identify that there is sufficient functionality in the system to provide immediate business value, they may choose to put this into production. Typically, a "release sprint" will follow when the sprint contains all the necessary sprint backlog tasks to put the system into production. These tasks shouldn't contain additional functionality but they may include:

- Full end-to-end system test, integration, performance, and regression test, if necessary
- Finishing required documentation
- Deploying the code to the production environment, production data population, setting up management and operational systems and processes, training and handover for support staff, and cutover and fallback planning

10.3.1.5 Scrum Artifacts

■ *Product backlog:* A product backlog is a prioritized list of project requirements with estimated times to turn them into completed product functionality. Estimates are in days and are more precise the higher the item is in product backlog priority. Priority should be assigned based on the items of most value to the business or that offer the earliest return on investment. This list should evolve, changing as the business conditions or technology changes.

■ *Sprint backlog:* The sprint backlog is a list of tasks that defines a team's work for a sprint. The list emerges during sprint planning. The tasks on the sprint backlog are what the team has defined as being required to turn committed product backlog items into system functionality. Each task identifies who is responsible for doing the work and the estimated amount of work remaining on the task on any given day during the sprint.

■ *Impediment list:* Anything around a Scrum project that impedes its productivity and quality is an impediment. It is the responsibility of the Scrum

Master to remove any impediment that is stopping the team from producing production-quality code. The impediment list is simply a set of tasks that the Scrum Master uses to track the impediments that need to be solved.

■ *Task list:* The task list is a list of tasks to turn product backlog into working product functionality. Tasks are estimated in hours, usually between 1 and 16. Tasks with more than 16 hours are broken down later. Team members sign up for tasks. Team members shouldn't sign up for work prematurely (until actually starting that task). Estimated work remaining is updated daily; any team member can add, delete, or change the sprint backlog. Work for the sprint emerges; if teams believe that this has become too much, they can meet again with the product owner.

■ *Product burndown chart:* The product burndown chart gives an indication of how quickly the team is "burning" through the work and delivering product backlog requirements. It is a useful tool for helping to plan when to release or when to remove requirements from a release if progress is not rapid enough.

10.3.1.6 Scaling Scrum in Big Projects

The primary way of scaling Scrum to work with large teams is to coordinate a "Scrum of Scrums." With this approach, each Scrum team proceeds as normal but each team also contributes one person who attends Scrum of Scrum meetings to coordinate the work of multiple Scrum teams. These meetings are analogous to the daily Scrum meeting, but do not necessarily happen every day. In many organizations, having a Scrum of Scrums meeting two or three times a week is sufficient.

Implementing Scrum is basically an organizational change. Implementing such a change in a large organization is a complex task that requires a field-tested methodology. The next subsection describes how EKD, a business process methodology, was integrated with Scrum at Systems, Inc.

10.3.1.7 EKD, Patterns, and Scrum at Systems, Inc.

In Chapter 2 we discussed the Enterprise Knowledge Development (EKD) methodology and how it can be used to describe processes. In this subsection we present how Scrum components were developed using the EKD description of patterns at Systems, Inc. After providing some examples, we will describe some basic elements of a Scrum patterns repository, and how it was populated, maintained, and used.

10.3.1.7.1 Scrum, EKD, and Pattern Description

Scrum patterns consist of Scrum activities, roles, and artifacts. Figure 10.3 lists key Scrum activities. After describing these elements at Systems, Inc. using EKD

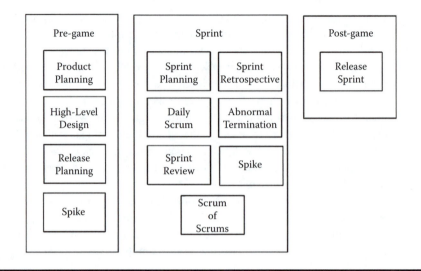

Figure 10.3   High-level view of Scrum activities.

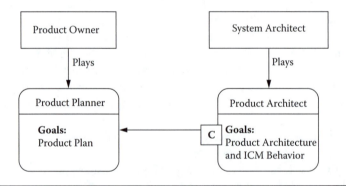

Figure 10.4   The product owner and system architect actor–role diagram.

notation (see Figures 10.4 and 10.5), we will discuss how they were used to elicit and organize patterns in a pattern repository.

10.3.1.7.2 A Sample Sprint Pattern

Context: The context for the sprint pattern is that of a Systems, Inc. "coach" managing a development team where there is a high percentage of discovery, creativity, and testing involved.

As noted previously, *sprints* are applicable for building systems, both new and existing, that allow partitioning of work, with clean interfacing, components, or objects. Every person on the team has to understand the problem fully and be aware of all the steps in development. This limits the size of the team and of the

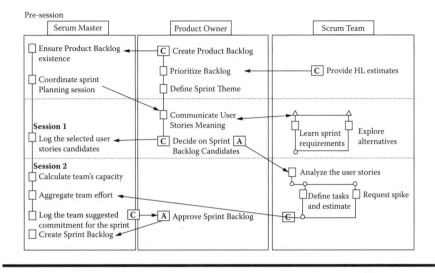

Figure 10.5 The *sprint* role–activity diagram.

system developed. Trust is a core value in Scrum, and especially important for the success of sprints, so *self-selecting teams* is a plus. During a sprint, communications are optimized, and maximized information sharing occurs in the daily Scrum. Each sprint takes a pre-allocated amount of work from the product backlog. The team commits to it. As a rule, nothing is added externally during a sprint. External additions are added to the product backlog. Issues resulting from the sprint can also be added to the product backlog. A sprint ends with a demonstration of new functionality.

Problem: The Systems, Inc. Scrum Masters want to balance the need of developers to work undisturbed and the need for management and the customer to see real progress.

Forces: For many people—project managers, customers—it is difficult to give up control and proof of progress as provided in traditional development. It feels risky to do so; there is no guarantee that the team will deliver. As previously noted, by the time systems are delivered, they are often obsolete or require major changes. The problem is that input from the environment is mostly collected at the start of the project, while the user learns most using the system or intermediate releases.

Solution: Give developers the space to be creative, and to learn by exploring the design space and doing their actual work undisturbed by outside interruptions, free to adapt their way of working using opportunities and insights. At the same time, keep the management and stakeholders confident that progress is being made by showing real progress instead of documents and reports produced as proof. Do this in short cycles, *sprints*, where part of the product backlog is allocated to a small team. In a sprint, during a period of approximately 30 days, an agreed-to amount of work will be performed to create a deliverable. During sprint planning, a backlog is assigned to sprints by priority and by approxima-tion of what can be accomplished during a month. Chunks of low cohesion and high coupling are selected. Focus is on enabling, rather than micromanagement. During the sprint, outside chaos is not allowed into the increment. The Scrum Master ensures this by managing the impediment list. The team, as it proceeds, may change course and its way of working. By buffering the team from the out-side, we allow the team to focus on the work at hand and on delivering the best it can and in the best way it can, using its skill, experience, and creativity. Each sprint produces a visible and usable deliverable. This is demonstrated in *demo after sprint*. An increment can be either intermediate or shippable, but it should stand on its own. The goal of a sprint is to complete as much quality software as possible and to ensure real progress, not paper milestones as alibis. Sprints set up a safe environment and time slots where developers can work undisturbed by outside requests or opportunities. They also offer a pre-allocated piece of work that the customer, management, and the user can trust to get a useful deliverable, such as a working piece of code at the end of the sprint. The team focuses on the right things to do, with management working on eliminating what stands in the way of doing it better.

Rationale: Developers need time to work undisturbed. They need support for logis-tics, and management and users need to stay convinced that real progress is made.

Examples: Systems, Inc. decided to use sprints on a number of end-user projects and for the development of a framework for database, document management, and workflow. The product backlog is divided into sprints that last about a month. At the end of each sprint, a working *Smalltalk* image is delivered with integration of all current applications. The team meets daily in daily Scrum meetings and the product backlog is allocated after the *demo after sprint* in a monthly meeting with the steering committee.

Applying the EKD methodology, Figure 10.4 presents the Systems, Inc. Actor diagram of the product owner and system architect. It indicates that product owners need to be coordinated by the system architect at the product architecture level.

Running a sprint requires coordination between the Product Owner and the Scrum Master. Figure 10.5 presents a role–activity diagram for the role of the Scrum Master, Product Owner, and Sprint Team, in the context of a sprint.

10.3.1.7.3 Defining Scrum Patterns

When defining patterns, we need to continually keep in mind two aspects: (1) how we propose to introduce and implement the pattern in the organization, and (2) the importance and relevance of issues being presented. Without clarifying these two points for each pattern, they will be of no use. The purpose of the process of defining patterns is to produce a description of reusable components. This process is mainly concerned with two things: identifying potential reusable knowledge and constructing the reusable component embedding this knowledge:

■ *Identifying reusable knowledge:* This activity is essential to the process of defining reusable components. There are two problems to be considered, namely: (1) identifying the sources of knowledge that will be used as the starting point for defining reusable knowledge, and (2) selecting the techniques that will be used for extracting the reusable knowledge. Two types of techniques have been proposed so far. One is based on the study of a wide variety of existing products, whereas the second is based on domain-oriented meta-modeling approaches. The latter focuses on the definition of specific concepts for describing domain knowledge.

■ *Constructing reusable components:* This activity mainly concerns the problems of specifying and organizing reusable components. The specification of reusable components is based on techniques that facilitate reuse, such as abstraction, genericity, and inheritance. These techniques try to balance the degree of reusability of the components with the effort that will be made while effectively reusing the component. The organization of the reusable components is driven toward facilitation of the search and retrieval of reusable components.

Scrum patterns elicitation. Systems, Inc. applied the following four-step procedure to define patterns (see Table 10.2). By define, we mean the discovery, selection, and demarcation of the relevant patterns. The stages are:

1. Collect candidates
2. Evaluate suitability
3. Derive structure
4. Validation

To minimize redundancy, it is not necessary to provide a full description of all patterns as provided for in the pattern template, already at Stage 1. The degree of detail should be increased as one moves through the stages. A scorecard for each candidate pattern should be completed. And for each stage, the various stakeholders play specific roles (see Table 10.3).

10.3.1.7.4 Scrum Pattern Repository and Reuse

To make the Scrum generic knowledge easy to organize and access for the benefit of the organization, one needs to systematize and structure the knowledge and experience gained in different parts of the company. The knowledge engineer's task is to provide a framework where potential patterns may be compared in a systematic way, according to commonly accepted criteria, so as to enable a satisfactorily informed decision.

A commonly recurring problem is the trade-off that is often made between having "easily organized knowledge," which is inherently incomplete, so as to be

Table 10.2 Stages in Defining Patterns

Part of Pattern Stage	Formal Signature	Informal Signature	Responsible Actor	Other
1. Collect candidates		Problem Solution Context	Project Management	Initial EKD Model
2. Evaluate suitability	Initial draft	Name Forces Rationale Consequences	Domain experts	Thesaurus, initial guidelines
3. Derive structure, context, and relationships	Verb Object Source Result Manner	Related patterns Related documents Contributing authors Hyperlinks Annotation Version	Knowledge engineer	
4. Validation	All attributes fully validated	All attributes fully validated	Project Management, Domain experts	Complete guidelines

Table 10.3 Roles, Tasks, and Actors for the Stages in Defining Patterns

Role	Tasks
Project Management	Overall responsibility: • Initiate definition procedure • Facilitate communication between analysts and domain experts
Knowledge engineer	Provide EKD and Pattern methodological support Ensure structure and consistency with objectives and method
Domain experts	Provide domain knowledge Ensure validity

structured, in comparison to knowledge that reflects reality, which is often not easily structured, or even understood. The choice is where to be on a knowledge continuum: between highly structured and unrealistic knowledge at one end, and unstructured but realistic knowledge at the other. Most developers tend to be somewhere in the middle of the continuum, leaning toward the unstructured side. In practical terms this means that when defining patterns, it is more important, for most people, to reflect real problems and solutions rather than flashy and technically brilliant presentations.

Knowledge, expressed by generic patterns, should facilitate the creativity process by reducing the need to "reinvent the wheel" when facing new problems and situations. The essence of the use of patterns is that they are applied to recurring problems. A pattern is of no use if it aims to solve a problem that is extremely unlikely to occur within the foreseeable future for those businesses that are envisaged to have access to the patterns.

To enhance reuse, patterns need to be evaluated and assessed periodically. Table 10.4 presents a set of criteria that can be used to classify a pattern on a High–Medium–Low scale. The criteria focus on usefulness, quality, and cost. Obviously each organization should develop its own criteria, in line with its strategy and organizational culture.

In using patterns, we advocate an approach to describing repeatable solutions to recognizable problems. In this context, both the problem and the solution must be uniquely identifiable and accessible. The pattern usage framework must therefore make the distinction between product or artifact patterns and process patterns, and includes an indexing schema for accessing them. The patterns' typology aims to distinguish between the way to solve a problem and the elements that will be used for the solution, while the indexing hierarchy characterizes each pattern by the problem that it addresses through the usage perspective and the knowledge perspective. The template in Table 10.4 represents the usage perspective, and the

Table 10.4 Pattern Evaluation Criteria

Criteria	Sub-criteria	High Value	Medium Value	Low Value
Usefulness	**Degree of triviality** The degree to which the pattern addresses a problem which is of little importance because the problem or solution is obvious.	The pattern is concerned with issues that are or most likely will be of concern to other parts of the company.	While the pattern deals with a pertinent problem, the solution is already well known.	The pattern is concerned with a problem which does warrant the creation of a pattern since it is so trivial with the proposed solution being obvious to domain experts.
	Grade of implementability Extent that pattern is thought to be practical and implementable. Is change compatible with business strategy. Have trade-offs been taken into account.	The pattern is useful in that it prescribes practical, easy to understand and implement solutions.	The pattern may be of some use despite some practical problems in implementation and some difficulty in understanding the solution.	The pattern is not usable. The solution is impractical and difficult to understand. The pattern only proposes "paper-based" change rather than real change.
	Degree of confidentiality	The pattern does not disclose any confidential business information.	Some information may be able to be used by other projects.	The pattern discloses sensitive project information.

Quality			
Degree of complexity The number of factors and their relationships.	The pattern addresses only a few manageable main concepts and ideas.	The pattern is complex but may still be useful in that the complexity is needed.	The large number of factors that affect the implementation of the solution minimizes the chances that the solution can be implemented.
Addition of value The local and global benefits accruing to the business with the implementation.	The consequences of a successful implementation are great value to the project directly affected as well as other projects.	The local and global benefits are unclear, difficult to determine or marginal.	There are no local or global benefits or there is a conflict between these so that in total no value is added.
Level of genericity Abstraction level of the problem that the pattern addresses.	The pattern addresses a problem that is general enough for all the company.	The pattern addresses a problem that applies only to part of the company.	The pattern addresses a problem that is only relevant to the project in which it was discovered.
Grade of understandability Visualizable and identifiable.	The pattern is easy for decision makers, domain experts and those to be affected, to comprehend.	The pattern is only partially understandable to decision makers, domain experts and those to be affected.	The pattern is incomprehensible to stakeholders.

Continued

Table 10.4 Pattern Evaluation Criteria

Criteria	Sub-criteria	High Value	Medium Value	Low Value
	External compatibility The extent to which the pattern could be used by other companies.	The pattern has taken into account differences in national and organizational cultures and ways of working among identified future external users of the pattern.	The pattern partially takes into account differences in national and organizational cultures and ways of working among identified future external users of the pattern.	The pattern does not take into account differences in national and organizational cultures and ways of working among identified future external users of the pattern.
Cost	**Level of experience in their use**	The pattern has been implemented within the company.	The pattern has been partially or sporadically used.	The pattern has never been implemented.
	Economic feasibility of the proposed solutions	Proposed solution is relatively easy to implement. Organizational support exists in terms of sufficient resources as well as managerial support. The solution is politically and socially acceptable.	Proposed solution is difficult but feasible to implement. Organizational support is lukewarm. Resources are available but may not be sufficient. There may exist political and social difficulties in making the pattern feasible.	Proposed solution is not feasible. Organizational support will be difficult to obtain. The resources will not be made available. The existing difficult social and/or political climate would make an implementation impossible.

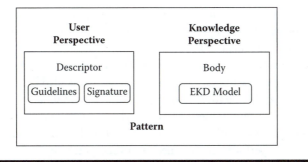

Figure 10.6   The Scrum pattern indexing organization.

EKD modeling presents the knowledge perspective. To enable reuse and pattern retrieved, a signature is required in either a formal or informal format. An example of how Scrum patterns can be indexed and used in practice by specific projects within the enterprise is illustrated in Figure 10.6. An example of an electronic patterns repository based on EKD is available at http://crinfo.univ-paris1.fr/EKD-CMMRoadMap/index.html.

10.3.2 The Systems, Inc. Measurement Program

Systems, Inc. recognized that the development cycle model currently in use was caus-ing problems with its customers, in that the development cycle was taking too long, and reliability was degraded. The changes to the development cycle that the organi-zation decided to implement were described in the discussion above. But change for the sake of change without knowing what the benefits of the change are is foolhardy. Accordingly, Systems, Inc. recognized that it needed to make some measurements of the current development cycle and process in use in order to compare that with the Scrum methodology that the organization intended to implement so as to deter-mine if the Scrum methodology was providing the hoped-for benefits.

Fortuitously, Systems, Inc. kept a crude record of development costs by devel-opment phase, and maintained a record of defects recorded during integration and validation testing (i.e., testing against requirements), as well as defects reported by its customers. Time spent in requirements analysis, detailed design, coding, and the two testing phases combined were recorded on separate charge numbers on each project. Because there were two individuals who developed architectures, it was possible to tease out the time spent in developing the architecture for each system by knowing who the architect was and when the work was done. The data was normalized as a function of the size of the software (in source lines of code) needed to implement the system, and control charts of the time expended in each phase were developed for the projects that had already been completed or were in progress at the time. The defect data was also normalized as a function of software size, and control charts were developed to characterize the number of defects per thousand lines of source code discovered during integration and validation testing, as well as reported by customers (control charts were introduced in Chapter 4 and further discussed in Chapters 6, 7, and 8).

Charge numbers were assigned for projects using the Scrum methodology to cover the requirements analysis, architectural design, detailed design, coding, integration testing, and validation testing phases. Ten new projects were selected as pilot projects for the Scrum methodology. Although using only ten projects to pilot the new methodology would introduce some uncertainty in the statistical validity of the evaluations performed, it was decided to accept this level of risk in evaluating the new methodology.

The intent was to develop a model of the development cycle that characterized, as a percent of the total development cycle, the time spent in each phase, normalized by the size of the software. By establishing control charts for each phase and establishing control charts of the defects discovered during each phase of testing, intelligent decisions concerning the efficacy of moving to a new development model could be made. A reliability model was also selected to determine if the MTBF had improved as a result of using a different development methodology (see the appendix to Chapter 6).

Typically, development organizations do not begin to implement statistical methods to evaluate processes until they have fully achieved Maturity Level 3. In this case, the organization already had a history of using statistical methods as part of its measurement program, and clearly had a business need to be able to make some quantitative comparisons of the present condition (current development methodology) to the new condition following the switch to a different development methodology. As pointed out in Chapter 5, in setting up a methodology for strategic planning for process improvement, there are times when process areas at higher maturity levels will have higher priority than process areas at the next level of maturity, simply because of business needs. That's what happened at Systems, Inc., although full implementation of the Maturity Level 4 process areas was not needed at this time.

Systems, Inc. evaluated the data it had on time charges for each phase of activity. Although, as noted previously in this chapter, developers did not keep accurate records of the time they spent on each task, some fairly consistent patterns emerged. Table 10.5 shows the mean percentage of time spent in each phase of the development cycle, as well as the standard deviations of these values over about a hundred projects. It also shows the same data for the data collected from the ten pilot projects of the new methodology, and the test of statistical significance between the two methodologies. Numbers in bracket indicate the P value of the test for equality of normalized time spent in task with the waterfall and the Scrum approach. A P value below 0.01 indicated that the difference is statistically significant. There is no statistically significant difference between the two methodologies for requirements analysis and architectural design, but the data does show that using Scrum significantly shortens the time for the rest of the development effort.

Table 10.5 Comparisons of Months/SLOC by Phase between Scrum and Waterfall Model

Methodology	Phase	Mean	Standard Deviation	Significant Difference?
Old	Requirements Analysis	0.00305	0.004016	No (P = 0.247)
New		0.00314	0.004134	
Old	Architectural Design	0.00221	0.003536	No (P = 0.964)
New		0.00217	0.003603	
Old	Detailed Design	0.00632	0.004760	Yes (P = 0.004)
New		0.00953	0.004022	
Old	Coding	0.00622	0.004639	Yes (P = 0.001)
New		0.00872	0.002419	
Old	Integration	0.00225	0.001194	Yes (P = 0.002)
New		0.00409	0.002268	
Old	Validation	0.00138	0.000785	Yes (P = 0.002)
New		0.00218	0.000945	

Similar Systems, Inc. results were found for the number of defects discovered. The number of defects discovered during detailed design increased significantly, while the number of defects discovered during the rest of the life cycle decreased significantly. It indicated that the Scrum methodology had a better capability to detect and trap defects earlier in the development cycle than the old waterfall methodology. A comparison was made of the number of defects reported by customers of the ten new systems for the 6-week period following the system going operational. A significant decrease in customer-reported defects was observed.

Calculations of the MTBF, based on defects found during integration and validation testing, showed an improvement in MTBF, but the systems still did not achieve the advertised value of 5,000 hours MTBF. This would remain an area for further work in process improvement. For more examples of quantitative control of agile development processes, see [2].

10.4 Summary and Conclusions

In this chapter we presented a case study that tied together the principles, tools, methodologies, and strategies discussed in this book. The experience of Systems,

Inc. demonstrates how the use of a process improvement model, like the CMMI, can help structure an organization's process improvement efforts in an organized, constructive, and effective manner. We showed how the appraisal methodology flushed out the significant process implementation problems in a development organization, and how it was followed by a structured method for planning and implementing process improvement. Such plans must take into account business objectives, CMMI implementation steps, and various risks. Implementation mainly consists of launching prioritized process improvement projects. We also showed how the use of the CMMI helped Systems, Inc. evaluate the efficacy of the process improvement projects it implemented.

One of the major process improvement projects initiated by Systems, Inc. was implementing a different development cycle model. The model that was selected was Scrum. Implementing *agile* methods for systems development such as Scrum in a large organization is a major challenge. The Systems, Inc. initiative in this area was considered by management a strategic initiative. We described in this chapter how Systems, Inc. formalized the approach by creating both a common language and a reuse repository of knowledge and experience. In doing this, we also illustrated the framework for applying EKD, a general *business process management* description language to map Scrum patterns. By creating an organizationwide accessible Scrum pattern repository, Systems, Inc. provided the infrastructure for companywide implementation of agile development. Such a repository provides both storage and updating features that are critical for expanding Scrum implementation beyond local islands of excellence.

Systems, Inc. still has some way to go before becoming recognized as CMMI Level 5. The various management initiatives described above indicate that it is committed to achieve improvements in process and products, and that such improvements are considered critical from a business perspective. The goal set by management was for Systems, Inc. to reach Level 5 in 2011. Our forecast is that with continued leadership by management and sound improvement methodology, Systems, Inc. has a good chance of achieving it. This will obviously depend on the vision, consistency, and perseverance of management.

Our final message to the top management of Systems, Inc. is to remember that there are:

> two approaches to improvement to avoid: systems without passion and passion without systems.

—Tom Peters, *Thriving on Chaos*, 1987

References

1. Boehm, B., Some Future Trends and Implications for Systems and Software Engineering Processes, *Systems Engineering*, 9(1), 1-19, Spring 2006.
2. Cangussu, J.W. and Karcich, R.M., 2005, A Control Approach for Agile Processes, *2nd International Workshop on Software Cybernetics – 29th Annual IEEE, International Computer Software and Applications Conference (COMPSAC 2005)*, p. 123–126 Edinburgh, Scotland, July 25–28, 2005.
3. Cohn M. and Ford D., Introducing an Agile Process to an Organization, IEEE Computer Society, http://www.mountaingoatsoftware.com/system/article/file/10/IntroducingAnAgileProcess.pdf, 2003.
4. Kenett, R. and Baker E., *Software Process Quality: Management and Control*, New York: Marcel Dekker, 1999.
5. Leffingwell, D. and Muirhead D., Tactical Management of Agile Development: Achieving Competitive Advantage, white paper, Boulder, CO: Rally Software Development Corporation, 2004.
6. Schwaber K. and Beedle M., *Agile Software Development with Scrum*, Upper Saddle River, NJ: Prentice Hall, 2001.
7. Software Engineering Institute, Process Maturity Profile, CMMI®, SCAMPI℠ Class A Appraisal Results, 2008 Mid-Year Update, Pittsburgh, PA: Software Engineering Institute, Carnegie Mellon University, September 2008.
8. Software Engineering Institute, CMMI Product Team. CMMI for Development, Version 1.2 (CMU/SEI-2006-TR-008), Pittsburgh, PA: Carnegie Mellon University, 2006.
9. Takeuchi, H. and Nonaka, I., The New New Product Development Game, *Harvard Business Review*, January–February 1986.
10. *The McKinsey Quarterly*, 2005, IT's Role in Media Convergence: An Interview with Verizon's CIO, www.mckinseyquarterly.com/article_abstract_visitor.aspx?ar=1596&L2=22&L3=78.
11. Wikipedia, http://en.wikipedia.org/wiki/Agile_software_development, (accessed August 19, 2008).

Index